# CARTESIAN GEOMETRY OF THE PLANE

# CARTESIAN GEOMETRY OF THE PLANE

BY

E. M. HARTLEY, M.A. Ph.D. (Cantab)

*Senior Lecturer in the University of Ghana*

CAMBRIDGE
AT THE UNIVERSITY PRESS
1966

CAMBRIDGE UNIVERSITY PRESS
Cambridge, New York, Melbourne, Madrid, Cape Town, Singapore, São Paulo, Delhi

Cambridge University Press
The Edinburgh Building, Cambridge CB2 8RU, UK

Published in the United States of America by Cambridge University Press, New York

www.cambridge.org
Information on this title: www.cambridge.org/9780521052221



First published 1960
Reprinted 1966
This digitally printed version 2008

*A catalogue record for this publication is available from the British Library*

ISBN 978-0-521-05222-1 hardback
ISBN 978-0-521-09871-7 paperback

# CONTENTS

## IV. CURVES DEFINED PARAMETRICALLY

### A. THE RECTANGULAR HYPERBOLA



### B. CUBIC CURVES



### C. TRIGONOMETRIC CURVES



## V. THE CIRCLE



## VI. THE ELLIPSE AND HYPERBOLA

# PREFACE

This book is intended as a first course in coordinate geometry, of the type envisaged in the Report on the Teaching of Higher Geometry in Schools, prepared for the Mathematical Association in 1953. I have had in mind those working for the Advanced and Scholarship levels of the General Certificate of Education, but there is sufficient material to meet also the needs of those reading for a General degree including Mathematics whose requirements stop short of the use of homogeneous coordinates. I have, as the title suggests, confined my attention to the Cartesian plane, feeling that most students benefit from gaining certainty and confidence in this field. Those who later specialize in Mathematics should be in a position to appreciate other geometries, after thoroughly mastering one, while those whose interest lies elsewhere will have certain necessary tools in their hands, and experience in using them.

I have tried to combine clarity in the exposition with more rigour than seems to be usual at this stage. It is probably unwise to insist on absolute standards of rigour from a pupil so early in a mathematical career, but the demands made later for logical thought will seem less unreasonable to the student who is accustomed to knowing exactly what has been proved. Although the chief emphasis is on the method of coordinates, I have not hesitated to use a combination of pure and analytical methods at every stage. References are made to other branches of mathematics, and to some methods of drawing the curves considered, which I hope will make the subject seem relevant and interesting.

For those working with little or no help I have given in the Introduction a summary of those results in Algebra, Pure Geometry, Trigonometry and Calculus which are used in the sequel. Proofs of the theorems in the Geometry section of this Introduction are outlined; for the other parts of the work the reader should consult the appropriate text-book. Some paragraphs are in smaller type; these contain comments which are included in the hope of answering some of the questions which may arise in the mind of the reader. These sections, and paragraphs marked with a star, may be omitted (and during a first reading probably should be), without the thread of the argument being lost.

The examples in each chapter are meant to provide immediate illustration of, and practice in the use of, the formulae (emphasized in bold type) which have just been obtained. Enough of these should be worked to make each formula thoroughly familiar; it is particularly important for those working alone not to omit such routine 'drill'. At the end of each chapter there is a set of miscellaneous examples which require more thought; the harder of these appear below a line. Some further examples, covering topics treated in different chapters, appear after the final chapter. In some of the cases where an example is not a straightforward application of bookwork, a hint as to an appropriate method of attack is given at the end of the book.

The number of those who have helped in different ways to make the book a reality is legion. In particular I must mention Dr E. A. Maxwell, of Queens' College, Cambridge, from whom I learned much as an undergraduate; he first suggested my writing on this subject, and with astonishing generosity read and criticized much of the manuscript. In this exercise he was later joined by Mr A. P. Rollett and Dr H. M. Cundy, and to each I owe gratitude for many useful suggestions. Among former pupils who gave me advice founded on their experience in the class-room, I am specially indebted to Miss E. A. Dickens, who also, with Miss I. L. Campbell and Miss P. E. Moss, provided the final list of answers. Many of the examples are taken from past papers of various examining bodies, and I wish to acknowledge my indebtedness to the University of Bristol; the University of Cambridge; the University of Sheffield; Girton College, Cambridge; Newnham College, Cambridge; Royal Holloway College, London; St Hugh's College, Oxford; Somerville College, Oxford; Cambridge Local Examinations Syndicate; Northern Universities Joint Matriculation Board; Oxford Local Examinations; Oxford and Cambridge Schools Examination Board; University of London Examinations Board for permission to use these questions from their papers. Finally, I am grateful for much encouragement and tolerance shown by the staff of the Cambridge University Press to a newcomer to the art of producing books.

Throughout, I have been aware of my indebtedness to a friend and former colleague, the late Dr Christine Hamill, of Newnham College, Cambridge, the University of Sheffield, and University College, Ibadan. I remember with pleasure and gratitude the many occasions on which together we evolved our methods of teaching, and in particular a few days' holiday in September 1955, during which we discussed the form and contents of this book; it was she who, with recent experience of the needs of West Africa in mind, urged me to remember those who would have to work largely alone. From these conversations, and from a subsequent six months correspondence, during which she criticized the first few chapters in merciless detail, I gained untold stimulation. I have tried to capture and communicate the interest, even fascination, of geometry for us; if in any measure I have succeeded, the praise is hers.

E. M. H.

CAMBRIDGE
*August 1959*

# INTRODUCTION

When a course of work is being planned, the perennial problem confronting a student or teacher of mathematics is one of *order*. The intricate way in which each part of the subject depends on each other part makes a final decision on what constitutes 'the right order' almost impossible to reach. In this book I have, naturally, arranged the material as I myself prefer. I have, however, attempted to make the problem less intractable for others, by collecting at the start a summary of those results in other branches of mathematics which are used later, and with each a note of the section in which it is first needed. Proofs of these results are not given; they can be found in the appropriate text-books.

## ALGEBRA

It is assumed throughout that the reader can simplify algebraic expressions and solve simple equations; in the interests of brevity the details of such computations are often omitted. For example, the steps between the statements

$$(x-3y+5)+\lambda(2x+y+3) = 0 \quad \text{where} \quad \lambda = -\tfrac{5}{3}$$

and

$$x+2y = 0$$

should be readily provided by the student with pencil and paper to hand.

**Determinants.** The theory of determinants is used in chapter II, but alternative proofs are given so that those who acquire the technique later will not be handicapped. A knowledge of both the evaluation and the properties of $3 \times 3$ determinants is essential for chapters VII and VIII. The results used earlier are the following:

[A 1]
$$\begin{vmatrix} a_1 & b_1 \\ a_2 & b_2 \end{vmatrix} = a_1b_2 - a_2b_1. \qquad \text{(II, §6)}$$

[A 2]
$$\begin{vmatrix} a_1 & b_1 & c_1 \\ a_2 & b_2 & c_2 \\ a_3 & b_3 & c_3 \end{vmatrix} = a_1(b_2c_3-b_3c_2) - a_2(b_1c_3-b_3c_1) + a_3(b_1c_2-b_2c_1). \qquad \text{(II, §9)}$$

[A 3] The equations,

$$a_1p+b_1q+c_1r = 0$$
$$a_2p+b_2q+c_2r = 0 \qquad (\text{II}, §9)$$
$$a_3p+b_3q+c_3r = 0$$

have a solution with $p$, $q$, $r$ not all zero if and only if the determinant of the coefficients is zero; that is

$$\begin{vmatrix} a_1 & b_1 & c_1 \\ a_2 & b_2 & c_2 \\ a_3 & b_3 & c_3 \end{vmatrix} = 0.$$

**Theory of Equations**

[A 4] The quadratic equation in $x$,

$$ax^2+bx+c = 0,$$

has (i) real roots if and only if $b^2-4ac \geqslant 0$; (III, §4)

(ii) no roots if and only if $b^2-4ac < 0$;

(iii) equal roots if and only if $b^2-4ac = 0$;

(iv) equal and opposite roots if and only if $b = 0$ and $ac \leqslant 0$. (III, §10)

(In the second case, the student who has heard of complex numbers will perhaps say that there are complex roots, but such numbers do not enter into the work of this book.)

[A 5] If $\alpha$ and $\beta$ are the roots of the quadratic equation

$$ax^2+bx+c=0, \qquad (\text{III}, §4)$$

then

$$\alpha+\beta = -b/a,$$
$$\alpha\beta = c/a.$$

(These are called the 'symmetric functions' of the roots.)

N.B. It is a good rule *never* to use the formula

$$x = \frac{-b\pm\sqrt{(b^2-4ac)}}{2a}$$

for the solutions of a quadratic equation when the coefficients are not numerical. In Cartesian geometry, a knowledge of the sum and product of the roots is almost always all that is needed.

[A 6] If $\alpha$, $\beta$, $\gamma$ are the roots of the cubic equation

$$ax^3+bx^2+cx+d = 0, \qquad (\text{III}, §11)$$

then
$$\alpha+\beta+\gamma = -b/a,$$
$$\beta\gamma+\gamma\alpha+\alpha\beta = c/a,$$
$$\alpha\beta\gamma = -d/a.$$

[A 7] If $\alpha$, $\beta$, $\gamma$, $\delta$ are the roots of the quartic equation
$$ax^4+bx^3+cx^2+dx+e = 0, \qquad \text{(IV, §5)}$$
then
$$\alpha+\beta+\gamma+\delta = -b/a,$$
$$\beta\gamma+\gamma\alpha+\alpha\beta+\alpha\delta+\beta\delta+\gamma\delta = c/a,$$
$$\beta\gamma\delta+\gamma\alpha\delta+\alpha\beta\delta+\alpha\beta\gamma = -d/a,$$
$$\alpha\beta\gamma\delta = e/a.$$

## PURE GEOMETRY

All the elementary pure geometry of lines, triangles and circles, a knowledge of which is needed for Ordinary Level of the G.C.E., is assumed.

**Properties of a triangle.** In a triangle $ABC$, $A'$, $B'$, $C'$ are the mid-points of the sides $BC$, $CA$, $AB$, and $D$, $E$, $F$ are the feet of the perpendiculars from $A$, $B$, $C$, to $BC$, $CA$, $AB$ respectively.

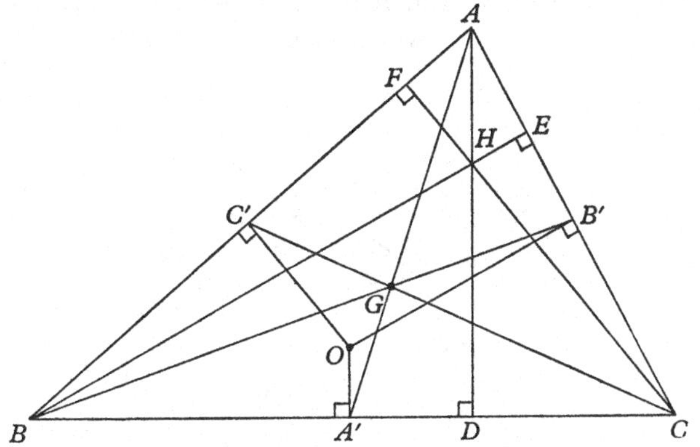


[G 1] The *circumcentre*, $O$, is the point of intersection *of perpendicular bisectors* of the sides; it is equidistant from $A$, $B$ and $C$ and so it is the centre of a circle passing through these points, the *circumcircle* of the triangle $ABC$. (I, Ex. I B.)

[G 2] The *centroid*, $G$, is the point of intersection of the *medians* $AA'$, $BB'$, $CC'$. $AG = 2GA'$, and $G$ is similarly a point of trisection of the other two medians. (I, §3.)

[G 3] The *orthocentre*, $H$, is the point of intersection of the *altitudes*, $AD$, $BE$, $CF$. (Ex. II C.)

[G 4] The *incentre*, $I$, is the point of intersection of the *internal* bisectors of the angles; it is equidistant from the sides, and so is the centre of a circle inscribed in the triangle, the *incircle* of the triangle. The *excentres* are the other three intersections of the

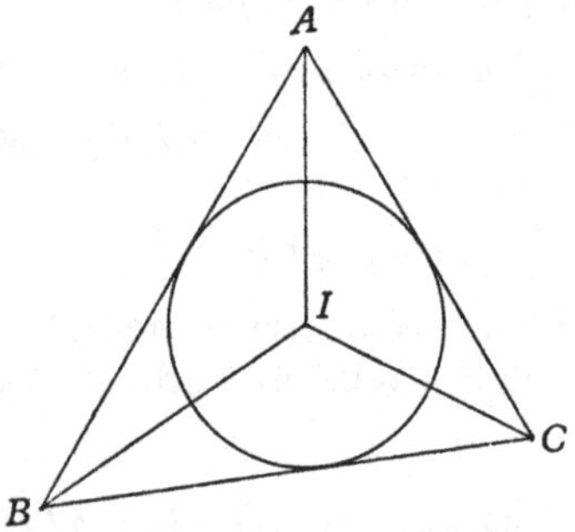


bisectors of the angles of the triangle. They are the centres of the *escribed circles*, each of which touches one side of the triangle internally, and the other two externally. (Ex. II E.)

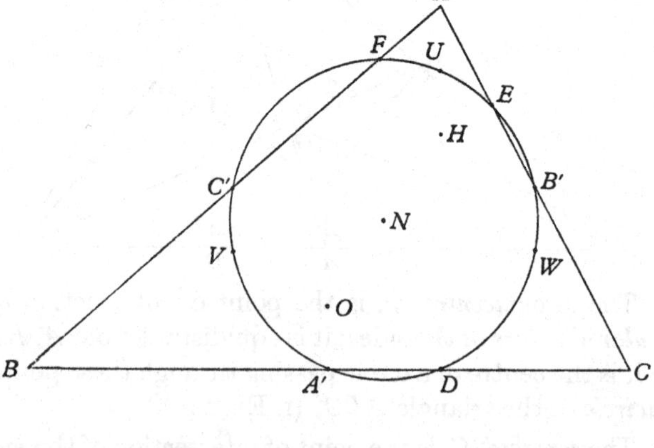


[G 5] Suppose the mid-points of $AH$, $BH$, $CH$ are $U, V, W$. The nine points $A', B', C', D, E, F, U, V, W$ all lie on a circle, the *nine-*

*point circle*, of the triangle, whose centre $N$ is the mid-point of $OH$. (Ex. II C.)

[G 6] *The Simson line.* Let $P$ be any point on the circumcircle of the triangle $ABC$, and let $L, M, N$ be the feet of the perpendiculars from $P$ to $BC$, $CA$, $AB$ respectively. Then $L$, $M$, $N$ are collinear, on the Simson line. Conversely, if $P$ is a point such that the feet of the perpendiculars from $P$ to the sides of a triangle are collinear, it lies on the circumcircle of the triangle. (III, § 8.)

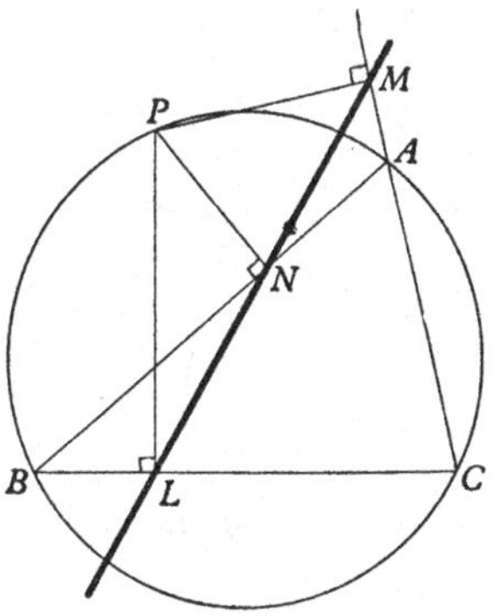


[G 7] *Theorem of Menelaus.* If a transversal $LMN$ meets the sides $BC$, $CA$, $AB$ of a triangle in $L$, $M$, $N$ respectively, then

$$\frac{\overrightarrow{BL}}{\overrightarrow{LC}} \cdot \frac{\overrightarrow{CM}}{\overrightarrow{MA}} \cdot \frac{\overrightarrow{AN}}{\overrightarrow{NB}} = -1. \qquad \text{(Ex. I C)}$$

The converse is also true. (For the meaning of $\overrightarrow{BL}$, see I, § 3.)

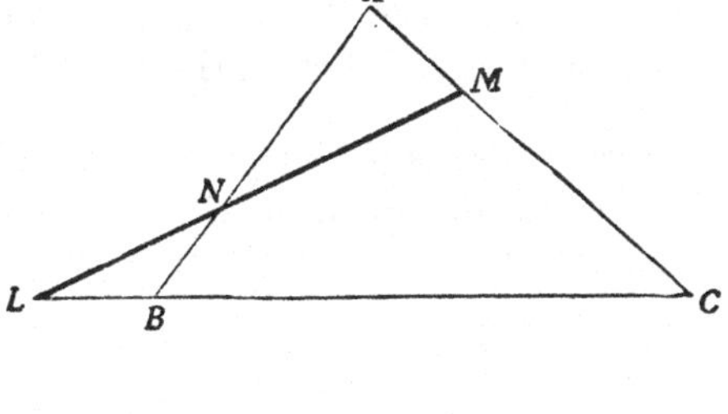

[G 8] *Theorem of Ceva.* If points $L, M, N$ are taken on the sides $BC, CA, AB$ of a triangle $ABC$ so that the lines $AL, BM, CN$ are concurrent in a point $O$, then

$$\frac{\overrightarrow{BL}}{\overrightarrow{LC}} \cdot \frac{\overrightarrow{CM}}{\overrightarrow{MA}} \cdot \frac{\overrightarrow{AN}}{\overrightarrow{NB}} = +1. \qquad \text{(Ex. 1c)}$$

The converse is also true.

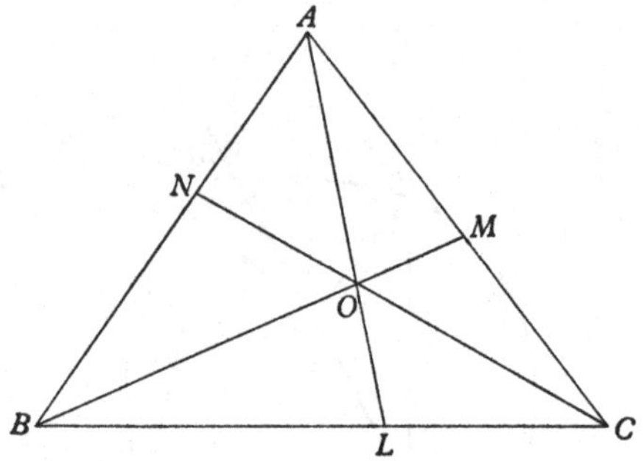


As a guide towards the proof of the results in this section, the following outlines are given.

[G 1] Let the lines through $B'$, $C'$ perpendicular to $CA$, $AB$ meet in $O$. Then $OC = OA$ and $OA = OB$, so that $OB = OC$ and $OA'$ is perpendicular to $BC$.

[G 2] Let $BB'$, $CC'$ meet in $G$, and produce $AG$ to $A''$ so that $G$ is the mid-point of $AA''$. Then $GBA''C$ has pairs of opposite sides parallel, so that it is a parallelogram and its diagonals bisect each other. It follows that $AGA''$ passes through $A'$, and that $AG = 2GA'$.

[G 3] Produce $OG$ to $H$ so that $GH = 2OG$. Then the triangles $GHA$, $GOA'$ are similar and so $AH$ is parallel to $OA'$ and perpendicular to $BC$. Similarly, $H$ lies on the other two altitudes.

[G 4] If the internal bisectors of two of the angles, say $B$, $C$ meet in $I$, then $I$ is equidistant from all of $BC$, $CA$, $AB$ and so $AI$ is the internal bisector of the third angle, $A$. A similar proof applies for each excentre.

[G 5] If $N$ is the mid-point of $OH$ and $A'N$ meets $AH$ in $U$, $A'O = UH$ and so $U$ is the mid-point of $AH$. The points $V$, $W$, defined similarly, are the mid-points of $BH$, $CH$. Since the triangles $A'ON$, $UHN$ are congruent, $A'N = NU$ and since angle $UDA'$ is a right angle, $N$ is the centre of a circle through $A', U, D$ whose radius is $NU$. Also $NU = \frac{1}{2}OA$, so that the radius of this circle is half the radius of the circumcircle. By symmetry the remaining six of the nine points listed lie also on the same circle.

[G 6] The points $P$, $N$, $A$, $M$ are concyclic, so that

$$\angle PNM = \angle PAM = \angle PBC = 180° - \angle PNL$$

(since the points $P$, $B$, $N$, $L$ are concyclic). It follows that $LMN$ is a straight line.

[G 7] Draw a line $AD$ parallel to $LMN$, to meet $BC$ in $D$ and use the ratio theorem for $\overrightarrow{CM}/\overrightarrow{MA}$ and for $\overrightarrow{AN}/\overrightarrow{NB}$.

[G 8] $\dfrac{\overrightarrow{BL}}{\overrightarrow{LC}} = \dfrac{\Delta BAL}{\Delta LAC} = \dfrac{\Delta BOL}{\Delta LOC} = \dfrac{\Delta BOA}{\Delta COA}$, and similarly for the other ratios.

## TRIGONOMETRY

The basic definitions of sine, cosine and tangent of an acute angle are assumed throughout, as are their inverses, cosecant, secant and cotangent.

[T 1] For any angle $\alpha$, $\tan\alpha = \overrightarrow{AB}/\overrightarrow{OA}$, as in the figure below. (For the definition of $\overrightarrow{AB}$, see chapter I, §2.) (II, §1)

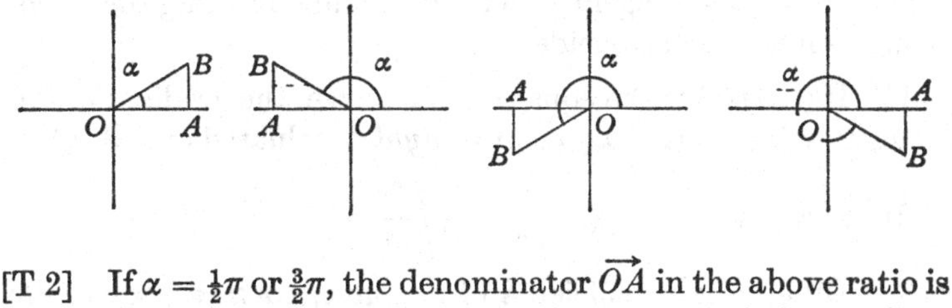


[T 2] If $\alpha = \frac{1}{2}\pi$ or $\frac{3}{2}\pi$, the denominator $\overrightarrow{OA}$ in the above ratio is zero. It is then customary to say that $\tan\alpha$ is infinite. Conversely, if $\tan\alpha = p/q$ and $q = 0$, then $\alpha = \frac{1}{2}\pi$ or $\frac{3}{2}\pi$. Similar results apply to $\sec\alpha$, $\operatorname{cosec}\alpha$ and $\cot\alpha$. (II, §2.)

[T 3] $$\tan(\alpha+\tfrac{1}{2}\pi) = -\cot\alpha. \qquad \text{(II, §2)}$$

[T 4] Sine and cosine of any angle, as for $\tan\alpha$. (II, §4)

[T 5] $$\tan(\alpha-\beta) = \frac{\tan\alpha-\tan\beta}{1+\tan\alpha\tan\beta}. \qquad \text{(II, §11)}$$

[T 6] If $\tan(\tfrac{1}{2}\theta) = t$, $\cos\theta = \dfrac{1-t^2}{1+t^2}$, $\sin\theta = \dfrac{2t}{1+t^2}$. (VI, §3)

[T 7]
$$\sin\alpha+\sin\beta = 2\sin\tfrac{1}{2}(\alpha+\beta)\cos\tfrac{1}{2}(\alpha-\beta),$$
$$\sin\alpha-\sin\beta = 2\cos\tfrac{1}{2}(\alpha+\beta)\sin\tfrac{1}{2}(\alpha-\beta),$$
$$\cos\alpha+\cos\beta = 2\cos\tfrac{1}{2}(\alpha+\beta)\cos\tfrac{1}{2}(\alpha-\beta),$$
$$\cos\alpha-\cos\beta = -2\sin\tfrac{1}{2}(\alpha+\beta)\sin\tfrac{1}{2}(\alpha-\beta). \qquad \text{(VI, §3)}$$

[T 8] Let $\tan\alpha_i = t_i$. For compactness we write

$$t_1+t_2+t_3+t_4 = \Sigma t_i,$$
$$t_1t_4+t_2t_4+t_3t_4+t_2t_3+t_3t_1+t_1t_2 = \Sigma t_it_j,$$
$$t_1t_2t_3+t_1t_2t_4+t_1t_3t_4+t_2t_3t_4 = \Sigma t_it_jt_k. \qquad \text{(VI, §3)}$$

Then $$\tan(\alpha_1+\alpha_2+\alpha_3+\alpha_4) = \frac{\Sigma t_i-\Sigma t_it_jt_k}{1-\Sigma t_it_j+t_1t_2t_3t_4}.$$

## CALCULUS

No calculus is essential for this book, except in chapter IV C, where it is indispensable. It is, however, in many cases so much quicker to use differentiation to find the gradient of a curve, that this is referred to at every stage from chapter III onwards, as an alternative. Until it is replaced by a 'calculus' definition of a tangent [C 1], the tangent is defined as a line meeting the curve in two points which coincide.

[C 1] If $y = f(x)$ is the equation of a curve, the gradient of the tangent at $(x_0, y_0)$ is $f'(x_0)$ (that is, $dy/dx$ evaluated at $x = x_0$).

[C 2] If $x = x(t)$, $y = y(t)$, $\dfrac{dy}{dx} = \dfrac{dy}{dt}\bigg/\dfrac{dx}{dt}$.

[C 3] Differential coefficients of $x^n$, $\sin x$, $\cos x$, $u+v$, $uv$, $u/v$ and $f\{g(t)\}$.

# I

# THE POINT

In this book we shall investigate geometrical properties of lines and curves in a plane, largely by means of coordinates. The points of the plane will be specified by their distances from two fixed perpendicular straight lines. We require of such a system of labelling the points of the plane, first that given any point, it has a 'label' which is determined precisely; and secondly that any 'label' corresponds to just one point. The points are not the same as their labels, and we shall find that there are ways (for example, in paragraph 5 of this chapter) in which the labelling can be altered; but this essential 'uniqueness' requirement of any labelling system will always be satisfied.

**1. The coordinates of a point.** Suppose $X'OX$, $Y'OY$ are two perpendicular straight lines in the plane intersecting in a point $O$. They are called *axes of coordinates*, or just *axes*, and $O$ is called the *origin*. If we wish to specify a point on $X'OX$ by its distance $d$ from $O$, we are faced with an ambiguity (unless $d = 0$), for the point may lie on either side of $O$. A convention of signs must therefore be introduced, and it is the usual mathematical practice to count distances from $O$ to points on the right of $O$ as positive, and distances from $O$ to points on the left of $O$ as negative. Similarly, points on $Y'OY$ are described by their distances from $O$, with a positive sign if they are above $X'OX$, and otherwise a negative one (see Fig. 1).

Once the points on the axes have been labelled in this way, we may label the other points of the plane. In Fig. 2 $P$ is any point, and $PM$, $PN$ are perpendiculars drawn to the axes from $P$. Then the (signed) distances $ON$, $OM$ are uniquely determined; they are denoted by $x, y$. Conversely, given two numbers $x, y$ the positions of $N$, $M$ on the axes can be found, and so the point $P$ can be located.

The two numbers $x, y$ will thus serve to label $P$ as we required; they are called the *coordinates* of $P$, and we say that $P$ is the point with coordinates $(x, y)$, or, more shortly, $P$ is the point $(x, y)$; we also refer to the point $P(x, y)$. The $x$-coordinate and $y$-coordinate of $P$ are called respectively the *abscissa* and *ordinate* of $P$. The whole system is called a system of *rectangular Cartesian* coordinates. The word 'Cartesian' recalls the French mathematician Descartes (1596–1650) who first devised this method of doing

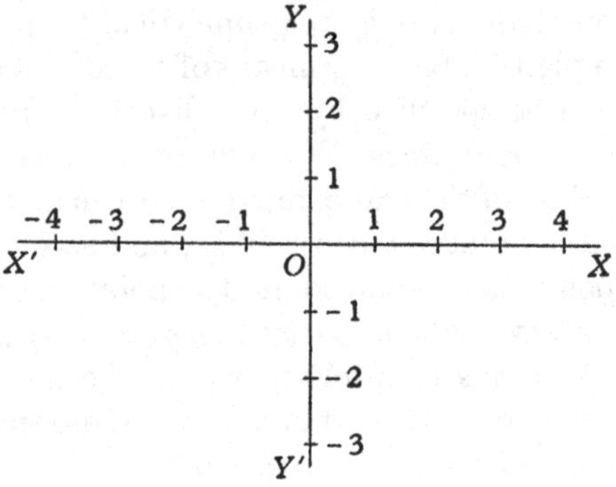


Fig. 1

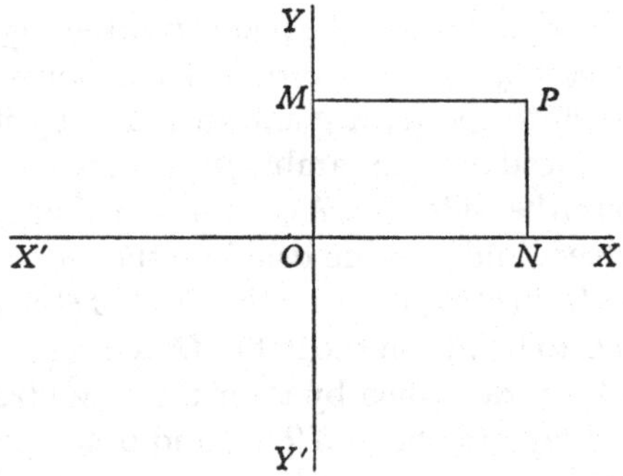


Fig. 2

geometry; 'rectangular' distinguishes this system from one in which the axes are inclined to each other at some angle $\omega$, not a right angle (the axes then being called *oblique*).

To make the definition precise, one other matter must be mentioned. We said that $x$ and $y$ are any two numbers, and we must make clear what type of numbers we mean. It would be possible, though not desirable, to restrict our attention to integers; the geometry we would then have would contain only points of the plane forming a rectangular pattern of dots. It is usual to allow the coordinates to be any real numbers; that is, including both rational numbers (like $\frac{1}{2}$, $-\frac{3}{7}$ and so on), and also irrational numbers (like $\sqrt{2}, \pi$). This is the geometry of experience; it carries the advantage that it can be illustrated by diagrams which will, with care, look credible.

Limiting the coordinates to the set of real numbers has one disadvantage. Quadratic equations like

$$x^2 = 4,$$

or

$$x^2+3x+2 = 0,$$

have two real roots, whereas others, like

$$x^2 = -1,$$

or

$$x^2+3x+4 = 0,$$

are not satisfied by any real numbers. Those who have met *complex numbers* will realize that, if we allow the numbers involved to be complex, the distinction between these two types of equation disappears. In this case, however, quite different definitions, not based on the idea of distance, would be needed, and at an elementary stage this is not desirable. In this book, therefore, we shall require the coordinates to be real numbers.

The reader who is not already accustomed to drawing graphs should make certain of the definitions, by plotting the positions

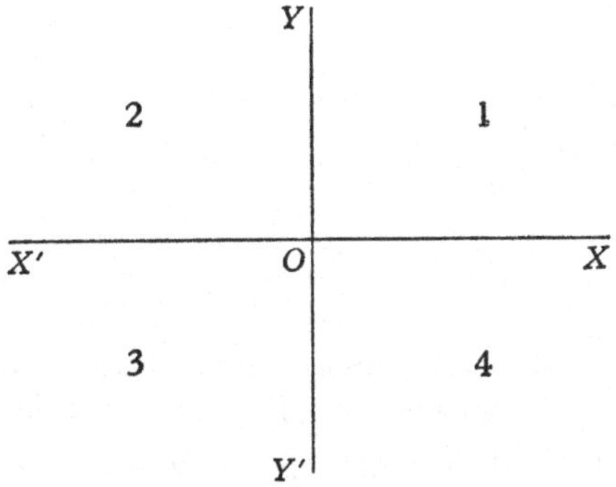


Fig. 3

of a number of points. Among other things, it will be seen that the origin $O$ is the point $(0, 0)$; that points $(x, 0)$ lie on $X'OX$, the $x$-axis, and that points $(0, y)$ lie on $Y'OY$, the $y$-axis. The four regions into which the plane is divided by the axes are called *quadrants*; the numbering of the quadrants is shown in Fig. 3.

### EXAMPLES I A

1. On a sheet of squared paper draw perpendicular axes and plot the positions of the following points: $(5, -1)$, $(2, 7)$, $(-4, -\frac{1}{2})$, $(3, 0)$, $(2, -6)$, $(-5, 3)$, $(0, -1)$, $(-\frac{3}{2}, 2)$, $(0, 5)$, $(2, 2)$. Which points are on the axes? Which are in the first quadrant, second quadrant, third quadrant, fourth quadrant?

2. The coordinates of a series of points are $(3+t, 2-t)$, where $t = -3, -2, -1, -\frac{1}{2}, 0, \frac{1}{2}, 1, 2, 3, 4$. Plot each point on a graph; what do you notice about the points? For what value of $t$ does the point lie on the $x$-axis? on the $y$-axis? For what range of values of $t$ is the point in the first quadrant?

3. Repeat Question 2 for

(i) $(-1+t, 2+t/2)$;

(ii) $(t^2, t)$;

(iii) $\left(\frac{1-t^2}{1+t^2}, \frac{3t}{1+t^2}\right)$;

in each case plotting the points for which $t = 0, \pm\frac{1}{2}, \pm 1, \pm 2, \pm 3, \pm 4$.

**2. Distance between two points.** The convention of signs which we introduced in the last section may be extended to include the distances between any two points on an axis. We shall say that

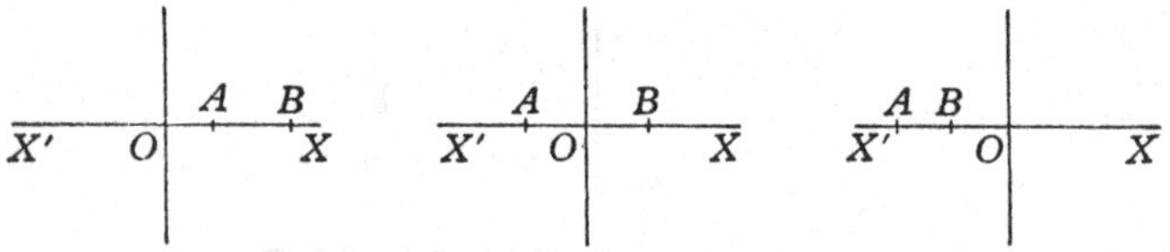


Fig. 4

the distance $\overrightarrow{AB}$ between $A$ and $B$ on the $x$-axis is positive if $B$ is on the right of $A$, and negative if $B$ is on the left of $A$. Thus $\overrightarrow{AB} = -\overrightarrow{BA}$. This fits in with the previous definition; if $A$ is $(\alpha, 0)$, then $\overrightarrow{OA} = \alpha$, and similarly if $B$ is $(\beta, 0)$, $\overrightarrow{OB} = \beta$.

To find an expression for $\overrightarrow{AB}$, we must consider several cases. If $B$ is on the right of $A$, the points may be in any one of the three positions in Fig. 4, $O$ being in each case the origin. In the first case

$$\overrightarrow{AB} = \overrightarrow{OB} - \overrightarrow{OA} = \beta - \alpha.$$

In the second case

$$\overrightarrow{AB} = \overrightarrow{AO} + \overrightarrow{OB} = \overrightarrow{OB} - \overrightarrow{OA} = \beta - \alpha,$$

and in the last case

$$\overrightarrow{AB} = \overrightarrow{AO} - \overrightarrow{BO} = \overrightarrow{OB} - \overrightarrow{OA} = \beta - \alpha.$$

If $B$ is on the left of $A$, then $A$ is on the right of $B$, and $\overrightarrow{BA} = \alpha - \beta$; thus $\overrightarrow{AB} = -\overrightarrow{BA} = -(\alpha - \beta) = \beta - \alpha$. We thus have the general formula, true for any positions of $A$ and $B$ on the $x$-axis,

$$\overrightarrow{\mathbf{AB}} = \boldsymbol{\beta} - \boldsymbol{\alpha}.$$

A similar formula

$$\overrightarrow{\mathbf{CD}} = \boldsymbol{\delta} - \boldsymbol{\gamma}$$

holds for the signed distances between any two points $C(0, \gamma)$, $D(0, \delta)$ on the $y$-axis. These ideas of signed distances can then be extended to distances on any lines parallel to the axes.

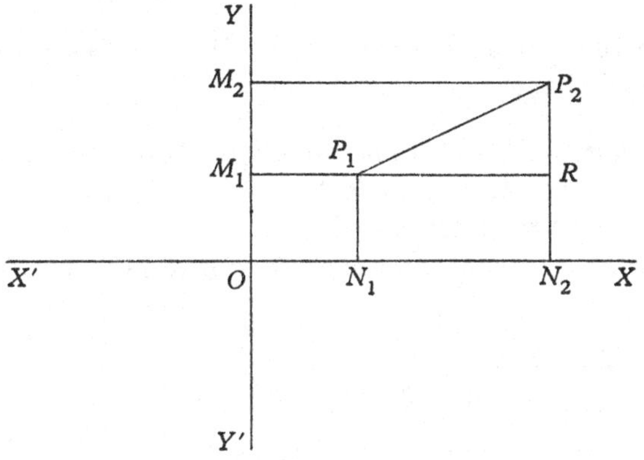


Fig. 5

Now suppose $P_1(x_1, y_1)$ and $P_2(x_2, y_2)$ are any two points in the plane. Let the perpendiculars from $P_1, P_2$ meet the axes in $N_1, N_2$ and $M_1, M_2$ respectively (Fig. 5) and let $R$ be the foot of the perpendicular from $P_1$ to $P_2N_2$. Then

$$\overrightarrow{P_1R} = \overrightarrow{N_1N_2} = x_2 - x_1 \quad \text{and} \quad \overrightarrow{RP_2} = \overrightarrow{M_1M_2} = y_2 - y_1;$$

since

$$P_1P_2^2 = P_1R^2 + RP_2^2 = \overrightarrow{P_1R}^2 + \overrightarrow{RP_2}^2$$

it follows that

$$\mathbf{P_1P_2^2 = (x_2 - x_1)^2 + (y_2 - y_1)^2}.$$

This formula is true for all positions of $P_1$, $P_2$ in the plane, since it depends on the expressions for $N_1N_2$, $M_1M_2$ which have been proved to be true for all positions on the axes.

In particular, if $P$ is $(x, y)$ and $O$ is the origin $(0, 0)$,

$$OP^2 = (x-0)^2 + (y-0)^2,$$

and so

$$\mathbf{OP^2 = x^2 + y^2.}$$

***Illustration 1.*** *Show that the triangle whose vertices are the points* $A(6, -1)$, $B(-3, -4)$ *and* $C(4, 5)$ *is right angled at* $A$.

Using the formula which has just been obtained

$$AB^2 = \{-3-6\}^2 + \{-4-(-1)\}^2 = 9^2 + 3^2 = 90,$$

$$AC^2 = \{4-6\}^2 + \{5-(-1)\}^2 = 2^2 + 6^2 = 40$$

and

$$BC^2 = \{4-(-3)\}^2 + \{5-(-4)\}^2 = 7^2 + 9^2 = 130.$$

Thus

$$AB^2 + AC^2 = 90 + 40 = 130 = BC^2,$$

and so, by the converse of the theorem of Pythagoras, the triangle is right angled at $A$.

***Illustration 2.*** *Find the equation satisfied by the coordinates of all points on a circle whose centre is* $(3, 1)$ *and whose radius is* 2. *Find also where this circle meets the axes.*

If $P(x, y)$ is any point on the circle, centre $C(3, 1)$, then

$$PC^2 = (x-3)^2 + (y-1)^2.$$

But $PC = 2$, and so

$$(x-3)^2 + (y-1)^2 = 4,$$

that is

$$x^2 + y^2 - 6x - 2y + 6 = 0.$$

This is the equation satisfied by the coordinates of every point on the circle.

On the $x$-axis, $y = 0$ and so $x^2 - 6x + 6 = 0$, giving two values of $x$, $3 \pm \sqrt{3}$. The circle therefore meets the $x$-axis in the points $(3 + \sqrt{3}, 0)$ and $(3 - \sqrt{3}, 0)$. On the $y$-axis, $x = 0$ and so

$$y^2 - 2y + 6 = 0.$$

This equation has no real roots, and so the circle does not meet the $y$-axis.

The reader should illustrate these examples, and some of those which follow, by drawing diagrams on squared paper.

### EXAMPLES I B

1. Find the distances between the following pairs of points: $(4, -3)$ and $(-1, 5)$; $(0, -1)$ and $(-1, 2)$; $(3, 1)$ and $(-5, -7)$; $(\frac{1}{2}, \frac{1}{3})$ and $(\frac{1}{3}, \frac{1}{2})$; $(\frac{1}{4}, 0)$ and $(\frac{3}{4}, -1)$; $(-1\frac{1}{3}, 0)$ and $(\frac{2}{3}, 3)$; $(a, b)$ and $(a+b, a-b)$.

2. If the distance between $(a, 3)$ and $(2, 4)$ is $\sqrt{2}$, what is $a$?

3. Show that the triangle whose vertices are $(3, 0)$, $(2, \sqrt{3})$ and $(4, \sqrt{3})$ is equilateral.

4. Show that the points $(6, 5)$, $(11, -7)$, $(-1, -12)$, $(-6, 0)$ are the vertices of a square.

5. Find the distance from $(2, 3)$ of the points $(-2, 0)$, $(-1, 7)$, $(6, 6)$, $(7, 3)$, $(5, 7)$ and $(2, -2)$. What can be deduced about the points?

6. Show that the points $(4, 3)$, $(7, 6)$, $(6, 9)$ and $(3, 6)$ are the vertices of a parallelogram.

7. Given $A(a, b)$ and $B(3a, 3b)$, show that if $P(x, y)$ is a point such that $AP = BP$, then $ax + by = 2(a^2 + b^2)$.

8. Find the circumcentre [G 1] of the triangle whose vertices are $(-2, -3)$, $(30, 7)$ and $(40, -3)$.

9. Find the radius of the circle with centre $(4, -3)$ which passes through the origin.

10. Find the equation satisfied by the coordinates of the points on a circle with centre $(1, 2)$ and radius 5. This circle meets the $x$-axis in $A$ and $B$ and the $y$-axis in $C$ and $D$. Find the coordinates of $A$, $B$, $C$ and $D$, and check the result by verifying the 'intersecting chord' property

$$OA \, . \, OB = OC \, . \, OD.$$

**3. Dividing a line in a given ratio.** Let $P_1(x_1, y_1)$ and $P_2(x_2, y_2)$ be two points. We shall, for this section, introduce a signed distance on $P_1P_2$, so that $\overrightarrow{P_1P_2}$ is positive and $\overrightarrow{P_2P_1}$ is negative.

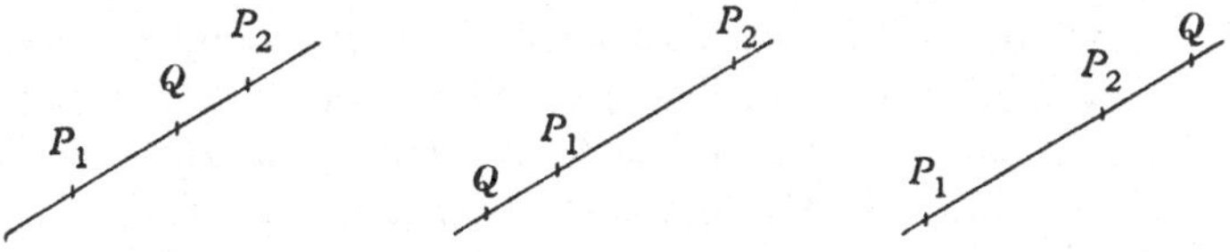


(a) (b) (c)

Fig. 6

Let $Q(X, Y)$ be any point on $P_1P_2$ and suppose that $\overrightarrow{P_1Q} = \lambda$, $\overrightarrow{QP_2} = \mu$. From Fig. 6 it is clear that $Q$ may lie, relative to $P_1$ and $P_2$, in one of three positions illustrated as $(a)$, $(b)$ and $(c)$.

We notice that:

(i) if $Q$ is between $P_1$, and $P_2$, as in (*a*), $\lambda$ and $\mu$ are both positive;

(ii) if $Q$ is outside $P_1P_2$ and beyond $P_1$, as in (*b*), $\lambda$ is negative and $\mu$ is positive, while $\mu > -\lambda$;

(iii) if $Q$ is outside $P_1P_2$ and beyond $P_2$, as in (*c*), $\lambda$ is positive and $\mu$ is negative, while $\lambda > -\mu$.

In each case we say that $P_1P_2$ is divided by $Q$ in the ratio $\lambda : \mu$. When $Q$ lies between $P_1$ and $P_2$, it is said to divide $P_1P_2$ *internally* (and the ratio is positive); otherwise it divides $P_1P_2$ *externally* (and

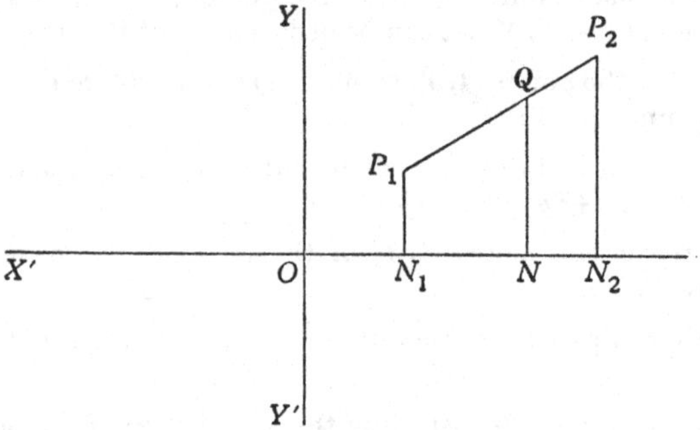


Fig. 7

the ratio is negative). It is customary for the word 'externally' to replace the minus sign in the statement about ratio. For example, the two statements

'$Q$ divides $P_1P_2$ in the ratio $-2:1$'

and '$Q$ divides $P_1P_2$ externally in the ratio $2:1$'

are regarded as equivalent.

We now obtain the coordinates $(X, Y)$ of the point $Q$ which divides $P_1P_2$ in the ratio $k_2 : k_1$. (This is the centre of gravity of a particle of mass $k_1$ at $P_1$ and a particle of mass $k_2$ at $P_2$.)

Let $N_1$, $N_2$ and $N$ be the feet of the perpendiculars from $P_1$, $P_2$ and $Q$ to $OX$ (Fig 7). By the intercept theorem

$$\frac{\overrightarrow{P_1Q}}{\overrightarrow{QP_2}} = \frac{\overrightarrow{N_1N}}{\overrightarrow{NN_2}} = \frac{X - x_1}{x_2 - X} = \frac{k_2}{k_1},$$

so that
$$\mathbf{X} = \frac{\mathbf{k_1 x_1 + k_2 x_2}}{\mathbf{k_1 + k_2}}.$$

In a similar way, dropping perpendiculars on to $Y'OY$,

$$\mathbf{Y} = \frac{\mathbf{k_1 y_1 + k_2 y_2}}{\mathbf{k_1 + k_2}}.$$

These results depend on the formulae for $\overrightarrow{N_1 N}$ and $\overrightarrow{NN_2}$, which are true for all positions of $P_1$ and $P_2$, so that the results hold for any positions of $P_1$ and $P_2$, and for all values of the ratio $k_2:k_1$.

From this, we can deduce the coordinates of the *mid-point* of $P_1P_2$. In this case, $k_1 = k_2 = 1$, so that the mid-point is

$$\{\tfrac{1}{2}(\mathbf{x_1 + x_2}), \quad \tfrac{1}{2}(\mathbf{y_1 + y_2})\}$$

The reader should write out a proof of this special case, on the lines of the one given above for the more general case.

***Illustration 3.*** *$P_1(x_1, y_1)$, $P_2(x_2, y_2)$, $P_3(x_3, y_3)$ are the vertices of a triangle. Find the coordinates of the centroid of the triangle* [G 2].

The mid-point $L$ of $P_1P_2$ has coordinates $\{\tfrac{1}{2}(x_1+x_2), \tfrac{1}{2}(y_1+y_2)\}$. If $G$ is the centroid, $LG:GP_3 = 1:2$, and so

$$X = \frac{2\{\tfrac{1}{2}(x_1+x_2)\}+x_3}{2+1} = \tfrac{1}{3}(x_1+x_2+x_3)$$

and similarly $\qquad Y = \tfrac{1}{3}(y_1+y_2+y_3)$.

(Notice the *symmetry* of this result. $G$ is similarly situated relative to each of $P_1$, $P_2$ and $P_3$, so that we should expect the coordinates of each of the points to enter on the same footing into the results, as is indeed the case.)

## EXAMPLES IC

1. Find the coordinates of the mid-point of $P_1P_2$, when (i) $P_1$ is (3, 6), $P_2$ is (5, 10); (ii) $P_1$ is $(-3, 7)$, $P_2$ is $(-6, -4)$; (iii) $P_1$ is $(a, b)$, $P_2$ is $(b, a)$.

2. Show that the lines joining $(-3, 3)$ and $(1, 5)$; $(4, 2)$ and $(-6, 6)$; $(0, 1)$ and $(-2, 7)$ all have the same mid-point.

3. Find the coordinates of the centre of a circle which has the points $(-6, 5)$ and $(3, -7)$ at the ends of a diameter.

4. $A(3, 2)$, $B(-2, 3)$ and $C(1, -5)$ are the vertices of a triangle. $A'$ is the mid-point of $BC$, $B'$ is the mid-point of $CA$ and $C'$ is the mid-point of $AB$. Find the coordinates of (i) the centroid of the triangle $ABC$; (ii) the centroid of the triangle $A'B'C'$. Is the result you obtain about these points true for *any* choice of the triangle $ABC$?

5. If $p+q+r = 0$, show that the centroid of the triangle with vertices $(ap^2, 2ap)$, $(aq^2, 2aq)$, $(ar^2, 2ar)$ lies on the $x$-axis.

6. If $A$ is the point $(4, -1)$ and the point $(3, -4)$ is the mid-point of $AB$, find the coordinates of $B$.

7. Find the fourth vertex of the parallelogram $ABCD$, when $A$ is $(4, -11)$, $B$ is $(5, 3)$ and $C$ is $(2, 15)$.

8. Find the coordinates of the point which divides $P_1(3, 2)$, $P_2(-6, 8)$ (i) internally in the ratio $1:2$; (ii) internally in the ratio $2:1$; (iii) externally in the ratio $1:2$.

9. Find the distance between the points which divide $A(4, 4)$ and $B(-3, 11)$ internally and externally in the ratio $3:4$.

10. For what value of the ratio $\lambda:\mu$ does the point which divides $A(7, 2)$, $B(3, 6)$ in that ratio lie on (i) the $x$-axis, (ii) the $y$-axis?

11. $A, B, C, D$ are any four points in the plane. $P, Q, R$ are the mid-points of $BC, CA, AB$ respectively, and $L, M, N$ are the mid-points of $AD, BD, CD$ respectively. Show that $PL$, $QM$, $RN$ all have the same mid-point.

12. A triangle $A(3, 2)$, $B(-2, 3)$, $C(1, 5)$ meets the $x$-axis in the points $L$ on $BC$, $M$ on $CA$ and $N$ on $AB$. Prove that

$$\frac{\overrightarrow{BL}}{\overrightarrow{LC}} \cdot \frac{\overrightarrow{CM}}{\overrightarrow{MA}} \cdot \frac{\overrightarrow{AN}}{\overrightarrow{NB}} = -1.$$

13. Prove the result of Example 12 for a general triangle whose vertices are $A(x_1, y_1)$, $B(x_2, y_2)$, $C(x_3, y_3)$. (This is the theorem of Menelaus [G 7].)

14. $ABC$ is the triangle of Example 12, and $O$ is the origin. $AO$ meets $BC$ in $P$. By considering $\tan\alpha$, where $\alpha$ is the angle between $AOP$ and the $x$-axis, show that $\overrightarrow{BP}/\overrightarrow{PC} = 13/17$. If $BO$ meets $CA$ in $Q$ and $CO$ meets $AB$ in $R$, show that

$$\frac{\overrightarrow{BP}}{\overrightarrow{PC}} \cdot \frac{\overrightarrow{CQ}}{\overrightarrow{QA}} \cdot \frac{\overrightarrow{AR}}{\overrightarrow{RB}} = +1.$$

15. Prove the result of Example 14 for a general triangle $ABC$. (This is Ceva's theorem [G 8].)

**4. Equation of the locus of a point.** A point $P(x, y)$ may, with suitable choice of $x$ and $y$, lie anywhere in the plane. If we impose some rule on $x$ and $y$, we limit the region of the plane where $P$ may be. For example, we have already seen that if $x$ and $y$ are both positive, $P$ lies in the first quadrant. Again, if $x = 0$, $P$ lies on $Y'OY$; or, if $y = 0$, $P$ lies on $X'OX$. In Illustration 2, there is another case of such a rule; it is given that $P$ lies on a circle with centre (3, 1) and radius 2, and it is found that $x$ and $y$ satisfy the equation $x^2+y^2-6x-2y+6 = 0$. Such an equation is called the equation of the locus of $P$; in this case it is the equation of a circle.

There are here two problems. We may be told some geometrical fact which limits the positions of $P$, and asked to find what relation is then satisfied by $x$ and $y$, the coordinates of $P$. For example, $P$ may be given as lying on a circle, that is, a fixed distance from a fixed point, or $P$ may be equidistant from two fixed points. This problem of finding an equation does not usually present many difficulties. The converse is more complicated. An equation connecting $x$ and $y$ is given, and the problem is to find out on what geometrical locus $P$ lies. This is a question of identification.

It is with one or other of these problems that coordinate geometry often deals, and the following examples illustrate both.

***Illustration 4.*** *$P(x,y)$ is equidistant from the points $A(4, -1)$ and $B(-3, 6)$. Find the equation of the locus of $P$.*

Using the distance formula (p. 13)

$$PA^2 = (x-4)^2+(y+1)^2$$

and

$$PB^2 = (x+3)^2+(y-6)^2.$$

It is given that $PA = PB$, so that $PA^2 = PB^2$, and so

$$(x-4)^2+(y+1)^2 = (x+3)^2+(y-6)^2.$$

Thus

$$x^2-8x+16+y^2+2y+1 = x^2+6x+9+y^2-12y+36,$$

which reduces to

$$14x-14y+28 = 0,$$

or

$$x-y+2 = 0.$$

This then is the equation of the locus of $P$.

As a check, we notice that the point $(\frac{1}{2}, \frac{5}{2})$, which is the mid-point of $AB$, lies on this line.

***Illustration 5.*** *$P(x,y)$ moves subject to the relation*

$$x^2+y^2-2x-3 = 0.$$

*Find what geometrical locus this represents.*

Using the experience gained when working Illustration 2, we write the given equation in the form $(x-1)^2+y^2 = 4$. Let $A$ be the point $(1,0)$; then $PA^2 = (x-1)^2+y^2$, and so $PA^2 = 4$, and $PA = 2$. The locus is therefore a circle, whose centre is $(1,0)$ and whose radius is 2.

Both problems, of finding an equation and of identification, are illustrated in the next example.

***Illustration 6.*** *A and B are two fixed points, and a point P moves so that $AP:PB = 1:k$, where $k \neq 1$. Find the locus of P.*

The problem is simplified if the axes and origin are chosen carefully, and it must be emphasized that we can choose axes to suit our convenience. Take the $x$-axis along $AB$ (thus making the

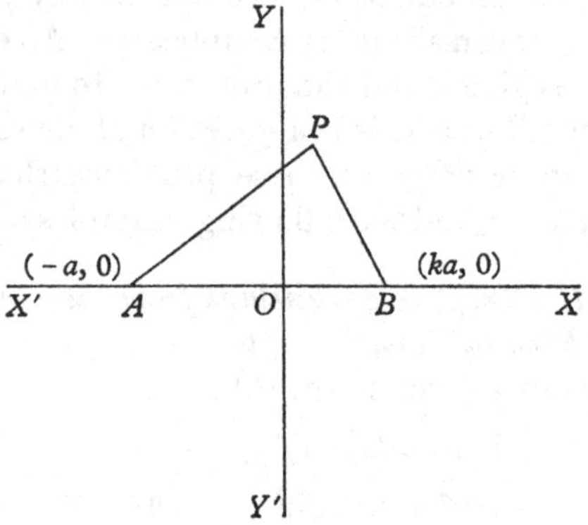


Fig. 8

coordinates of $A$ and $B$ simple), and choose as origin, $O$, the point on $AB$ and on the locus of $P$, between $A$ and $B$. The point $O$ then divides $AB$ internally in the ratio $1:k$, so that if $A$ is $(-a, 0)$, $B$ is $(ka, 0)$ (Fig. 8). Then, if $P$ is $(x, y)$,

$$AP^2 = (x+a)^2+y^2,$$
$$PB^2 = (x-ka)^2+y^2.$$

Since $k.AP = PB$, $k^2.AP^2 = PB^2$, and so

$$k^2\{(x+a)^2+y^2\} = \{(x-ka)^2+y^2\}.$$

This reduces to

$$(1-k^2)(x^2+y^2)-2kax(1+k) = 0.$$

Since $k$ is positive, $1+k \neq 0$, and we are given that $1-k \neq 0$, so the equation becomes

$$x^2+y^2-\frac{2ka}{1-k}x = 0.$$

This is the equation of the locus of $P$.

To identify it, we notice that the equation is similar in type to the one in Illustration 5. We therefore write it in the form

$$\left(x-\frac{ka}{1-k}\right)^2+y^2=\frac{k^2a^2}{(1-k)^2},$$

and this expresses the fact that the distance of $P(x, y)$ from $(ka/(1-k), 0)$ is $ka/(1-k)$. The locus of $P$ is therefore a circle with centre $(ka/(1-k), 0)$ and radius $ka/|1-k|$.

We may compare the method just used with one which does not use coordinates.

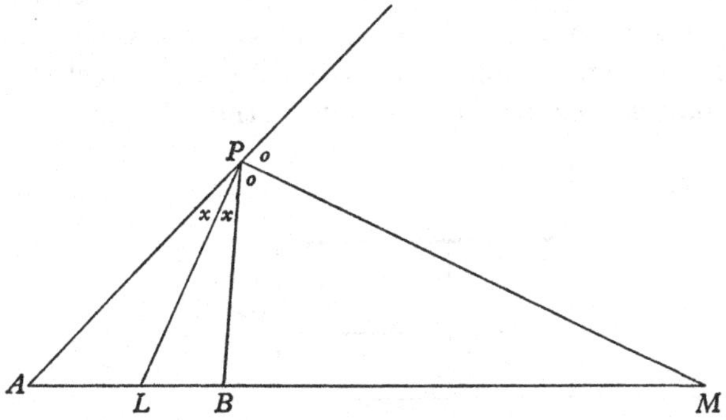


Fig. 9

Let $L$, $M$ be the points on $AB$ such that $AL:LB = 1:k$ and $AM:MB = 1:k$. $L$ and $M$ therefore divide $AB$ internally and externally in the ratio $1:k$ (Fig. 9). Then $AP:PB = AL:LB$ and $AP:PB = AM:MB$. It follows that $PL$ and $PM$ are the internal and external bisectors of the angle $APB$, so that the angle $LPM$ is a right angle, and so $P$ lies on a circle whose diameter is $LM$.

This locus is called the Circle of Apollonius, recalling a Greek geometer who lived in the third century B.C. and contributed largely to the study of geometrical conics.

## EXAMPLES I D

$A$ is the point $(1, 0)$ and $B$ is the point $(-1, 0)$. Find the equation of the locus of $P$ in each of the following cases:

1. $AP = PB$.
2. $AP^2+PB^2 = 3$.
3. $AP = 2PB$.

4. $AP+PB = 3$.

5. $AP-PB = 1$.

6–10. Repeat these examples with $A$ as $(3, -2)$ and $B$ as $(4, -1)$.

11. Identify the locus whose equation is

$$\text{(i) } x^2+y^2 = 9; \quad \text{(ii) } (x+1)^2+y^2 = 4.$$

12. Show that the locus whose equation is $x^2+y^2-12x+4y-9 = 0$ represents a circle with radius 7 and centre $(6, -2)$.

13. Consider what happens to the locus of $P$ in Illustration 6, if $k = 1$.

**5*. Change of origin.** It is sometimes convenient, particularly in later work, to make a change in the coordinate system, measuring distances from a new pair of axes. One change that can be made is to use new axes parallel to the old ones but through

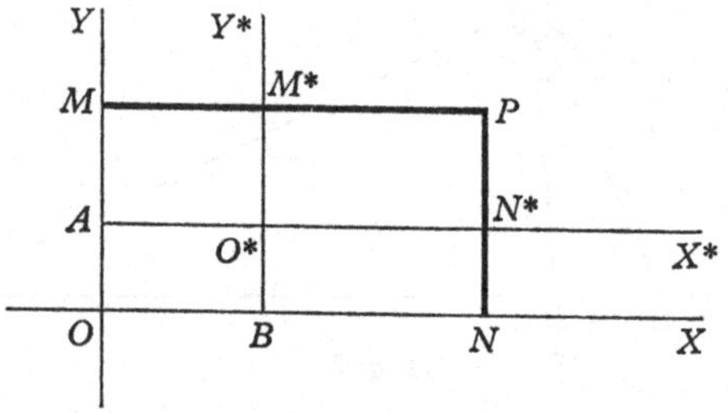


Fig. 10

a different point $O^*$. The formulae relating the old and the new coordinates are very simple.

Suppose $P$ is the point $(x, y)$ referred to axes $OX$, $OY$, and let $O^*$ be the point $(\lambda, \mu)$ referred to these axes. Take new axes $O^*X^*$, $O^*Y^*$ through $O^*$ parallel to $OX$, $OY$, meeting $OY$, $OX$ in $A$, $B$ respectively (Fig. 10). Draw perpendiculars from $P$ to $OX$, $OY$ meeting then in $N$, $M$ and meeting $O^*X^*$, $O^*Y^*$ in $N^*$, $M^*$. Let the coordinates of $P$ referred to the new axes be $(x^*, y^*)$; that is, $\overrightarrow{O^*N^*} = x^*$ and $\overrightarrow{O^*M^*} = y^*$. Then

$$x = \overrightarrow{ON} = \overrightarrow{BN}+\overrightarrow{OB} = \overrightarrow{O^*N^*}+\overrightarrow{OB}$$
$$= x^*+\lambda,$$

and similarly $y = y^*+\mu$.

If, therefore, an equation in $x, y$ is given, and if it is desirable to refer to new axes through $(\lambda, \mu)$, we substitute $x^*+\lambda$ for $x$ and

$y^* + \mu$ for $y$ and the resulting equation between $x^*$ and $y^*$ will be the new equation of the locus.

For example, the equation $(x-4)^2+(y-2)^2 = 1$ can be simplified by putting $x = x^*+4, y = y^*+2$, so changing the origin to $O^*$ (4, 2). The new equation is $x^{*2}+y^{*2} = 1$, representing a circle of radius 1 with centre $O^*$.

## EXAMPLES I E

Simplify the following equations by changing to a new origin in each case:

1. $x^2+y^2-6x-10y-2 = 0$; new origin (3, 5).
2. $x^2+y^2+2x-4y+1 = 0$; new origin $(-1, 2)$.
3. $y^2-4ax+2ay+5a^2 = 0$; new origin $(a, -a)$.
4. $x^2+2y^2-6x+16y+39 = 0$; new origin $(3, -4)$.
5. $x^3+3x^2+3x-y+5 = 0$; new origin $(-1, 4)$.

## MISCELLANEOUS EXAMPLES I

1. Prove that the points (0, 4), (3, 7), (4, 4) and (1, 1) are the four vertices of a parallelogram.

2. A triangle $ABC$ is formed by joining the points $(-2, 1)$, (2, 3), $(2, -2)$. Sketch the figure and find the coordinates of its centroid.

3. A triangle has vertices at the points (1, 1), (4, 1) and (3, 4). Find the coordinates of its circumcentre.

4. $A$ is the point (2, 3) and $B$ is the point (0, 1). The angle $BAC$ is a right angle and $BC$ is 5 units of length. Find the coordinates of the two possible positions of $C$.

5. Two vertices of a triangle are the points (25, 2), $(10, -10)$ and the centroid of the triangle is the point (7, 4). Find the coordinates of the third vertex and show that the triangle is right-angled.

6. The coordinates of the points $A$, $B$, $C$ are $(-2, 1)$, (2, 7), (5, 5) respectively. Prove that these points form three corners of a rectangle, and that $AB = 2BC$. If $D$ is the fourth corner of the rectangle, calculate the distance of $C$ from the diagonal $BD$.

7. Prove that the line joining the origin $O$ to the point $P$ lying between $A(11, 2)$ and $B(3, 6)$, such that $AP/PB = 5/3$, bisects the angle $AOB$. Prove also that the line through $O$ perpendicular to $OP$ cuts $AB$ externally in the same ratio.

8. Find the equation of the locus of a point which moves so that its distance from the point (2, 0) is three times its distance from the axis of $y$.

9. A point moves so that its distance from the point $(-1, -1)$ is equal to its distance from the point $(3, 5)$. Find the equation of its locus.

10. $O$ is the origin and $B$ the point $(2, 0)$. $P$ is a variable point which moves so that the projection of $OP$ on the $y$-axis is three times the projection of $BP$ on the $x$-axis. Find the equation of the locus of $P$.

11. $A$ is the point $(1, 1)$ and $B$ is the point $(3, 5)$. Find the equation of the locus of a point $P$ which moves so that:

(i) $PA$ is equal to $PB$;

(ii) the angle $APB$ is a right angle;

(iii) $AP^2 + BP^2 = 60$.

12. A point $P$ moves so that its distance from the point $(2, 3)$ is equal to the length of a tangent from $P$ to a circle with centre the origin and radius 4. Find the equation of the locus of $P$.

---

13. $E$ and $F$ are points of the sides $AD$, $BC$ of a quadrilateral $ABCD$ such that $AE = k.ED$ and $BF = k.FC$. If $P$, $Q$, $R$ are the mid-points of $AB$, $EF$, $DC$ respectively, prove that $P$, $Q$, $R$ are collinear and that $PQ = k.QR$.

14. $A, B, C, D$ are four points $(x_1, y_1)$, $(x_2, y_2)$, $(x_3, y_3)$, $(x_4, y_4)$. $F$ is the mid-point of $AB$, $G$ lies on $CF$ so that $CG : GF = 2 : 1$, and $H$ lies on $DG$ so that $DH : HG = 3 : 1$. Find the coordinates of $H$, and show that it is the mid-point of the line joining $F$ to the mid-point of $CD$.

# II

## THE LINE

The simplest locus of a point in a plane is a straight line; in future in this book we shall call it simply a line. In this chapter we obtain the equation of a line, in various forms, and prove a number of basic results.

**1. Gradient of a line.** If any two points on a line are selected, their join makes a constant angle with a fixed direction, and the angle is independent of the two particular points selected on the line. This is a precise way of saying that any line has a constant slope. It is customary to measure the angle $\alpha$ which a line makes with the $x$-axis, and then $\tan\alpha$ is called the *gradient* of the line.

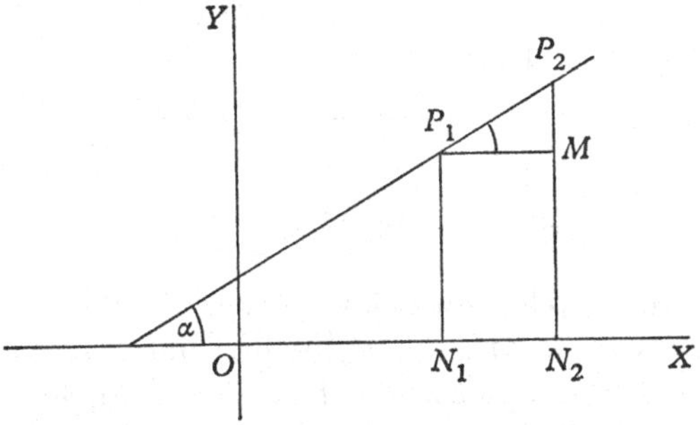


Fig. 11

To find the gradient of the line joining $P_1(x_1, y_1)$ and $P_2(x_2, y_2)$, draw lines $P_1N_1$, $P_2N_2$ perpendicular to $OX$, and let the perpendicular from $P_1$ to $P_2N_2$ meet it in $M$ (Fig. 11). If $P_1P_2$ makes an angle $\alpha$ with $OX$, angle $P_2P_1M = \alpha$, and

$$\tan\alpha = \frac{\overrightarrow{MP_2}}{\overrightarrow{P_1M}} = \frac{y_2-y_1}{x_2-x_1},$$

using the result of p. 13, and [T 1]. This then is the gradient of $P_1P_2$; it will hold for any positions of $P_1$, $P_2$ in the plane.

We notice that lines which make an acute angle with the positive direction of $OX$ have a positive gradient, while lines which make an obtuse angle with $OX$ have a negative gradient.

**2. Parallel and perpendicular lines.** A set of parallel lines all make the same angle $\alpha$ with the $x$-axis, and so all the lines have the same gradient. Conversely, if two lines have the same gradient, they are parallel. The condition for lines with gradients $m$ and $m'$ to be parallel is therefore

$$\mathbf{m = m'}.$$

In particular, all lines parallel to $OX$ have zero gradient, for $\alpha = 0$ and so $\tan\alpha = 0$. For lines parallel to $OY$, $\alpha = \frac{1}{2}\pi$, and so $\tan\alpha$ is infinite [T 2].

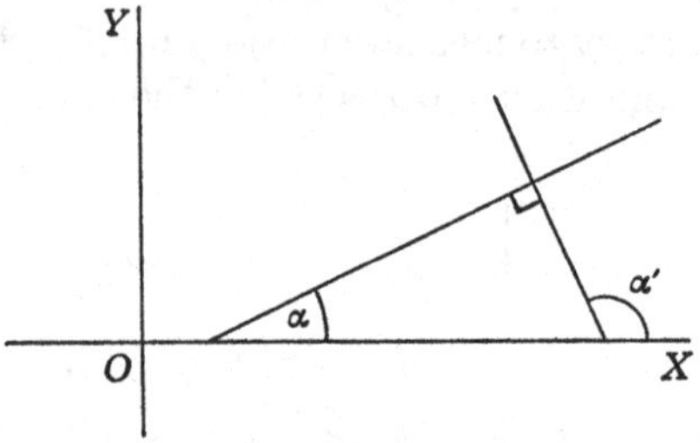


Fig. 12

Now consider two lines which are perpendicular, and suppose their gradients are $m$ and $m'$. If the first line makes an angle $\alpha$ with $OX$ (Fig. 12), and the second line makes an angle $\alpha'$ with $OX$, then $\alpha' = \alpha + \frac{1}{2}\pi$. So

$$\tan\alpha' = \tan(\alpha + \tfrac{1}{2}\pi)$$

$$= -\cot\alpha \quad [\text{T } 3],$$

$$= -1/\tan\alpha.$$

Therefore since $\tan\alpha = m$ and $\tan\alpha' = m'$, the relation between the gradients is

$$m' = -1/m,$$

or

$$\mathbf{mm' = -1}.$$

Conversely, if this relation is satisfied, the lines with gradients $m$ amd $m'$ are perpendicular.

**3. Equation of a line.** If a line passes through $A(h, k)$ and has gradient $m$ it is uniquely determined, and so the equation satisfied by the coordinates of any point $P(x, y)$ on it can be found. Using the result of §1, the gradient of $AP$ is

$$(y-k)/(x-h),$$

and this is given as $m$. Therefore

$$\frac{y-k}{x-h} = m;$$

that is
$$y-k = m(x-h).$$

This is the equation satisfied by the coordinates of any point $P$ on the line; it is called the *equation of the line.*

The *form* of this equation is important. It can be written

$$mx-y+(k-mh) = 0,$$

and we see that it contains a term in $x$ (namely, $mx$), a term in $y$, $(-y)$, and a constant term, $(k-mh)$, which does not involve $x$ or $y$. There are no other terms, like $x^2$ or $xy$. An equation of this type is called a *linear* equation.

Conversely, we consider the general linear equation

$$ax+by+c = 0,$$

and prove that (as its name suggests) it always represents a line. Since the equation gives a relation satisfied by some, but not all, values of $x$ and $y$, not both of $a$ and $b$ can be zero. If $a = 0$, and $b \neq 0$, the equation is

$$y = -c/b;$$

this is satisfied by all the points with a fixed ordinate, $-c/b$, on a line parallel to $OX$. Similarly, if $b = 0$ and $a \neq 0$, the equation is

$$x = -c/a.$$

and represents a line parallel to $OY$.

In the general case, when $a \neq 0$ and $b \neq 0$, suppose $(x_1, y_1)$ and $(x_2, y_2)$ are two points whose coordinates satisfy the equation, so that

$$ax_1+by_1+c = 0$$

and
$$ax_2+by_2+c = 0.$$

Subtracting,
$$a(x_2-x_1)+b(y_2-y_1) = 0,$$

and so
$$\frac{y_2-y_1}{x_2-x_1} = -\frac{a}{b}.$$

It follows that the gradient of the line joining any two points on the locus has the constant value $-a/b$, and so the locus is that of a line.

We shall often speak briefly of 'the line $ax+by+c = 0$' instead of 'the line whose equation is $ax+by+c = 0$'.

## EXAMPLES II A

1. Find the gradients of the lines joining the following pairs of points; $(4, 2)$ and $(-3, 7)$, $(6, 1)$ and $(5, 3)$, $(5, -4)$ and $(-7, -4)$, $(-1, 4)$ and $(8, 3)$, $(a, b)$ and $(b, a)$.

2. Show that the line joining $(1, 2)$ to $(3, -2)$ is perpendicular to the line joining $(-2, 0)$ to $(0, 1)$.

3. Which of the following points lie on the line $3x+2y-6 = 0$: $(3, -1)$, $(2, 0)$, $(1, 1)$, $(0, 3)$, $(-2, 6)$, $(-6, -6)$?

4. Find the point on the line $4x-y+2 = 0$, where (i) $x = 0$, (ii) $y = 6$, (iii) $x = -3$, (iv) $x = 1$, (v) $y = -9$, (vi) $x = \frac{3}{4}$, (vii) $y = \frac{4}{5}$.

5. If the line $ax+by+c = 0$ passes through the origin, what algebraic condition must be satisfied?

6. What is the gradient of the line $ax+by+c = 0$? Show that the lines $ax+by+c = 0$ and $a'x+b'y+c' = 0$ are (i) parallel, if and only if $ab' = a'b$; (ii) perpendicular, if and only if $aa'+bb' = 0$.

7. In each of the following cases, find the equation of the line through the point $A$ with gradient $m$:

(i) $A$ is $(0, 4)$, $m = \frac{1}{2}$; (ii) $A$ is $(7, 4)$, $m = 3$;
(iii) $A$ is $(-1, 6)$, $m = 2$; (iv) $A$ is $(4, -3)$, $m = -1$;
(v) $A$ is $(4, 0)$, $m = -\frac{1}{3}$; (vi) $A$ is $(a, 0)$, $m = 1$.

8. In each of the following cases, find the equation of the line joining $A$, $B$:

(i) $A$ is $(1, 5)$ and $B$ is $(2, -3)$;
(ii) $A$ is $(7, -1)$ and $B$ is $(-1, -4)$;
(iii) $A$ is $(0, 5)$ and $B$ is $(-2, 1)$;
(iv) $A$ is $(8, 11)$ and $B$ is $(-3, 11)$;
(v) $A$ is $(-1, -1)$ and $B$ is $(-1, 7)$;
(vi) $A$ is $(a, b)$ and $B$ is $(b, a)$.

9. Prove that the points $(-1, -2)$, $(2, 1)$, $(1, 4)$ and $(-2, 1)$ form a parallelogram, and find the equations of its diagonals.

10. Find the equation of the line joining $(4, 1)$ and $(7, -1)$. Prove that the point $(1, 3)$ lies on it, and find where it meets the axes.

11. What is the gradient of the line $4x-3y+5 = 0$? Find the equations of the lines through $(4, -1)$ parallel to it, and perpendicular to it.

12. Find the equations of the medians of the triangle whose vertices are $(3, 1)$, $(-1, 7)$ and $(1, -3)$.

13. Find the equation of the line with gradient 2 which meets the $x$-axis in the same point as the line $2x+3y+4=0$.

14. Three vertices of a parallelogram $ABCD$ are $A(a, 0)$, $B(-a, 0)$ and $C(b, c)$. Find the coordinates of $D$.

15. The vertices of a triangle are $(0, -2)$, $(2, 0)$ and $(3, -3)$. Prove that it is isosceles, and find the equation of the median which bisects the shortest side. Verify that this line is perpendicular to the side it bisects.

16. Show that the quadrilateral with vertices $(3, 4)$, $(-2, -1)$, $(1, -4)$ and $(6, 1)$ is a rectangle. Find the equations of the diagonals, and verify that the mid-point of either lies on the other.

17. If the line $ax+5y+3=0$ is parallel to the line $7x-10y+2=0$, what is the value of $a$?

18. Show that the points $(a, 0)$, $(at^2, 2at)$, $(-a, 2at)$ and $(-at^2, 0)$ form a rhombus, and verify that its diagonals bisect each other at right angles.

19. Find the equation of the tangent at the point $(3, -4)$ to the circle with centre $(0, 0)$ and radius 5.

20. Show that if $x = 3-t$, $y = 1+2t$, the point $(x, y)$ lies on the line $2x+y=7$.

**4. Special forms of the equation of the line.** There are a number of useful alternative forms in which the equation of a line may be written. The form chosen in any particular case will depend either on what information is given about the line, or on what is to be found out about it. For convenience of reference, all the alternative forms are gathered together in this paragraph; a beginner may prefer to acquire the information more slowly, and if so, sections VI and VII may be deferred until practice has familiarized the others.

The first two forms have already been found in § 3.

I. *The standard form*, $\mathbf{ax+by+c=0}$.

The equation of any line can be written in this form.

II. *The gradient form*, $\mathbf{y-k=m(x-h)}$.

This form is useful because it can be written down at once if the gradient $m$ and one point $(h, k)$ on the line are known. Conversely, from this form the gradient of the line and the coordinates of a particular point on it can be seen at a glance. The equation of a line parallel to the $y$-axis, however, cannot be put into this form.

III. *The gradient—intercept form.*

This standard form is obtained when the point given on the line is $(0, c)$. That is, the *intercept* $c$ which the line makes on the $y$-axis is given, and also the gradient $m$. Then, as in II, the equation takes the form

$$y - c = m(x - 0),$$

or

$$\mathbf{y = mx + c.}$$

When the equation is written in this form (which is possible for all lines except those parallel to $OY$), the gradient $m$ and the intercept $c$ on $OY$ can be found by inspection.

IV. *The line joining two points.*

If $P_1(x_1, y_1)$ and $P_2(x_2, y_2)$ lie on the line, its gradient is

$$\frac{y_2 - y_1}{x_2 - x_1} \quad (\S 1).$$

Substituting this value for $m$ in the gradient form, we obtain the equation of the line as

$$\mathbf{y - y_1 = \frac{y_2 - y_1}{x_2 - x_1}(x - x_1).}$$

Some people prefer to remember this in the more symmetrical form

$$\frac{x - x_1}{x_2 - x_1} = \frac{y - y_1}{y_2 - y_1}.$$

V. *The intercept form.*

The equation of the line through the points $A(a, 0)$ and $B(0, b)$ is (applying the result just obtained)

$$y - 0 = \frac{b - 0}{0 - a}(x - a).$$

This reduces to

$$\mathbf{\frac{x}{a} + \frac{y}{b} = 1.}$$

This is called the *intercept form*, because it can be written down at once if the lengths of the intercepts $a$ on the $x$-axis and $b$ on the $y$-axis are known. Conversely, from the equation in this form the values of $a$ and $b$ can be seen at once. The equations of all lines other than those through the origin can be written in this form.

VI. *The normal form.*

In this form, the perpendicular distance $p$ from the origin to the line is given, and also the angle $\alpha$ which this perpendicular, or *normal*, makes with $OX$ (Fig. 13). For preciseness we take $p$ to be positive, and measure $\alpha$ as shown in the diagram, so that $\alpha$ lies between $0°$ and $360°$. (This corresponds to the usual convention for measuring angles in trigonometry.) The intercepts on the

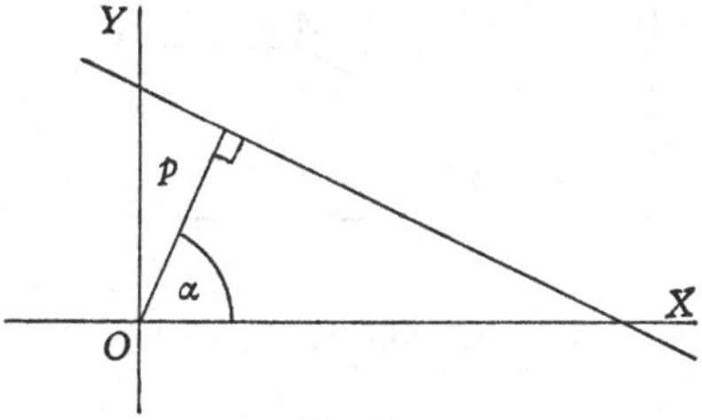


Fig. 13

$x$-axis and the $y$-axis are $p/\cos\alpha$ and $p/\sin\alpha$, and substituting these values in the intercept form we obtain the equation of the line as

$$\frac{x\cos\alpha}{p}+\frac{y\sin\alpha}{p}=1,$$

or

$$\mathbf{x\cos\alpha+y\sin\alpha=p.}$$

This form of the equation is found useful when calculating the distance of a point from a line (see p. 47). The equation of any line not through the origin can be put into this form, as we shall see in Illustration 1.

VII. *The parametric form.*

It is frequently useful in coordinate geometry to express the coordinates of a variable point on a line or on a curve in terms of a single variable called a *parameter*. Two properties must be satisfied:

(i) when a point on the locus is given, the corresponding unique value of the parameter can be found;

(ii) when a value of the parameter is given, there is corresponding to it just one point on the locus.

Consider a line through the point $A(h, k)$ whose gradient is $\tan\alpha$. Let $P(x, y)$ be a point on the line and suppose that $AP = r$

(Fig. 14). If we allow $r$ to be either positive or negative, $P$ will take any position on the line, and conversely when $P$ is given, the value of $r$ (which is its distance from $A$ with the appropriate sign) can be found. It follows that $r$ serves as the *parameter* of $P$, in accordance with the definition.

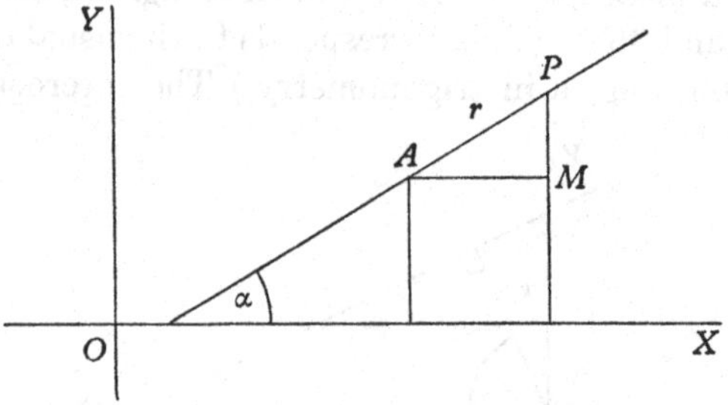


Fig. 14

To find the coordinates of $P$ in terms of $r$, suppose the perpendicular from $A$ to the ordinate through $P$ meets it in $M$. Then in the right-angled triangle $PAM$,

$$\overrightarrow{AM} = r\cos\alpha, \quad \overrightarrow{MP} = r\sin\alpha, \quad [\text{T } 4].$$

But $$\overrightarrow{AM} = x-h, \quad \overrightarrow{MP} = y-k,$$

and so $$x-h = r\cos\alpha, \quad y-k = r\sin\alpha;$$

or $$\mathbf{x = h + r\cos\alpha, \quad y = k + r\sin\alpha.}$$

These equations serve to determine the coordinates of any point $P$ on the line, in terms of the variable parameter $r$. They are often called the 'freedom equations' of the line. The uses of this form will be apparent later, particularly in the last chapters of this book.

## VIII. *The determinantal form.*

Those who are familiar with determinants may use the equation of the line joining $(x_1, y_1)$ and $(x_2, y_2)$ in the form

$$\begin{vmatrix} x & y & 1 \\ x_1 & y_1 & 1 \\ x_2 & y_2 & 1 \end{vmatrix} = 0.$$

This is a linear equation in $x$ and $y$, and is evidently satisfied when $(x, y)$ is $(x_1, y_1)$ or $(x_2, y_2)$, so that it represents the line through these two points.

***Illustration 1.*** *The equation of a line is* $3x-4y+7=0$. *Write this in (i) the gradient form* $y=mx+c$, *(ii) the intercept form, (iii) the normal form. Hence give the gradient of the line, the intercepts it makes on the axes, and its perpendicular distance from the origin.*

(i) To put the equation in the gradient form, we arrange the terms so that the term in $y$ is alone on the left-hand side of the equation and the other terms are on the right. This gives

$$4y = 3x+7,$$

or

$$y = \tfrac{3}{4}x+\tfrac{7}{4}.$$

From this form we see that the gradient of the line is $\frac{3}{4}$, and that it cuts an intercept $\frac{7}{4}$ on the $y$-axis.

(ii) For the equation in the intercept form, it must be expressed with the constant term alone on the right-hand side of the equation. We have then

$$-3x+4y = 7;$$

that is

$$\frac{x}{-\frac{7}{3}}+\frac{y}{\frac{7}{4}} = 1.$$

This is the intercept form; comparing it with $x/a+y/b=1$ we see that the line cuts intercepts $-\frac{7}{3}$ and $\frac{7}{4}$ on the $x$-axis and $y$-axis respectively.

(iii) Lastly, to put the equation in normal form, we again arrange the terms with the constant term alone on one side of the equation, and since $p$ is to be positive, we write it

$$-3x+4y = 7.$$

For any value of a constant $k$, other than zero, this is the same as

$$-3kx+4ky = 7k.$$

We now choose $k$ so that

$$\cos\alpha = -3k$$

and

$$\sin\alpha = 4k.$$

Now

$$\cos^2\alpha+\sin^2\alpha = 1,$$

so that

$$(-3k)^2+(4k)^2 = 1,$$

which gives

$$25k^2 = 1$$

and

$$k = \pm\tfrac{1}{5}.$$

Taking $k = +\frac{1}{5}$ (in order to keep $p$ positive), we obtain

$$(-\tfrac{3}{5})x + (\tfrac{4}{5})y = \tfrac{7}{5}$$

as the equation in the required form.

From this we deduce that the line is at a perpendicular distance of $\frac{7}{5}$ from the origin.

(The device used to obtain the coefficients of $x$ and $y$ as the cosine and sine of an angle $\alpha$ will, with experience, be shortened. The reader may have recognized that the right-angled triangle with sides of length 3 and 4 will have a hypotenuse of length 5, so that $\cos\alpha$ and $\sin\alpha$ can be written down at sight. Any short cut of this kind (provided it is accurate) is desirable).

***Illustration 2.*** *A line has equation $ax+by+c = 0$. Show that any line parallel to it has equation $ax+by+k = 0$, and that any line perpendicular to it has equation $bx-ay+k' = 0$, where $k$ and $k'$ are any constants.*

The equation of the given line is

$$ax+by+c = 0,$$

or, in gradient form,

$$y = -\frac{a}{b}.x - \frac{c}{b},$$

so that the gradient of the line, and also of any line parallel to it, is

$$-\frac{a}{b}.$$

It follows that the equation of a line parallel to the given one is

$$y = -\frac{a}{b}.x + c_1,$$

where $c_1$ is any constant; this is

$$ax+by+k = 0,$$

where $k = -bc_1$ and is also an arbitrary constant.

If the gradient of a line perpendicular to the given one is $m$, the condition for perpendicularity (p. 26) gives

$$-\frac{a}{b}.m = -1,$$

or

$$m = \frac{b}{a}.$$

The equation of a perpendicular line is then

$$y = \frac{b}{a}x + c_2,$$

where $c_2$ is any constant; this is

$$bx - ay + k' = 0,$$

where $k' = ac_2$, and again is an arbitrary constant.

These two results are extremely useful and are worth remembering; in each case it should be noticed that the particular value of the constant in any example is found by expressing the fact that the line passes through a particular point.

***Illustration 3.*** *ABC is a triangle of which AD, BE, CF are the altitudes. Prove by the use of coordinate geometry that the altitudes are concurrent.*

(A proof of this result by coordinates has no particular merit, except that it serves as an illustration of the methods available.)

In this problem we are at liberty to choose the origin and axes so as to simplify the problem. After the diagram (Fig. 15) is drawn, an obvious choice appears, and we take $D$ as origin with $BC$ and $AD$ as axes of $x$ and $y$. Suppose then that $A$ is $(0, a)$, $B$ is $(b, 0)$ and $C$ is $(c, 0)$. (As the diagram is drawn, $b$ will be negative, but this does not matter.)

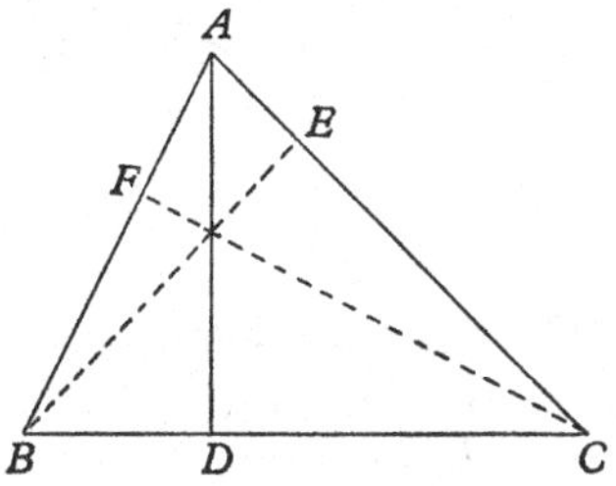

Fig. 15

The equation of $AC$ (intercept form) is

$$\frac{x}{c} + \frac{y}{a} = 1.$$

The equation of $BE$, perpendicular to $AC$ and through $B$, is then (using the result of Illustration 2),

$$\frac{x}{a} - \frac{y}{c} = \frac{b}{a},$$

or

$$cx - ay = bc.$$

Similarly, the equation of $CF$ (replacing $B$ by $C$, and so $b$ by $c$ in the above result) is

$$bx - ay = bc.$$

These both meet $AD$, $x = 0$, in the point $H(0, -bc/a)$, and so the three altitudes are concurrent.

**5. The sign of the expression $ax+by+c$.** We already know that the line $OX$, $y = 0$, divides the plane into two parts, one above $OX$ where $y > 0$ and one below $OX$ where $y < 0$. The boundary between these two regions is the line $y = 0$. In the same way, any line $ax+by+c = 0$ divides the plane into two regions, in one of which the expression $ax+by+c$ is positive and in the other of which $ax+by+c$ is negative.

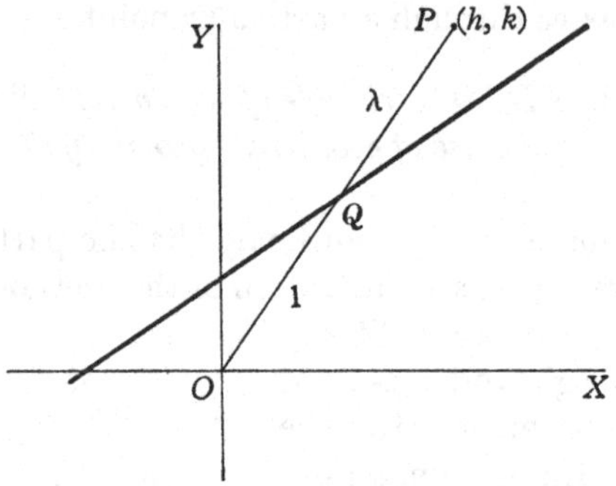


Fig. 16

Consider for definiteness the line $2x-3y+1 = 0$. When $(x, y)$ is at the origin $(0, 0)$, the expression $2x-3y+1$ is positive. Let $P(h, k)$ be another point of the plane, and let $OP$ meet $2x-3y+1 = 0$ in a point $Q$ such that $OQ:QP = 1:\lambda$ (Fig. 16). Then $Q$ is the point $\left(\frac{h}{1+\lambda}, \frac{k}{1+\lambda}\right)$ and since $Q$ lies on the line,

$$2h-3k+(1+\lambda) = 0,$$

$$\lambda = -(2h-3k+1).$$

Now $O$ and $P$ will be on *opposite* sides of the line (as in the diagram) if $Q$ lies between them and $\lambda$ is positive. For this, $2h-3k+1$ must be negative. Similarly, $O$ and $P$ are on the *same* side of the line if $Q$ divides $OP$ externally, and so $\lambda$ is negative and $2h-3k+1$ is positive.

This proves that the line $2x-3y+1 = 0$ divides the plane into two regions, and any other line can be treated in the same way. It should be noticed that we cannot attach the terms 'positive' and 'negative' infallibly to the two sides of the line. In the case considered above, for example, the equation could be written $-2x+3y-1 = 0$, and then the positive and negative regions of the plane would be interchanged.

## EXAMPLES II B

1. Find the gradient of each of the following lines, and the intercepts they make on the axes:

(i) $3x+2y = 6$; (ii) $2x-7y+3 = 0$;

(iii) $x-y+1 = 0$; (iv) $x+3y-k = 0$;

(v) $-3x+5y+2 = 0$; (vi) $x+y+3 = 0$.

2. The line $lx+my = 1$ meets the axes in the points $(3, 0)$ and $(0, -1)$. Find the values of $l$ and $m$.

3. From the following lines, pick out (i) any that are parallel, (ii) any that are perpendicular:

(*a*) $2x-y+4 = 0$; (*b*) $x+3y+7 = 0$;

(*c*) $4x-2y+3 = 0$; (*d*) $x+2y-1 = 0$;

(*e*) $ax+2ay-4 = 0$; (*f*) $3x-y+8 = 0$;

(*g*) $4x+12y+1 = 0$; (*h*) $y = 2x-1$;

(*k*) $y = 3x$.

4. Derive the equation of section IV (p. 30), by using the fact that if $P_1(x_1, y_1)$, $P_2(x_2, y_2)$ and $P(x, y)$ are collinear, the gradient of $P_1P$ is the same as the gradient of $P_1P_2$.

5. Write down the equations of the lines whose intercepts on the axes are $\frac{3}{2}, \frac{3}{4}$; $2, -1$; $-1, -1$; $-3, \frac{4}{3}$; $1/l, 1/m$. What are their gradients?

6. Find the equation of the line through the point $(3, -2)$ such that the intercept on the $x$-axis is equal to the intercept on the $y$-axis but opposite in sign.

7. Find the equation of the line through the origin perpendicular to the line joining the points $(3, -6)$ and $(4, 5)$.

8. A square is drawn on the line joining $(4, 0)$, $(0, 3)$ on the side away from the origin. Find the equations of the lines forming the square.

9. What is the geometrical relationship of the line $ax+by = 1$ to each of the following lines: (i) $ax+by = 0$; (ii) $bx-ay = 0$; (iii) $ax-by = 0$?

10. The coordinates of two points $A$ and $B$ are $(3, 7)$ and $(-1, -5)$. Find the equation of $AB$, and of the perpendicular bisector of $AB$.

11. Find the equation of the line which is parallel to the line $3x+4y = 12$ and which makes an intercept of 5 units on the $x$-axis. Find also the

equation of the line which is perpendicular to the given line and which passes through the point $(4, 5)$.

**12.** Find the equation of the locus of a point which moves so that it is equidistant from $A(2, -3)$ and $B(-1, 4)$. Show analytically that the locus is a line bisecting $AB$ at right angles.

**13.** Derive the equation of a line in normal form, by using the fact that its gradient is $-1/\tan\alpha$ and it contains the point $(p\cos\alpha, p\sin\alpha)$.

**14.** Write down the equations of the lines for which the perpendicular distance $p$ from the origin and the angle $\alpha$ which this normal makes with $OX$ are given as follows:

(i) $p = 3, \alpha = 30°$; (ii) $p = 2, \alpha = 210°$;

(iii) $p = 4, \alpha = 315°$; (iv) $p = 1, \alpha = 135°$;

(v) $p = \frac{3}{4}, \tan\alpha = -\frac{4}{3}$; (vi) $p = 2, \tan\alpha = \frac{1}{2}$.

Find the gradient $m$ of each of these lines, and verify that in each case $m.\tan\alpha = -1$.

**15.** Write each of the following equations in normal form:

(i) $3x+4y-5 = 0$; (ii) $2x-y-4 = 0$;

(iii) $5x-7y+12 = 0$; (iv) $x+2y+3 = 0$;

(v) $x+y+1 = 0$; (vi) $ax+by+c = 0$.

Hence find their distances from the origin.

**16.** $P$ is the point $(3+2t, 1-t)$. Find the equation of the line on which $P$ lies as $t$ varies, and its gradient.

**17.** Find the coordinates of the two points which lie on the line $x+2y = 5a$ and whose coordinates are $x = at^2$, $y = 2at$.

**18.** The vertices of a triangle are $A(-3, 2)$, $B(7, 6)$, $C(1, -2)$, and $E$, $F$ are the mid-points of $AC$, $AB$ respectively. Prove that $EF$ is parallel to $BC$. Prove also that if $E'$ divides $AC$ in the ratio $\lambda:\mu$ and $F'$ divides $AB$ in the ratio $\lambda:\mu$, then $E'F'$ is parallel to $BC$ for all values of $\lambda:\mu$.

**19.** A line through $(1, -3)$ has gradient $\frac{3}{4}$. Write down the general point $P$ on the line, in terms of the distance $r$ of $P$ from $(1, -3)$. For what value of $r$ does $P$ lie on (i) $OX$, (ii) $OY$, (iii) the line $x = y$? Give the coordinates of $P$ in each case.

**20.** What is the equation of the line whose general point is

$$\left(3-\tfrac{1}{2}r, 4+\frac{\sqrt{3}}{2}r\right)?$$

Find the coordinates of the points where the line meets the circle whose equation is $x^2+y^2 = 25$.

**6. Point of intersection of two lines.** We have already found where a given line meets the axes, namely the lines whose equations are $x = 0$ and $y = 0$. More generally, given the equations of any two lines, we can find the coordinates of the point which lies on both, by solving the two equations simultaneously. This is always possible, unless either the two lines are parallel, or the two equations which are given represent the same line.

For example, if the lines are

$$x+2y = 5,$$

and

$$2x+y = 7,$$

to find the point of intersection we solve these two equations. Eliminating $x$,

$$3y = 3, \quad y = 1;$$

and so

$$x+2 = 5, \quad x = 3.$$

It follows that the two lines meet in the point (3, 1). It is prudent always to check that the point found does lie on both lines.

Typical examples in which a solution cannot be found are $x+y-2 = 0$, $3x+3y-6 = 0$; and $x-y-1 = 0$, $x-y-2 = 0$. The reader should in each case decide what makes the unique solution impossible.

More generally, the same method may be applied to the two lines

$$a_1x+b_1y+c_1 = 0$$

and

$$a_2x+b_2y+c_2 = 0.$$

Elimination of $y$ gives

$$x = \frac{b_1c_2-b_2c_1}{a_1b_2-a_2b_1}$$

and elimination of $x$ gives

$$y = -\frac{a_1c_2-a_2c_1}{a_1b_2-a_2b_1}.$$

(Here $a_1b_2-a_2b_1$ must not be zero; the vanishing of this expression is the condition for the two lines to be either parallel or coincident.)

The results may alternatively be expressed in the form

$$\frac{x}{b_1c_2-b_2c_1} = \frac{-y}{a_1c_2-a_2c_1} = \frac{1}{a_1b_2-a_2b_1}.$$

Those familiar with the notation of determinants [A 1] will recognize that this may be written

$$\frac{x}{\begin{vmatrix} b_1 & c_1 \\ b_2 & c_2 \end{vmatrix}} = \frac{-y}{\begin{vmatrix} a_1 & c_1 \\ a_2 & c_2 \end{vmatrix}} = \frac{1}{\begin{vmatrix} a_1 & b_1 \\ a_2 & b_2 \end{vmatrix}}.$$

A knowledge of this result, applied systematically and accurately in all cases except when the coefficients are so small that the solution of the equations is easy to guess, makes it easier to avoid slips. In the example above, the equations are

$$x + 2y - 5 = 0,$$

$$2x + y - 7 = 0.$$

The solution is

$$\frac{x}{\begin{vmatrix} 2 & -5 \\ 1 & -7 \end{vmatrix}} = \frac{-y}{\begin{vmatrix} 1 & -5 \\ 2 & -7 \end{vmatrix}} = \frac{1}{\begin{vmatrix} 1 & 2 \\ 2 & 1 \end{vmatrix}},$$

that is

$$\frac{x}{-14+5} = \frac{-y}{-7+10} = \frac{1}{1-4},$$

so that

$$x = 3, \quad y = 1.$$

**7. Line through the point of intersection of two lines.** Consider the problem of finding the equation of the line joining the point $(1, -1)$ to the point of intersection of the lines $x + 2y - 5 = 0$ and $2x + y - 7 = 0$. We could find the coordinates of the point of intersection $(3, 1)$, as was done in the last paragraph, and then write down the equation of the line joining $(1, -1)$ and $(3, 1)$, as

$$y + 1 = \frac{1-(-1)}{3-1}(x-1),$$

or

$$x - y - 2 = 0.$$

The disadvantage of this method is that it involves two stages; first the coordinates of the point of intersection must be found, and then the equation of the line must be obtained. In cases where the point of intersection of the two lines is not itself needed, it is more direct to use a method based on the following result:

*If $lx + my + n = 0$ and $l'x + m'y + n' = 0$ are the equations of two given lines, then the equation*

$$(lx + my + n) + \lambda(l'x + m'y + n') = 0,$$

*where $\lambda$ is any constant number, represents a line passing through the point of intersection of these two lines.*

This important fact is established by the method of *inspection*. First, we notice that the equation

$$(lx+my+n)+\lambda(l'x+m'y+n') = 0$$

consists just of linear terms; there is a term in $x$, $[(l+\lambda l')x]$, a term in $y$, $[(m+\lambda m')y]$, and a constant term $[n+\lambda n']$. The equation is therefore a linear equation, and so (§3) it represents a line.

Secondly, we notice that at the point of intersection of the two given lines, both the expressions

$$lx+my+n \quad \text{and} \quad l'x+m'y+n'$$

are zero, and so

$$(lx+my+n)+\lambda(l'x+m'y+n')$$

is also zero, whatever value $\lambda$ may have. This means that the line contains the point of intersection of the two given lines. The result is thus established.

We notice that in any particular problem the number $\lambda$ is still at our disposal; it can be chosen in order to make the line satisfy some other condition, to pass through a point, or be parallel to a given line, for example.

The original problem can now be solved more neatly. The equation

$$(x+2y-5)+\lambda(2x+y-7) = 0$$

represents any line through the point of intersection of the two given lines. If it is also to contain the point $(1, -1)$ then

$$(1+2.-1-5)+\lambda(2.1-1-7) = 0;$$

that is

$$\lambda = -1.$$

The equation of the required line is thus

$$(x+2y-5)-1(2x+y-7) = 0,$$

or

$$x-y-2 = 0.$$

**8. Abridged notation.** It becomes desirable, as the expressions used in coordinate geometry become more cumbersome, to adopt some technique for shortening what must be written down. In the last section we have had an example where some abridgement results both in less manual labour and in increased clarity.

Let us write $L$ to represent $(lx+my+n)$, and $L'$ to represent

$(l'x+m'y+n')$. Then the result we proved may be restated as follows:

*If $L = 0$ and $L' = 0$ are the equations of two given lines, then the equation*

$$L+\lambda L' = 0,$$

*where $\lambda$ is any constant number, represents a line passing through the point of intersection of these two lines.*

The advantages of this 'shorthand' notation are obvious, provided always that the *meaning* of the expressions $L = 0$ $L' = 0$ is kept in mind. Naturally, any abridged notation of this type becomes useless if the reader does not or cannot remember what the symbols stand for.

It should be noticed that it is prudent to aid the memory by choosing the shorthand notation intelligently. For example, one would denote $l_1x+m_1y+n_1$ by $L_1$ and $l_2x+m_2y+n_2$ by $L_2$.

## EXAMPLES II C

**1.** Find the coordinates of the points of intersection of the following pairs of lines:

(i) $2x-5y+1 = 0$ and $x+y+4 = 0$;

(ii) $7x-4y+1 = 0$ and $x-y+1 = 0$;

(iii) $x/a+y/b = 1$ and $x/b+y/a = 1$.

**2.** What is represented by the following equations?

(i) $(x+y+2)+\lambda(3x-2y+1) = 0$;

(ii) $(2x-3y-4)+\lambda(5x+y+1) = 0$;

(iii) $(x+2y-3)+\lambda(x+2y-5) = 0$.

**3.** Find the equation of the line through the point of intersection of the lines $x-4y-7 = 0$ and $2x+y-1 = 0$ which (i) passes through the origin; (ii) passes through the point $(-1,-1)$; (iii) is parallel to the line $5x-2y = 0$; (iv) is perpendicular to the line $10x+7y+3 = 0$; (v) cuts off equal intercepts on the axes.

**4.** Find the coordinates of the vertices of the triangle whose sides are $x-y+1 = 0$, $x-2y+4 = 0$ and $x+3y+9 = 0$. What are the coordinates of the orthocentre [G 3] of this triangle?

**5.** For what value of $k$ do the lines $3x+y-4 = 0$ and $kx+2y-6 = 0$ meet in a point on the line $2x-y = 6$?

**6.** Show that the lines $x = 3$, $x = y\sqrt{3}+1$, $x = -y\sqrt{3}-1$ form an equilateral triangle, and find its area.

7. Show that the lines

$$3x-y+4 = 0, \quad 24x+7y-3 = 0 \quad \text{and} \quad 3x-4y-15 = 0$$

form an isosceles triangle, and find the lengths of its altitudes.

8. Find the equation of the line through the point of intersection of the lines $6x+7y+4 = 0$ and $5x-21y+2 = 0$ which is also (i) parallel to the $x$-axis; (ii) parallel to the $y$-axis; (iii) through the point $(2, 1)$; (iv) perpendicular to $x+2y = 0$.

9. A parallelogram $ABCD$ has edges $4x-y+1 = 0$, $2x+3y-10 = 0$ meeting in $B$, and $D$ is the point $(-1, -1)$. Find the equations of (i) the sides $AD$, $CD$; (ii) the diagonal $BD$; (iii) the diagonal $AC$. Verify that the diagonals bisect each other.

10. A triangle has vertices $P(1, 5)$, $Q(-7, -3)$, $R(5, -3)$. Find (i) the mid-points of the sides; (ii) the equations of the medians; (iii) the coordinates of the centroid $G$; (iv) the equations of the perpendicular bisectors of the sides; (v) the coordinates of the circumcentre $C$; (vi) the equations of the altitudes; (vii) the coordinates of the orthocentre $H$. Verify that $G$, $C$ and $H$ lie on a line (called the Euler line of the triangle $PQR$) and that $CG : GH = 1 : 2$. List the nine points which give the name to the nine-point circle [G 5], and show that they are all equidistant from the mid-point of $CH$.

11. Find the equation of the line joining the point $(1, -1)$ to the point of intersection of the lines $3x-y = 2$, $2x+5y = 2$.

12. $ABCD$ is a parallelogram in which $A$ is the point $(4, 8)$, $B$ is the point $(3, 6)$ and $C$ is the point $(-5, -2)$. Calculate (i) the gradients of $AB$ and $BC$; (ii) the equations of $AD$ and $CD$; (iii) the coordinates of $D$; (iv) the coordinates of the point of intersection of $AC$ and $BD$.

13. Find the equation of the line joining the origin to the point of intersection of the lines $3x+2y = 12$ and $2x-3y = 5$. Find the length of this line.

14. Find in its simplest form the equation of the line which is parallel to the line $4y+x = 1$ and which passes through the point of intersection of the lines $y = 2x$, $y+x = 3$.

15. One side of a rhombus is the line $2y = x$ and two opposite vertices are the points $(0, 0)$ and $(6, 6)$. Find the equations of the diagonals, the coordinates of the other two vertices, and the length of the side.

16. The points $A, B, C, D$ have coordinates $(0, 3)$, $(6, 0)$, $(0, 2)$, $(3, 0)$ respectively. The lines $AD$, $BC$ meet in $P$, and the lines $AB$, $CD$ meet in $Q$. Find the point $R$ where the line joining $P$ to the origin meets $AB$. If $C'$ is the point $(0, -2)$ and $QC'$ meets $y = 0$ in $D'$, show that the lines $AD'$ and $BC'$ meet in a point on the line $PR$.

17. $O(0, 0)$, $A(3, 0)$, $B(0, 4)$ are the vertices of a triangle, and squares $OADE$, $OBFG$ are described externally on the sides $OA$ and $OB$. If $T$ is

the point of intersection of $AF$ and $BD$, prove that $OT$ is perpendicular to $AB$.

18. The points $(1, 2)$, $(2, 9)$ are the ends of a diagonal of a rectangle and the other diagonal is parallel to the line $x = 7y$. Find the coordinates of the other two vertices and the area of the rectangle.

**9. Condition for three points to be collinear.** The condition for the points $P_1(x_1, y_1)$, $P_2(x_2, y_2)$ and $P_3(x_3, y_3)$ to lie on a line can be found in an elementary way. The gradients of $P_1P_2$ and $P_1P_3$ are

$$\frac{y_2-y_1}{x_2-x_1} \quad \text{and} \quad \frac{y_3-y_1}{x_3-x_1},$$

and if $P_1P_2P_3$ is a line, these are equal. The condition is therefore

$$\frac{y_2-y_1}{x_2-x_1} = \frac{y_3-y_1}{x_3-x_1},$$

which reduces to

$$x_2y_3 - x_3y_2 + x_3y_1 - x_1y_3 + x_1y_2 - x_2y_1 = 0.$$

(Such a result should not be memorized, but the *method* should be noted.)

This result also follows directly from the result on determinants [A 3]. If the three points lie on the line $ax + by + c = 0$, then

$$ax_1 + by_1 + c = 0,$$
$$ax_2 + by_2 + c = 0,$$
$$ax_3 + by_3 + c = 0.$$

These three equations have a non-zero solution for $a:b:c$ if and only if

$$\begin{vmatrix} x_1 & y_1 & 1 \\ x_2 & y_2 & 1 \\ x_3 & y_3 & 1 \end{vmatrix} = 0,$$

and this when evaluated, [A 2], is the condition for collinearity already found.

This result also leads to the *determinantal form* for the equation of the line joining two points $(x_1, y_1)$ and $(x_2, y_2)$. The three points $(x_1, y_1)$, $(x_2, y_2)$ and $(x, y)$ are collinear if and only if

$$\begin{vmatrix} x & y & 1 \\ x_1 & y_1 & 1 \\ x_2 & y_2 & 1 \end{vmatrix} = 0,$$

which is the expression given in § 4, section VIII.

**10. Condition for three lines to be concurrent.** Let

$$a_1x+b_1y+c_1 = 0,$$

$$a_2x+b_2y+c_2 = 0,$$

$$a_3x+b_3y+c_3 = 0,$$

be the equations of three lines. The last two meet in a point

$$\left(\frac{b_2c_3-b_3c_2}{a_2b_3-a_3b_2}, \quad \frac{c_2a_3-c_3a_2}{a_2b_3-a_3b_2}\right) \quad (\S 6),$$

and this lies on the first line, if and only if

$$a_1\left(\frac{b_2c_3-b_3c_2}{a_2b_3-a_3b_2}\right)+b_1\left(\frac{c_2a_3-c_3a_2}{a_2b_3-a_3b_2}\right)+c_1 = 0.$$

This reduces to the condition

$$a_1b_2c_3-a_1b_3c_2+a_2b_3c_1-a_2b_1c_3+a_3b_1c_2-a_3b_2c_1 = 0,$$

which must hold if the three lines meet in a point.

This is the determinant

$$\begin{vmatrix} a_1 & b_1 & c_1 \\ a_2 & b_2 & c_2 \\ a_3 & b_3 & c_3 \end{vmatrix} = 0 \quad [\text{A } 2].$$

It may be obtained from [A 3]. The three lines meet in a point, if and only if there is a solution $x:y:1$ which satisfies simultaneously all three equations

$$a_1x+b_1y+c_1 = 0,$$

$$a_2x+b_2y+c_2 = 0,$$

$$a_3x+b_3y+c_3 = 0,$$

and the condition for this is the vanishing of the determinant formed by the coefficients.

### EXAMPLES II D

1. Show that the points $(1, -3)$, $(-1, -5)$ and $(2, -2)$ are collinear.

2. Show that the lines $3x+5y = 7$, $4x-3y+10 = 0$ and $2x+y = 0$ meet in a point, and find the distance of this point from the origin.

3. Show that the three points $(3, 1)$, $(-1, 2)$ and $(19, -3)$ lie on a line.

4. Show that the points $(4, 3)$, $(-2, 1)$ and $(1, 2)$ are collinear, and that the lines $4x+3y = 1$, $-2x+y = 1$ and $x+2y = 1$ are concurrent.

5. Find for what value of $p$ the points $(3, 1)$, $(5, 2)$ and $(p, -3)$ are collinear.

6. The lines $-x+3y = 3$, $x+5y = 7$ and $2x-2y = k$ are concurrent. Find the value of $k$.

7. Show that in each of the following cases the points are collinear:

(i) $(1, 10)$, $(3, 2)$, $(4, -2)$;

(ii) $(2, 3)$, $(-4, 1)$, $(14, 7)$, $(-13, -2)$;

(iii) $(a, 2b)$, $(3a, 0)$, $(2a, b)$, $(0, 3b)$.

8. Show that in each of the following cases the lines are concurrent:

(i) $3x-y-2 = 0$, $5x-2y-3 = 0$, $2x+y-3 = 0$;

(ii) $2x-5y+1 = 0$, $x+y+4 = 0$, $x-3y = 0$;

(iii) $7x-4y+1 = 0$, $x-y+1 = 0$, $y = 2x$.

**11. The angle between two lines.** Suppose two lines, $p$ and $q$, have gradients $m$ and $m'$. Then, if the lines make angles $\alpha$ and $\beta$ with the $x$-axis, $\tan\alpha = m$ and $\tan\beta = m'$ (Fig. 17). One of the

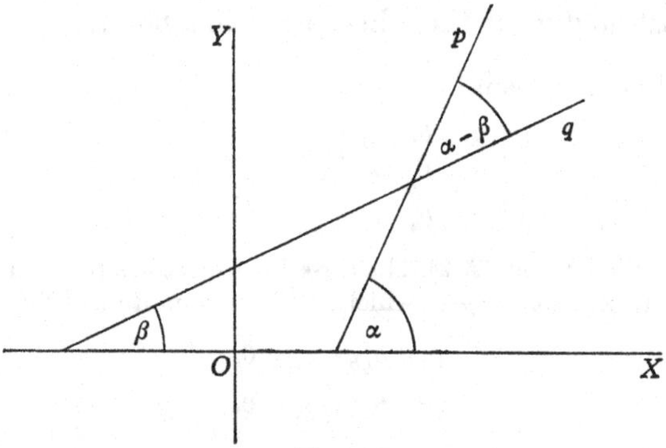


Fig. 17

angles between $p$ and $q$ is then $\alpha-\beta$ (the other being the supplement of this). Now the trigonometrical formula [T 5] gives

$$\tan(\alpha-\beta) = \frac{\tan\alpha - \tan\beta}{1+\tan\alpha\tan\beta}$$

$$= \frac{m-m'}{1+mm'}.$$

The expression $$\frac{\mathbf{m-m'}}{\mathbf{1+mm'}}$$

therefore gives the tangent of one of the angles between the two lines; if the modulus of this (that is, the numerical value, omitting the minus sign if the number is negative) is taken, we obtain the tangent of the acute angle between the two lines.

If one or other of the lines is parallel to $OY$, then either $m$ or $m'$ is infinite, and the expression cannot be used. In either of these cases, however, it is easy to calculate the angle between the lines from the gradient of the other line.

Two results which are already familiar can be deduced from this formula.

(i) If the lines are parallel, $\alpha - \beta = 0$ and $\tan(\alpha-\beta) = 0$. This gives $m - m' = 0$ as the condition for parallel lines.

(ii) If the lines are perpendicular, $\alpha - \beta = \frac{1}{2}\pi$ and $\tan(\alpha-\beta)$ is infinite. This gives $1 + mm' = 0$ as the condition for perpendicular lines.

**12. The perpendicular distance of a point from a line.** To find the perpendicular distance $d$ of the point $P(h, k)$ from a line, we first take the equation of the line in the normal form

$$x\cos\alpha + y\sin\alpha = p.$$

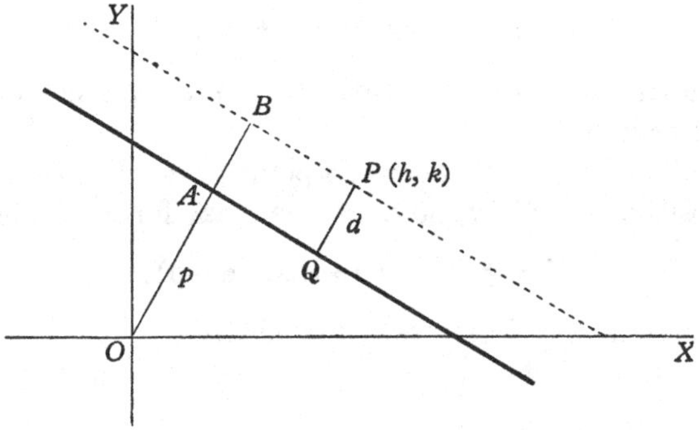


Fig. 18

Then (p. 31) $p$ is the perpendicular distance $OA$ from the origin to the line (Fig. 18). Let $Q$ be the foot of the perpendicular from $P$ to the line, and let $B$ be the point where $OA$ (produced if necessary) meets the line through $P$ parallel to the given line.

The equation of any line parallel to the given line is

$$x\cos\alpha + y\sin\alpha = \lambda$$

for any value of the constant $\lambda$. If this line passes through $(h, k)$,

$$h\cos\alpha + k\sin\alpha = \lambda;$$

this gives the value of $\lambda$ for the particular parallel line through $P$, and $\lambda = OB$, the perpendicular distance from $O$ to the line. From the figure,

$$\begin{aligned} d &= QP = AB \\ &= OB - OA \\ &= \lambda - p, \end{aligned}$$

and substituting the value of $\lambda$, we have

$$d = h\cos\alpha + k\sin\alpha - p.$$

This expression has been obtained from a diagram in which $P$ is on the opposite side of the line from $O$. If the diagram were drawn with $P$ on the same side of the line as $O$, we should have $d = p - \lambda$, and the resulting expression for $d$ would be the one obtained above, but with the opposite sign. (This corresponds to the fact, discussed in § 5, that any line divides the plane into a 'positive region' and a 'negative region'.) As the diagram is drawn here, and taking the equation of the line as

$$x\cos\alpha + y\sin\alpha - p = 0$$

with $p$ positive, $O$ is in the negative half of the plane and $P$ is in the positive half.

We shall define the distance of a point from a line as an essentially positive quantity, so that the final form of the result obtained is

$$\mathbf{d = \pm (h\cos\alpha + k\sin\alpha - p)},$$

where the sign is so chosen that $d$ is positive.

We notice that the expression may be written down in the following way.

(1) Express the equation of the line in normal form

$$x\cos\alpha + y\sin\alpha - p = 0.$$

(2) In the left-hand side of this equation, replace $x$ by $h$ and $y$ by $k$; attach a sign so that the resulting expression is positive.

In the general case, finding the perpendicular distance of the point $(h, k)$ from the line $ax + by + c = 0$, we carry out the same routine.

(1) The given line in normal form is

$$\frac{a}{\sqrt{(a^2+b^2)}}x+\frac{b}{\sqrt{(a^2+b^2)}}y+\frac{c}{\sqrt{(a^2+b^2)}}=0$$

(using the method of Illustration 1, p. 33).

(2) Replacing $x$ by $h$ and $y$ by $k$ in the left-hand side of this, and attaching a suitable sign, gives

$$\mathbf{d=\pm\frac{ah+bk+c}{\sqrt{(a^2+b^2)}}}.$$

***Illustration 4.*** *Find the locus of a point which moves so that it is equidistant from the lines* $x-y=1$, $7x+y=3$.

Applying the result just proved, the perpendicular distances of $P(h,k)$ from the two lines are found to be

$$\pm\left(\frac{h-k-1}{\sqrt{2}}\right) \quad \text{and} \quad \pm\left(\frac{7h+k-3}{\sqrt{50}}\right).$$

These distances are equal, so either

$$\left(\frac{h-k-1}{\sqrt{2}}\right)=+\left(\frac{7h+k-3}{\sqrt{50}}\right),$$

or

$$\left(\frac{h-k-1}{\sqrt{2}}\right)=-\left(\frac{7h+k-3}{\sqrt{50}}\right).$$

These equations reduce to

$$2h+6k+2=0$$

and

$$12h-4k-8=0.$$

The locus of $P$ is therefore in two parts

$$x+3y+1=0$$

and

$$3x-y-2=0;$$

that is, the locus is a pair of lines.

As a check, we notice that this result is in accord with what we should have expected. A point equidistant from two lines can be shown by pure geometry to lie on one or other of the two bisectors of the angles between the lines, and these bisectors, internal and external, are themselves perpendicular. The lines obtained above are evidently perpendicular and pass through the point of intersection $(\frac{1}{2}, -\frac{1}{2})$ of the given lines.

Bisectors of the angles between any two lines can be found in this way. To discriminate in a particular example between the internal and external bisector it is best to draw a rough diagram.

## EXAMPLES II E

1. Find the acute angles between the following pairs of lines:

(i) $y = x+3$ and $2x+y = 5$;

(ii) $x-y+2 = 0$ and $3x+2y-1 = 0$;

(iii) $x+1 = 0$ and $x+y = 0$;

(iv) $y = 3$ and $2x+7y-6 = 0$;

(v) $3x+4y-7 = 0$ and $x-y-1 = 0$.

2. Find the equations of the lines through $(1, 3)$ which make angles $\tan^{-1}\frac{4}{3}$ with the line $2x+y = 7$.

3. Find the perpendicular distances of

(i) $(3, 4)$ from $3x+4y = 10$;

(ii) $(1, -1)$ from $8x-3y+6 = 0$;

(iii) $(0, 5)$ from $y = 3x+7$;

(iv) $(b, a)$ from $ax+by = 0$;

(v) $(4, 2)$ from $\frac{1}{3}x-\frac{1}{4}y = 1$.

4. Find the equations of the bisectors of the angles between the lines $2x+y+4 = 0$ and $2x-4y-7 = 0$. (Use the method of Illustration 4.)

5. By calculating the angles between each bisector and each given line in Example 4, verify that the lines found are in fact bisectors.

6. A triangle is formed by joining the points $(0, 0)$, $(2, 2)$ and $(3, 0)$. Find the lengths of the perpendiculars from the point $(1, 2)$ on to the sides.

7. Find the equations of the bisectors of the angles of the triangle whose vertices are $(2, 5)$, $(18, -7)$, $(-7, -7)$, and hence find the coordinates of the incentre and of the three excentres of the triangle [G 4].

8. Find the equations of the two lines which make angles of 45° with the line $3x+4y = 21$ and which pass through the point $(3, 1)$.

9. The distances of a point $P$ from the lines $x+y = 2$, $x+7y+1 = 0$ and $x-y+1 = 0$ are in the ratio $1:2:3$. Find the coordinates of $P$.

10. Find the coordinates of the four points whose distances from the lines in the previous example are all equal. Which of these points lies inside the triangle?

11. The line $p$ has equation $ax+by+c = 0$, and meets the axes in $A$, $B$; the origin is $O$ and $P$ is $(h, k)$. Express the area of the quadrilateral

$OAPB$: (i) in terms of the areas of the triangles $OPA$ and $OPB$; (ii) in terms of the areas of the triangles $OAB$ and $PAB$. Hence establish the formula for the perpendicular distance $d$ from $P$ to $p$.

12. With the notation of Example 11, (i) find where the line $y = k$ meets $p$; (ii) hence find the horizontal distance of $P$ from $p$. Deduce the formula for $d$.

13. With the notation of Example 11, show that the equation of the line through $P$ perpendicular to $p$ is $bx - ay + (ak - bh) = 0$. Find the coordinates of the point where this line meets $p$, and hence the value of $d$.

14. With the notation of Example 11, show that if the line perpendicular to $p$ has gradient $\tan\alpha$, then

$$\cos\alpha = a/\sqrt{(a^2+b^2)} \quad \text{and} \quad \sin\alpha = b/\sqrt{(a^2+b^2)}.$$

A point $Q(h + r\cos\alpha,\ k + r\sin\alpha)$ lies on $p$, and $PQ$ is perpendicular to $p$. Find the value of $r$, and deduce the formula for $d$.

15. Find the equation of the locus of the point which moves so that its distance from the line $3x + 4y - 5 = 0$ is three times its distance from the line $12x - 5y + 2 = 0$.

16. Find the equations of the two lines through the point $(5, 4)$ which each make an angle of 45° with the line $3y = 2x - 1$.

**13. Area of a triangle.** To find the area of the triangle (Fig. 19) whose vertices are $A(x_1, y_1)$, $B(x_2, y_2)$ and $C(x_3, y_3)$, we apply the formula for the perpendicular distance which was found in the last section.

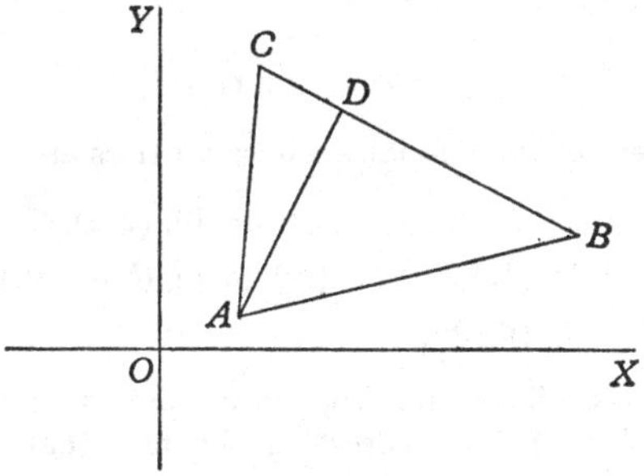


Fig. 19

The equation of $BC$ is

$$y - y_2 = \frac{y_2 - y_3}{x_2 - x_3}(x - x_2),$$

or $$x(y_2 - y_3) - y(x_2 - x_3) + (x_2y_3 - x_3y_2) = 0.$$

The perpendicular distance, $AD$, from $A$ to $BC$ is therefore

$$\frac{x_1(y_2-y_3)-y_1(x_2-x_3)+(x_2y_3-x_3y_2)}{\sqrt{\{(x_2-x_3)^2+(y_2-y_3)^2\}}}.$$

The denominator of this fraction is the length of $BC$. The area of the triangle $ABC$ is then

$$\begin{aligned}\tfrac{1}{2}AD.BC &= \tfrac{1}{2}\{x_1(y_2-y_3)-y_1(x_2-x_3)+(x_2y_3-x_3y_2)\}\\ &= \tfrac{1}{2}\{(x_2y_3-x_3y_2)+(x_3y_1-x_1y_3)+(x_1y_2-x_2y_1)\}.\end{aligned}$$

This formula is more memorable in the determinantal form

$$\frac{1}{2}\begin{vmatrix} x_1 & y_1 & 1 \\ x_2 & y_2 & 1 \\ x_3 & y_3 & 1 \end{vmatrix}.$$

Those capable of doing so should repeat the above proof, using the determinant notation throughout, starting from the line in determinant form (§ 4, section VIII).

The expression found may, for particular values of the coordinates, be either positive or negative. An area is generally regarded as positive, and so the expression must be taken with the appropriate sign. When finding areas of plane figures, it is prudent to draw and examine a diagram, as there are often particular geometrical features, such as right angles, which make it easier to use a special method rather than the general formula.

## EXAMPLES II F

1. Find the areas of the triangles whose vertices are:

(i) $(3, 1), (2, -4), (-5, 6)$; (ii) $(0, 0), (2, 2), (3, -4)$;

(iii) $(2, -1), (-2, 1), (1, 4)$; (iv) $(a, b), (b, a), (0, 0)$;

(v) $(-1, 1), (3, -5), (6, 2)$.

2. Obtain the condition for the three points $A(x_1, y_1)$, $B(x_2, y_2)$, $C(x_3, y_3)$ to be collinear, by expressing the fact that the area of the triangle $ABC$ is zero.

3. Find the area of the triangle formed by the lines

$$x-y=0, \quad x+y+10=0 \quad \text{and} \quad x-5y=8.$$

4. The lines $2x+y=2$ and $2x+4y=5$ cut the axes of $x$ and $y$ at $A$, $B$ and $C$, $D$ respectively. If $O$ is the origin, find the area common to the two triangles $AOB$ and $COD$.

5. The points (3, 1), ( − 1, − 2) and ( − 4, − 6) are three vertices of a rhombus. Find the coordinates of the fourth vertex, the equations of the diagonals, and the area of the rhombus.

6. Show that the triangle formed by the axes and the line $x + t^2y = 2ct$ has an area which is independent of the value of $t$.

7. A triangle is formed by the lines $y = x - 8$, $y = 2x - 4$ and $y = 4x - 2$. Find its area and the coordinates of its orthocentre.

8. Show that the points ( − 2, 4), (3, − 4), (4, − 6) and ( − 1, 2) are the vertices of a parallelogram, and calculate its area.

**14*. Rotation of axes.** We have already shown in the previous chapter how to obtain the coordinates of a point relative to new axes which are drawn parallel to the original ones. We now investigate the result of rotating the axes through an angle $\alpha$, keeping the origin $O$ fixed.

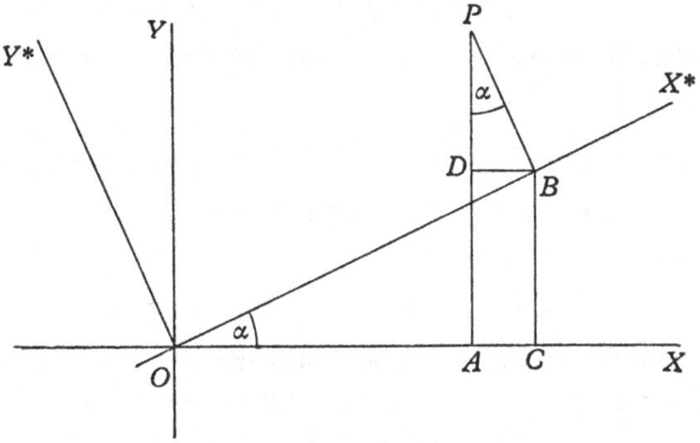


Fig. 20

Let the old axes be $OX$, $OY$, and the new ones $OX^*$, $OY^*$, where the angle $XOX^* = \alpha$. Let $P$ be a point whose coordinates referred to the old axes are $(x, y)$ and referred to the new axes are $(x^*, y^*)$. In Fig. 20, $A$ is the foot of the perpendicular from $P$ to $OX$, $B$ the foot of the perpendicular from $P$ to $OX^*$, $C$ the foot of the perpendicular from $B$ to $OX$ and $D$ the foot of the perpendicular from $B$ to $AP$. Then angle $DPB = \alpha$, and $\overrightarrow{OA} = x$, $\overrightarrow{AP} = y$, $\overrightarrow{OB} = x^*$, $\overrightarrow{BP} = y^*$. Now

$$\begin{aligned} x = \overrightarrow{OA} &= \overrightarrow{OC} - \overrightarrow{AC} \\ &= \overrightarrow{OC} - \overrightarrow{DB} \\ &= \overrightarrow{OB}\cos\alpha - \overrightarrow{BP}\sin\alpha \\ &= x^*\cos\alpha - y^*\sin\alpha, \end{aligned}$$

and
$$y = \overrightarrow{AP} = \overrightarrow{AD} + \overrightarrow{DP}$$
$$= \overrightarrow{CB} + \overrightarrow{DP}$$
$$= \overrightarrow{OB}\sin\alpha + \overrightarrow{BP}\cos\alpha$$
$$= x^*\sin\alpha + y^*\cos\alpha.$$

The formulae connecting the old and new coordinates are therefore

$$\mathbf{x = x^* \cos \alpha - y^* \sin \alpha,}$$
$$\mathbf{y = x^* \sin \alpha + y^* \cos \alpha.}$$

The reverse equations, found by solving for $x^*$ and $y^*$, are

$$\mathbf{x^* = x \cos \alpha + y \sin \alpha,}$$
$$\mathbf{y^* = -x \sin \alpha + y \cos \alpha.}$$

As a check of accuracy, we notice that the axes $x = 0$ and $y = 0$ in the new coordinates are

$$x^*\cos\alpha - y^*\sin\alpha = 0 \quad \text{and} \quad x^*\sin\alpha + y^*\cos\alpha = 0,$$

which are, as they should be, perpendicular lines.

## MISCELLANEOUS EXAMPLES II

1. The coordinates of the vertices of a triangle $ABC$ are $A(-4, 2)$, $B(4, 6)$, $C(8, -2)$. Show by calculation that $AB = BC$ and that $AB$ is perpendicular to $BC$. Find the equation of the perpendicular from $B$ to $AC$, and the area of the triangle $ABC$.

2. $A, B$ and $C$ are three points whose coordinates are $(3, -3)$, $(8, 7)$ and $(2, 5)$ respectively. The perpendicular from $C$ to $AB$ meets $AB$ at $D$. Find the equation of the line $CD$, the coordinates of $D$, and the ratio of the lengths of $AD$ and $DB$.

3. $OABC$ is a trapezium with one corner $O$ at the origin, $A$ at $(5, 3)$, and $OC$ of length 2 units measured in the positive direction of the $y$-axis. $OA$ and $CB$ are parallel, and $BA$, when produced, makes an angle of 45° with the positive direction of the $x$-axis. Find the equations of the lines $AB$ and $CB$, and the coordinates of the mid-point of $AB$.

4. The coordinates of the vertices $ABC$ of a triangle are $A(0, 0)$, $B(3, 5)$ and $C(6, 2)$. Calculate the equations of the perpendiculars from $B$ to $AC$, and from $C$ to $AB$. Find the coordinates of their point of intersection $H$, and prove that $AH$ is perpendicular to $BC$.

5. The equation $2x^2 - xy - 3y^2 = 0$ represents two straight lines. Find the equation of each line and the angle between them.

6. Find the equation of the line passing through the origin and perpendicular to $x-2y = 3$. Find also the length of this perpendicular.

7. Sketch the trapezium formed by the lines $x+2y = 2$; $x = 0$; $x-2y = 4$; $x = 2$. Find the area of the figure so formed, and the equations of its diagonals.

8. A line is drawn through the point $(3, 0)$ parallel to the line $y = 2x$; also a line is drawn through the point $(1, 2)$ parallel to the line $2y = x$. Sketch the figure and find the equations of the diagonals of the parallelogram thus formed.

9. Two points $A$, $B$ have coordinates $(3, 4)$ and $(7, 10)$. Find in its simplest form the equation of the perpendicular bisector of $AB$.

10. The equations of the sides $BC$, $CA$ and $AB$ of a triangle are $y = 2$, $2x-y = 1$ and $3x+y = 14$, respectively. Sketch the figure, and find the coordinates of $A$, $B$ and $C$. If the point $D$ completes a parallelogram $ABCD$, find the equations of $AD$ and $CD$.

11. Find the equation of the perpendicular bisector of the line joining the points $(-4, -3)$ and $(2, 6)$.

12. The vertices $O$, $A$, $B$, $C$ of a rectangle are $(0, 0)$, $(3, 0)$, $(3, 5)$, $(0, 5)$ respectively. $P$, $Q$ are points on $AB$ such that $P$ bisects $AB$ and $CQ$ is perpendicular to $OP$. Find the ratio in which $Q$ divides $AB$, the equation of the line $BD$ parallel to $OP$, and the length of the intercept made on $CQ$ by $OP$ and $BD$.

13. $ABCD$ is a rectangle; $A$ is $(0, 4)$, $B$ is $(3, 0)$ and $BC$ is of length 3. If $D$ and the origin are on opposite sides of $AB$, find the coordinates of $D$, the equation of $CD$, and the angle between $CA$ and the $x$-axis.

14. $P$ and $Q$ are the feet of the perpendiculars from $A(0, 5)$ and $B(-6, 0)$ to the line $2y-x = 4$. Find the equations of $AP$ and $BQ$, the distance $PQ$, and the point of intersection of $AQ$ with the $x$-axis.

15. Find the coordinates of the centroid of the triangle formed by the lines $x-y-2 = 0$, $3x+y-10 = 0$, $7x-3y-2 = 0$.

16. $ABCD$ is a parallelogram. The equation of $AB$ is $3y = x+4$, $DC$ passes through the point $(-2, 1)$, and $AD$ and $BC$ are parallel to the $y$-axis and pass through the points $(-3, 4)$ and $(2, 1)$ respectively. Obtain the coordinates of $A$, $B$, $C$ and $D$, and of the point of intersection of the diagonals of the parallelogram.

17. $A$ is the point $(4, 4)$, $B$ is $(5, 3)$ and $C$ is $(6, 0)$. Find the equations of the perpendicular bisectors of $AB$ and $BC$. Hence calculate the coordinates of the circumcentre and the length of the circumradius of the triangle $ABC$.

18. The vertices of the triangle $ABC$ are the points $A(-3, -1)$, $B(11, 13)$, $C(-1, -3)$. Find the length of the median through $A$ and the

coordinates of the centroid of the triangle. Find also the tangent of the angle between the median and the perpendicular from $A$ on to $BC$.

19. Two straight lines pass respectively through the points $(2, 2)$, $(-6, -1)$ and $(0, 2)$, $(-2, 6)$. Calculate the tangent of the angle between the two lines and the coordinates of their point of intersection. Find the area of the triangle bounded by these lines and the axis of $x$.

20. Prove that the lines $2x+y+3 = 0$, and $x-2y-1 = 0$ are perpendicular. If these two lines are taken as sides of a rectangle whose other sides meet in the point $(3, 4)$, find the equations of these other sides and the area of the rectangle.

21. The sides $BC$, $CA$, $AB$ of a triangle $ABC$ lie along the lines

$$3x+4y = 1, \quad 5x+y = 13, \quad 2x-3y+5 = 0,$$

respectively. Find the coordinates of the orthocentre of the triangle. Show that the locus of a point $P$ such that $CP^2-BP^2 = 13$ is the altitude through $A$.

22. The point $(4, 5)$ is a vertex of a parallelogram, two of whose adjacent sides are $y-x = 0$, $4x-3y+1 = 0$. Find the equations of the diagonals.

23. A triangle is formed by the three straight lines

$$y = m_1x+a/m_1, \quad y = m_2x+a/m_2, \quad y = m_3x+a/m_3.$$

Prove that its orthocentre always lies on the line $x+a = 0$.

24. Calculate the acute angle between the pair of straight lines represented by the equation $3x^2-10xy+7y^2 = 0$, giving your answer correct to the nearest minute.

25. The equations of the sides $BC$, $CA$, $AB$ of a triangle are respectively $y = 3x$, $x = 2y$, $2x+y = 5$. $P$ is the point $(0, 3)$. Find the coordinates of $L$, $M$, $N$, the feet of the perpendiculars drawn from $P$ to $BC$, $CA$, $AB$ respectively. Show that $L$, $M$, $N$ lie on a straight line and find its equation.

26. Find, to the nearest minute, the acute angle between the pair of lines given by the equation $4x^2-24xy+11y^2 = 0$. If another pair of lines is drawn through the point $(1, 1)$ to form a parallelogram with the original pair, calculate the area of the parallelogram.

27. The line whose equation is $2x-y+2 = 0$ intersects the $y$-axis in $A$. $B$ is the point $(9, 0)$. Find the coordinates of a point $C$ on the line such that $BC = BA$. Find the coordinates of $D$, where $ABCD$ is a parallelogram, and the area of the parallelogram.

28. $P$, $Q$, $R$ are three points with coordinates $(1, 0)$, $(2, -4)$, $(-5, -2)$ respectively. Determine the equations of the line through $P$ perpendicular to $QR$ and of the line through $Q$ perpendicular to $PR$, the point of intersection of these two lines, and the area of the triangle $PQR$.

29. The coordinates of the extremities of one diagonal of a square are $(-1, 1)$, $(2, 5)$. Find the coordinates of the extremities of the other diagonal, and the area of the square.

30. Find the coordinates of the centre of the parallelogram formed by the four lines $ax+by = 1$, $ax+by = 2$, $bx+ay = 1$, $bx+ay = 2$. Prove that the parallelogram is a rhombus.

31. The equations of the sides of a parallelogram are

$$2x-y+3 = 0, \quad 10x+3y-25 = 0, \quad 2x-y-5 = 0, \quad 10x+3y+15 = 0.$$

Find the area of the parallelogram and the equations of its diagonals.

32. Show that the lines whose equations are

$$x-2y+1 = 0, \quad 9x+2y-11 = 0 \quad \text{and} \quad 7x+6y-53 = 0$$

form the sides of an isosceles triangle. Find the equation of the line parallel to the line $x-2y+1 = 0$ which forms, with the other sides produced, a triangle equal in area to that of the given triangle.

33. A triangle is formed by the three straight lines $x+y+1 = 0$, $y = 2x$ and $2x+3y-1 = 0$. Find the equations of the three lines drawn through the vertices perpendicular to the opposite sides, and find the coordinates of their common point of intersection.

34. Find the equations of the bisectors of the angles between the lines

$$24x+7y = 20 \quad \text{and} \quad 4x-3y = 2.$$

35. Find the angle between the lines $2x-y+7 = 0$ and $5x+2y-11 = 0$. Find also the equation of the line passing through the point $(3, -5)$ and the point of intersection of these lines. Which of the three lines cuts the $y$-axis at the greatest distance from the origin?

36. If $A$ is the point $(4, 3)$, find the coordinates of the point $A'$ such that $AA'$ is bisected at right angles by the line $2x+3y = 4$. If $B$ is the point $(-2, 7)$, find the coordinates of the point $C$ on the given line such that $AC$ and $BC$ are equally inclined to the line, and find the angle between $AC$ and the line.

37. $A, B, C$ are the points $(-2, -2)$, $(6, -4)$, $(8, 4)$ respectively. Find the equations of $AB$ and $AC$. $R$ is the foot of the perpendicular from $P(5, 9)$ to $AB$, and $PR$ meets $AC$ at $Q$. Prove that $P$ divides $QR$ externally in the ratio $2:3$.

38. The coordinates of four points $O$, $A$, $B$, $C$ are $(0, 0,)$, $(3, 0)$, $(0, 4)$, $(-2, 1)$ respectively. $BC, AO$ produced meet in $D$; $BA, CO$ produced meet in $E$. Find the equations of the lines $AB, BC, CO$, and the coordinates of $D$ and $E$.

39. The equations of the sides $BC$, $CA$, $AB$ of a triangle are, respectively, $y = 0$, $2x+y-8 = 0$, $x-y-1 = 0$. Find the equations of the lines through $A$ perpendicular to $AB$ and $AC$. If these lines cut $BC$ (or $BC$ produced) in $D$ and $E$, show that $DE = 2BC$.

40. The coordinates of the points $A$, $B$, $C$, $D$ are $(-4, -2)$, $(8, -5)$, $(4, 2)$ and $(0, 3)$ respectively, the unit on each axis being 1 cm. $AC$ and $BD$ meet in $H$. Prove that $AH/HC = BH/HD$. Calculate the area of the triangle $CHD$, and hence write down the area of each of the triangles $HAB$, $HAD$ and $HBC$.

41. Write down the equations of the lines through the points $A(1, 0)$ and $B(3, 0)$ parallel to $y = x$, and the line through the point $(1, 4)$ perpendicular to $y = x$. If the last line meets the lines through $A$ and $B$ in the points $C$ and $D$, calculate the area of the trapezium $ABDC$. Find the equations of the diagonals of $ABDC$ and the coordinates of their common point.

42. Referred to rectangular axes through $O$, the point $B$ is $(6, 8)$, $A$ is a point on the $y$-axis and $OABC$ is a convex quadrilateral of area 50 square units in which $AO = AB$ and $CO = CB$. Calculate the coordinates of $A$ and $C$. If $OB$ and $AC$ intersect in $X$ and the ordinate of $B$ intersects $AC$ in $Y$, calculate $AX:XY:YC$.

43. Two straight lines, perpendicular to each other, intersect in a point $P$. One passes through the point $A(0, 4)$ and the other passes through the point $B(8, 0)$. If $PB = 2PA$, find the coordinates of $P$ and the equations of the two lines for each of the two possible positions of $P$.

44. The sides of a triangle have equations

$$y = 0, \quad x + y = 4, \quad x - 3y + 4 = 0.$$

Find the equation of the locus of a point which moves so that the sum of the squares of its distances from the sides of the triangle is 12.

45. Calculate the area of the triangle formed by the line $2x + 3y = 7$ with the axes. Find the equation of the line through the origin which divides the triangle into two parts equal in area.

46. Find the ratio in which the join of $A(12, -5)$ and $B(3, 4)$ is cut by the line $3x - 11y = 0$. Prove that the line joining the origin to the point dividing $AB$ externally in the same ratio is perpendicular to the given one.

47. If $Q$ is the foot of the perpendicular from $(8, 7)$ to the line joining $A(3, 6)$ to $B(9, 2)$, find the coordinates of $Q$, and the ratio in which $Q$ divides $AB$.

48. The line $x/a + y/b = 1$ (where $a$ and $b$ are positive and $a > b$) meets the $x$-axis at $A$ and the $y$-axis at $B$. Find the equation of the line perpendicular to $AB$ and passing through the mid-point $C$ of $AB$. This line meets the $x$-axis at $D$. Find the area of the quadrilateral $ODCB$, where $O$ is the origin.

49. The sides of a triangle are given by

$$5x + 4y = 2, \quad x = 4, \quad x + 2y = 4.$$

Find the equations of the altitudes and medians of the triangle. Find also the coordinates of its orthocentre and centroid.

50. A parallelogram $ABCD$ has vertices $A(1, -2)$, $B(6, 10)$, $C(26, 25)$. Find the coordinates of $D$, the area of the figure, and the tangent of the acute angle between the diagonals $AC$, $BD$.

51. $A, B, C$ are the feet of the perpendiculars from the point $(0, -4)$ to the lines $x+y=0$, $y+2x=1$, $3y-9x+32=0$ respectively. Find the equation of $AB$ and prove that it passes through $C$.

52. One vertex of a rhombus is at the point $(0, 4)$. One diagonal lies along the line $3y-x=2$ and two of the sides are parallel to the line $y=x$. Find the coordinates of the remaining three vertices. Find the area of that part of the rhombus which lies in the first quadrant.

53. Two sides of a parallelogram lie along the lines $2x-3y+10=0$, $5x-y+12=0$ and one vertex is at the point $(3, 1)$. Show that one diagonal lies along the line $9x-7y+6=0$ and calculate the area of the parallelogram.

54. From the point $P(1, 3)$ a line is drawn perpendicular to the line $8x-14y-31=0$ to meet it in $Q$ and $PQ$ is produced to $R$ so that $PQ=QR$. Find the coordinates of $R$ and the equation of the line through $R$ which is parallel to the given line.

55. Two sides of a parallelogram lie along the lines $x-y+1=0$, $2x+3y=6$, and the diagonals meet at the point $(1, \frac{1}{2})$. Find the coordinates of the vertices of the parallelogram and the equations of the remaining sides.

56. Find the coordinates of the centroid of the triangle with sides along the lines $x+y=1$, $x-y=1$, $x-3y+3=0$, and hence, or otherwise, prove that the point $(2{\cdot}8, 1{\cdot}9)$ lies inside the triangle.

57. The vertices $B$, $C$ of a triangle $ABC$ lie along the lines $3y=4x$, $y=0$ respectively, and the side $BC$ passes through the point $(\frac{2}{3}, \frac{2}{3})$. If $ABOC$ is a rhombus, where $O$ is the origin of coordinates, find the equation of the line $BC$ and prove that the coordinates of $A$ are $(\frac{8}{5}, \frac{4}{5})$.

58. The sides $BC$, $CA$ and $AB$ of a triangle $ABC$ lie along the lines $x+y=1$, $2x-3y=1$ and $x-2y=1$ respectively. Find the equations of the altitudes of the triangle through the vertices $A$ and $B$ and prove that the distance of the orthocentre of the triangle from the side $AB$ is $8/5\sqrt{5}$.

59. The sides $OA$, $OC$ of a parallelogram $OABC$ lie along the lines $3y=x$, $y=3x$ respectively, and $B$ is the point $(4, 3)$. Find, in their simplest forms, the equations of the lines which contain the sides $AB$, $BC$ and the diagonal $AC$.

60. A square $ABCD$ has its opposite vertices $A$ and $C$ at $(2, -5)$ and $(-4, 3)$. Find the coordinates of $B$ and $D$, and the equations of the lines which contain the sides of the square.

---

61. A straight line cuts the $x$- and $y$-axes at $A$ and $B$, and the coordinates of the mid-point of $AB$ are $(3, 2)$. Find the coordinates of the point of intersection of this straight line with the straight line $x+3y+3=0$.

A variable straight line passes through the fixed point $(h, k)$ and meets the $x$- and $y$-axes at $P$ and $Q$; prove that the locus of the mid-point of $PQ$ is $h/x+k/y = 2$.

62. Find the locus of a point which moves so that (i) the sum of its distances from the lines $x+y = 1$ and $x-y = 2$ is a constant $k$; (ii) the sum of the *squares* of its distances from the same two lines is a constant $p^2$. For what values of $k$ will the loci obtained in (i) touch the locus obtained in (ii)?

63. Show that $x-my-1 = 0$ and $x-ny+1 = 0$ represent lines through the point $(1, 0)$ and $(-1, 0)$ respectively, and find the equations of the lines joining their point of intersection to the points $(0, 1)$ and $(0, -1)$. If this latter pair of lines includes an angle of 45°, prove that

$$m^2-n^2 = \pm 2(mn+1).$$

64. The equations of two lines $PQ$ and $RS$ are $2x-y+3 = 0$ and $9x-13y-26 = 0$. Show that the lines $3x-y-3 = 0$ and $x-2y = 0$ cut $PQ$ and $RS$ in concyclic points.

65. $OPQR$ is a square with one vertex $O$ at the origin. The coordinates of the opposite vertex $Q$ are $(2a, 2b)$. Prove that the coordinates of the other vertices are $(a+b, b-a)$ and $(a-b, a+b)$. Prove also that, if the square varies in size and position with $O$ always at the origin and $Q$ moving on a line parallel to the $x$-axis, then the corners $P$ and $R$ move on straight lines which are perpendicular to each other.

66. $A(x_1, y_1)$ and $B(x_2, y_2)$ are two fixed points. $C$ is the point on the $y$-axis such that the slopes of the lines $CA$, $CB$ are equal and opposite. Prove that $C$ is the point $\{0, (x_1y_2+x_2y_1)/(x_1+x_2)\}$. $D$ is the point on the $x$-axis such that the slopes of $DA$, $DB$ are equal and opposite. Prove that the slope of $CD$ is equal and opposite to that of the line joining the origin to the mid-point of $AB$.

67. Show that the equation of a line through the fixed point $(a, 0)$ may be written $y = m(x-a)$. This line meets the line $y = x$ at $Q$ and the line $y = -x$ at $R$. Find the coordinates of $Q$ and $R$. The line through $Q$ perpendicular to $y = x$ meets the line through $R$ perpendicular to $y = -x$ at $P$. Find the coordinates of $P$ and show that $P$ lies on the curve

$$x^2-y^2 = 2ax.$$

68. The vertices of a triangle are $A(-2, 3)$, $B(3, 7)$ and $C(4, 0)$. Find the coordinates of the point $D$ on the same side of $AC$ as $B$ such that the triangle $ACD$ is right-angled at $C$ and equal in area to the triangle $ABC$. Calculate the area of the triangle $ACD$.

69. The vertices of a triangle are $A(4, 4)$, $B(-4, -2)$ and $C(4, -4)$ respectively. Calculate the coordinates of the point which is equidistant from $B$ and $C$ and lies on the internal bisector of the angle $BAC$.

70. The equations of two straight lines are $2x-y=0$ and $x-2y+2=0$ respectively. If $P$ is the point $(4, 5)$, find by calculation the equations of the two straight lines which pass through $P$ and are each equally inclined to the two given lines.

71. Calculate the coordinates of the centres of the two circles of radius 5 which pass through the point $(4, 4)$ and touch the straight line

$$3x-4y-28=0.$$

72. Sketch the triangle formed by the lines

$$3x-4y-4=0, \quad 12x-5y+6=0 \quad \text{and} \quad 7x+24y-56=0,$$

and verify by calculation and reference to your sketch that the point $(1, 1)$ is the centre of the inscribed circle.

73. Show that, as $r$ varies, the locus of the point $P(a+r\cos\beta,\ b+r\sin\beta)$ is the straight line through the point $Q(a, b)$ inclined at an angle $\beta$ with the positive direction of the $x$-axis. What is the geometrical meaning of $r$? If $Q$ is the point $(1, 1)$, find the coordinates of the point $R$ such that the line $3x+4y=12$ bisects $QR$ at right angles.

74. If $O$ is the origin and $P$, $Q$ the points $(x_1, y_1)$, $(x_2, y_2)$ and if $OPJQ$ is a parallelogram, find the coordinates of $J$, and the area of the parallelogram. Find also the conditions that $OPJQ$ should be (i) a cyclic quadrilateral, (ii) a rhombus.

75. Show that any straight line through $A$, the intersection of the lines $ax+by+c=0$, $a'x+b'y+c'=0$, can be written in the form

$$(ax+by+c)+\lambda(a'x+b'y+c')=0.$$

Show that the line through $A$ which is perpendicular to the line $y=mx$ has the equation $(x+my)(b'a-a'b)+(b'c-c'b)+m(c'a-a'c)=0$.

76. $F$ is the point $(a, 0)$, $P$ the point $(at^2, 2at)$ and $N$ the point $(0, 2at)$. Find the equations of the bisectors of the angle $FPN$. Find the points where these bisectors meet the $y$-axis, and prove that one of these points, say $Q$, is such that $FQP$ is a right angle.

77. Show that the area of the triangle formed by the three straight lines whose equations are $y=m_1x+c_1$, $y=m_2x+c_2$ $(m_1>m_2)$ and $x=0$, is $(c_1-c_2)^2/2(m_1-m_2)$. Hence, or otherwise, find the area of the triangle formed by the lines $y=2x+3$, $y=-2x+7$, $y=6x+2$.

78. A triangle is formed by the lines $3x-4y=0$, $4x+3y=0$ and $5x+12y=14$. Prove that the incentre is the point $(1/11, 7/11)$ and find the radius of the incircle.

79. The point $P$ has coordinates $(a, b)$; $Q$ and $R$ are the feet of the perpendiculars from $P$ to the lines $y=2x$, $x=-2y$, respectively. Find the equation of the line $QR$. Find also the locus of $P$ if $QR$ is perpendicular to the line joining $P$ to the origin.

80. Prove that, if $\lambda$ is any constant, the equation

$$lx+my+n+\lambda(l'x+m'y+n') = 0$$

represents a line through the point of intersection of the lines

$$lx+my+n = 0 \quad \text{and} \quad l'x+m'y+n' = 0,$$

provided that these are not parallel. What does the equation represent if they are parallel? The sides $AB$, $AD$ of a parallelogram $ABCD$ have equations $7x+3y-13 = 0$, $x-2y+3 = 0$ respectively. The diagonal $AC$ is perpendicular to $AD$, and $BC$ passes through the point $(9, 1)$. Find the equations of $BC$ and $CD$.

# III

## THE PARABOLA

The most familiar curve, other than the straight line, is the circle. A study of this curve, however, does not so obviously need the tools provided by coordinate geometry, and many of its properties will already be well known. We therefore turn first to another curve, the parabola, which illustrates to the full the methods we are using. This curve may already have been met in other branches of mathematics. It is the path of a particle projected under gravity in a vacuum; it is also the form of the curve which, when rotated about its axis, forms a surface which reflects rays of light from a point source into a beam of parallel rays.

**1. Definition of a parabola.** The curve may be defined in a number of ways. We shall start from the 'focus directrix' definition:

The parabola is the locus of a point which varies so that its distance from a fixed point, the *focus*, is equal to its distance from a fixed line, the *directrix*, which does not contain the focus.

Let $S$ be the focus and $DD'$ the directrix. Let the perpendicular from $S$ to $DD'$ meet it in $Z$, and let $O$ be the mid-point of $SZ$, (Fig. 21). Take as coordinate axes $OX$, $OY$ the lines through $O$ perpendicular to and parallel to $DD'$, and let $OS = a$, so that $S$ is the point $(a, 0)$. Then if $P$ is a general point $(x, y)$ on the curve and the perpendicular from $P$ to $DD'$ meets it in $M$,

$$SP^2 = (x-a)^2+y^2;$$
$$PM^2 = (x+a)^2.$$

By definition $SP = PM$ and so $SP^2 = PM^2$; that is

$$(x-a)^2+y^2 = (x+a)^2,$$

whence $$\mathbf{y^2 = 4ax}.$$

This then is the equation satisfied by the coordinates of all points on the parabola, and conversely, any point whose coordinates satisfy this equation lies on the curve. We say, more shortly, that $y^2 = 4ax$ is the equation of the parabola (referred to the axes we have chosen).

It may happen that in some problem the axes are already chosen, and are not at our disposal. Suppose that in such a case the focus $S$ is the point $(h, k)$ and the equation of the directrix is

$$x \cos\alpha + y \sin\alpha = p.$$

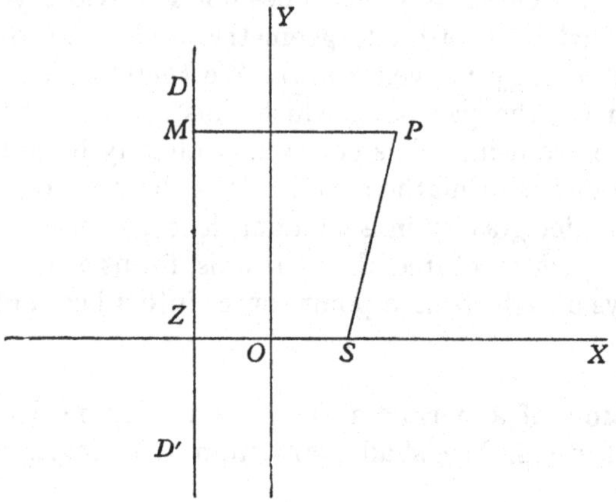


Fig. 21

Then if as before, $P$ is a general point $(x, y)$ on the curve and the foot of the perpendicular from $P$ to the directrix is $M$,

$$SP^2 = (x-h)^2 + (y-k)^2,$$

$$PM^2 = (x\cos\alpha + y\sin\alpha - p)^2 \quad \text{(from the formula on p. 48).}$$

Since $SP^2 = PM^2$, the locus of $P$ is

$$\{(x-h)^2 + (y-k)^2\} = (x\cos\alpha + y\sin\alpha - p)^2,$$

which reduces to

$$(x\sin\alpha - y\cos\alpha)^2 - 2x(h - p\cos\alpha) - 2y(k - p\sin\alpha) + (h^2 + k^2 - p^2) = 0.$$

The appearence of this equation will be sufficient to emphasize the desirability of using the much simpler form of equation of the parabola whenever that is justifiable, as it is in the large majority of cases. But it must be realized that this simplicity is due to the selection of convenient axes, and such selection can be made only *once* in any one problem.

**2. The shape of the parabola.** From the equation, $y^2 = 4ax$, in the standard form, we can deduce a number of facts about the curve.

(i) If in the equation we put $-y$ for $y$, the equation is unaltered. That is, if $(x, y)$ lies on the parabola, so does $(x, -y)$. The curve is thus symmetrical about the $x$-axis. This line, about which the parabola is symmetrical, is called the *axis* of the parabola.

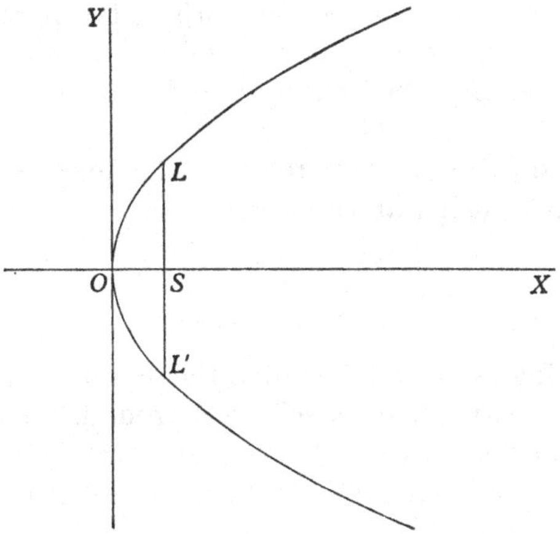


Fig. 22

(ii) Since in the diagram $a$ is positive, and $y^2 = 4ax$, $x$ must always be positive. This means that every point on the curve lies to the right of $OY$.

(iii) When $x = 0$, $y = 0$; the curve passes through $O$. This point, where the parabola meets its axis, is called the *vertex* of the parabola.

(iv) When $x$ is large, so is $y$.

(v) A line $x = h$ parallel to the directrix meets the curve where $y = \pm 2\sqrt{(ah)}$. When $h > 0$ these will be two distinct points; they coincide in the origin when $h = 0$.

(vi) A line $y = k$ parallel to the axis meets the curve in one point $(k^2/4a, k)$.

Those familiar with elementary calculus will complete the picture by finding the gradient of the curve at any point. From $y^2 = 4ax$, by differentiation,

$$2y\frac{dy}{dx} = 4a,$$

so

$$\frac{dy}{dx} = \frac{2a}{y}.$$

Thus for positive values of $y$, the gradient decreases as $y$ increases.

With the help of these results, the curve can now be sketched (Fig. 22). We notice that all parabolas have the same shape; varying $a$ merely involves a change of scale, since $y^2 = 4bx$ can be written $(ya/b)^2 = 4a(xa/b)$. The breadth of the parabola at the focus is called the *latus rectum*. It follows from the definition that $SL = 2a$, so that the latus rectum $LL' = 4a$.

**3. Parametric form of the parabola.** The line $y = mx$ through the origin meets the parabola where

$$(mx)^2 = 4ax,$$

$$x = 0, \quad 4a/m^2.$$

The point of intersection other than the origin is $(4a/m^2, 4a/m)$, and we see that, as the line $y = mx$ sweeps round, $m$ varies and the point on the parabola varies. A tidier form for this is obtained by putting $2/m = t$, so that the coordinates of the general point on the parabola are

$$(at^2, 2at).$$

This point lies on the parabola for all values of $t$; the reader should consider which points on the curve are given by $t = 0$, $t = \pm 1$, $t = \pm 2$, positive values of $t$, negative values of $t$.

We have here another example of the use of a parameter (see the definition on p. 31). Given any point $(x_1, y_1)$ on the parabola, the parameter $T = y_1/2a$ is uniquely determined, and conversely corresponding to any particular value $T$ of $t$ the unique point $(aT^2, 2aT)$ arises. We shall speak more shortly of the point with parameter $t$, or the point $t$, on the parabola, meaning the point with coordinates $(at^2, 2at)$.

In examples, it is usually simpler to take a point on the parabola in parametric form, instead of calling it $(x_1, y_1)$ with the condition $y_1^2 = 4ax_1$.

**4. Equations of chord, tangent and normal.** A line

$$lx+my+n=0$$

meets the parabola in points whose parameters satisfy the equation

$$lat^2+2mat+n=0.$$

This is a quadratic equation in $t$, so [A 4] it has either no real roots (in which case the line does not meet the parabola), or two roots $p$ and $q$, which may or may not coincide. Thus any line meets the parabola in at most two points.

Using the result [A 5] giving the sum and product of the roots of a quadratic, we have

$$p+q=-2m/l,$$

$$pq=n/al.$$

Thus the equation of a chord joining points $P$ and $Q$ with parameters $p$ and $q$ is (substituting for $m/l$ and $n/l$ in the equation $lx+my+n=0$)

$$\mathbf{x-\tfrac{1}{2}(p+q)\,y+apq=0.}$$

When the quadratic equation

$$lat^2+2mat+n=0$$

(whose roots are the parameters of the points on the parabola and the line) has *equal* roots, the line is said to meet the curve in *coincident* points, and to be a *tangent* there. Alternatively, we may think of one of the roots, $q$, say, as *tending to* coincidence with $p$, so that the tangent is the limiting position of the chord. Putting $q=p$ in the equation of the chord, we find that the equation of the tangent at the point $p$ is

$$\mathbf{x-py+ap^2=0.}$$

This may also be written as

$$y\,.\,2ap=2a(x+ap^2),$$

so that the tangent to the parabola at $(x_1, y_1)$ is

$$\mathbf{y\,.\,y_1=2a(x+x_1).}$$

The *normal* to a curve at a point is the line through the point on the curve perpendicular to the tangent there. The equation of the normal to the parabola at the point $(ap^2, 2ap)$ follows at once.

The gradient of the tangent is $1/p$, so the normal has gradient $-p$. Its equation is therefore

$$y-2ap = -p(x-ap^2),$$

which reduces to $\mathbf{px+y=ap^3+2ap.}$

***Illustration 1.*** *Show that the normal to the parabola $y^2 = 4ax$ at the point $(ap^2, 2ap)$ meets the parabola again at the point whose parameter is $-(p^2+2)/p$.*

The equation of the normal, as found above, is

$$px+y = ap^3+2ap.$$

This meets the parabola, whose general point is $(at^2, 2at)$, where

$$pat^2+2at = ap^3+2ap.$$

This is a quadratic equation

$$(ap)\,t^2+(2a)\,t-(ap^3+2ap) = 0$$

in $t$, so the sum of its roots [A 5] is

$$-2a/ap = -2/p.$$

One root is known to be $p$; so the other must be $-p-2/p$, or $-(p^2+2)/p$.

## EXAMPLES IIIA

1. Find from the focus directrix definition the equations of the parabolas with

(i) focus $(0,0)$, directrix $x = -2a$;
(ii) focus $(0,0)$, directrix $y = 2b$;
(iii) focus $(2,1)$, directrix $3x+4y = 0$.

2. Find the locus of $P(x,y)$, if $x = ut-\frac{1}{2}gt^2$, $y = vt$ and $u, v, g$ are constants, $t$ being a variable parameter. By changing the origin to $(u^2/2g, uv/g)$, (see p. 22), show that this is a parabola with latus rectum $2v^2/g$. [This gives the path of a particle projected under gravity with a velocity whose horizontal and vertical components are $u$ and $v$.]

3. What are the parameters of the points $(a, 2a)$, $(a, -2a)$, $(0,0)$, $(\frac{1}{4}a, -a)$, $(16a, 8a)$ on the parabola $y^2 = 4ax$?

4. Find the equation of the chord of the parabola joining $(ap^2, 2ap)$, $(aq^2, 2aq)$, by writing down the equation of the line joining these two points, and cancelling $p-q$.†

† To find the equation of the tangent, it is not now correct merely to put $q = p$, for in finding the equation of the chord we have when cancelling $p-q$ assumed that it is not zero. This step can be correctly taken by using a limiting argument, but a full discussion of this procedure belongs to the realm of the calculus.

5. Show by differentiation that the gradient of the tangent to the parabola $y^2 = 4ax$ at $(ap^2, 2ap)$ is $1/p$. Hence obtain the equation of the tangent.

6. $P$ and $Q$ are the points with parameters 3 and $-2$ on the parabola $y^2 = 4x$. Find the equation of $PQ$, and the equations of the tangents and normals at $P$, $Q$.

7. Find the equation of the normal to the parabola $y^2 = 4x$ at the point where $y = 2$. Find also the coordinates of the other point in which the normal cuts the curve.

8. One end of a chord through the focus of the parabola $y^2 = 8x$ is $(2, 4)$. What are the coordinates of the other end?

9. Find the equation of the normal to the curve $y = 4x^2$ at the point $(1, 4)$. If this normal cuts the axes at the points $P$ and $Q$ and if $O$ is the origin find the area of the triangle $POQ$.

10. Find the equations of the tangent and of the normal to the curve $y^2 = 4x$ at the point $(4, 4)$. Find also the area of the triangle formed by these lines and the $x$-axis.

11. Find the coordinates of the points in which the line $y = 4x - 2a$ meets the parabola $y^2 = 4ax$. Find the equations of the tangents to the parabola at these points and the coordinates of their point of intersection.

12. Show that the directrix is a tangent to any circle whose centre lies on a parabola and which passes through the focus.

13. Find the equation of the tangent to the parabola $y^2 = 12x$ which intercepts a length 2 on the $y$-axis.

14. A line with gradient 3 touches the parabola $y^2 = 4x$. Find the coordinates of its point of contact, and the equation of the normal at the point.

15. Find the equations of the tangents to the parabola $y^2 = 4x$ through the point $(5, -6)$.

16. Show that there are no tangents to the parabola $y^2 = 32x$ through the point $(1, 4)$.

17. Find the equation of the tangent to the parabola $y^2 = 12x$ at the point $(3t^2, 6t)$. Prove that the foot of the perpendicular from the point $(3, 0)$ to this tangent lies on the line $x = 0$ for all values of $t$.

18. If the chord $PQ$ of the parabola $y^2 = 4ax$ is normal at $P(a, 2a)$ find the length of $PQ$, the equations of the tangents $PT$ and $QT$, and the area of the triangle $PQT$.

**5. The point of intersection of tangents, and the equation of the chord of contact.** Suppose $(h, k)$ lies on the tangent at $(at^2, 2at)$, whose equation is

$$x - ty + at^2 = 0.$$

Then $$h-tk+at^2=0.$$

This is a quadratic equation in $t$, so that there are at most two tangents from a point to a parabola. If the roots of the equation are $p$ and $q$, the usual sum and product formulae [A 5] give

$$p+q=k/a, \quad pq=h/a.$$

Thus the tangents to the parabola at the points with parameters $p$ and $q$ meet in the point where

$$h=apq, \quad k=a(p+q),$$

that is, in the point

$$\{\mathbf{apq}, \quad \mathbf{a(p+q)}\}.$$

(It is worth noticing this method, which is easier and shorter than the more obvious method of writing down the equations of the two tangents and solving to find their common point.)

The same piece of algebra furnishes the equation of the chord joining the point of contact of tangents drawn to the parabola from the point $(h, k)$. This line is called the *chord of contact*; it is also called the *polar line* of $(h, k)$ with respect to the parabola. (A more complete definition of the polar line will be given later in the book; see p. 251.)

The equation of the chord joining the points with parameters $p$ and $q$ is

$$x-\tfrac{1}{2}(p+q)y+apq=0.$$

If the tangents at these points meet in $(h, k)$, we know already that

$$p+q=k/a, \quad pq=h/a.$$

Substituting these values in the equation of the chord, we find that the equation of the chord of contact is

$$x-(k/2a)y+h=0,$$

or

$$\mathbf{ky=2a(x+h)}.$$

From the same equation

$$h-kt+at^2=0,$$

which states that the tangent at $(at^2, 2at)$ passes through $(h, k)$, we can deduce the condition on a point in the plane for there to be a tangent through it. The equation has real and distinct roots in $t$ provided [A 4]

$$k^2>4ah.$$

We say that points of the plane through which there are two tangents to the parabola are *outside* the parabola, and points through which there are no tangents are *inside*. Thus we see that the parabola divides the plane into two regions:

*outside* the parabola when $y^2 > 4ax$

and *inside* the parabola when $y^2 < 4ax$.

Between these two regions there is the boundary, the parabola itself, where $y^2 = 4ax$.

**6. The condition for a line to touch the parabola.** The line

$$lx + my + n = 0$$

meets the parabola whose general point is $(at^2, 2at)$ where

$$lat^2 + 2mat + n = 0.$$

This is a quadratic in $t$, and it has equal roots if and only if

$$(am)^2 = n.al,$$

that is

$$am^2 = nl.$$

This then is the condition for the line to touch the parabola.

The same result may be obtained by an alternative method. If the line

$$lx + my + n = 0$$

is a tangent to the parabola at the point with parameter $t$, it represents the same locus as the equation

$$x - ty + at^2 = 0,$$

which is the equation of the tangent at the point $t$. The coefficients of $x$ and $y$ and the constant term must therefore be proportional; that is

$$\frac{l}{1} = \frac{m}{-t} = \frac{n}{at^2}.$$

(Notice that we cannot assert that the corresponding coefficients in the two equations are equal, but merely that the ratios are equal.) Thus

$$lt = -m \quad \text{and} \quad l^2t^2 = ln/a,$$

and these results hold for any value of $t$. Eliminating $t$, therefore, we obtain

$$(ln/a) = (-m)^2,$$

that is

$$am^2 = nl,$$

as the condition for the line to be a tangent to the parabola.

**7. Illustrations.** The results obtained so far, and particularly the methods used, are very important and will be used frequently. They are further illustrated in the following worked examples.

***Illustration 2.*** *The normal at a point $P$ of a parabola, focus $S$, meets the axis in $G$. Prove that there are two positions of $P$ such that the triangle $SPG$ is equilateral, and that the sides then have length $4a$.*

Take the equation of the parabola in the standard form $y^2 = 4ax$, and $P$ as the point $(at^2, 2at)$. Then the equation of the normal at $P$ is

$$tx + y = at^3 + 2at \quad (\S 4),$$

and this meets the axis of the parabola, $y = 0$, in the point

$$G(at^2 + 2a, 0).$$

The focus $S$ is the point $(a, 0)$, so that

$$SG^2 = (at^2 + a)^2,$$

$$SP^2 = PM^2 \text{ (with the usual notation, see p. 63)}$$

$$= (at^2 + a)^2$$

and $$PG^2 = (2a)^2 + (2at)^2.$$

We notice that $SP = SG$ for every value of $t$, that is for every position of $P$. If the triangle $SPG$ is to be equilateral, $SP^2 = PG^2$, and so

$$(2a)^2 + (2at)^2 = (at^2 + a)^2.$$

This reduces to

$$t^4 - 2t^2 - 3 = 0,$$

$$(t^2 - 3)(t^2 + 1) = 0.$$

Since $t$ is real, $t^2 + 1$ is not zero, so the roots of this equation are

$$t = \pm\sqrt{3}.$$

There are thus two positions of $P$ such that the triangle $SPG$ is equilateral, and in each the length of the side is $4a$, since

$$SG = at^2 + a = 4a.$$

***Illustration 3.*** *A triangle is formed by three tangents to the parabola $y^2 = 4ax$. Prove that its orthocentre lies on the directrix.*

Let the triangle be $LMN$, where $MN$, $NL$ and $LM$ touch the parabola at the points $P$, $Q$ and $R$ with parameters $p$, $q$ and

$r$ respectively. Then $L$ is the point of intersection of the tangents at $Q$ and $R$, so that it is the point

$$\{aqr,\, a(q+r)\} \quad (\S 5).$$

The line $MN$, which is the tangent at $P$, has gradient $1/p$, so that the gradient of a line perpendicular to $MN$ is $-p$. The equation of the altitude of the triangle through $L$ is therefore

$$y-a(q+r) = -p(x-aqr).$$

Similarly, the equation of the altitude through $M$ is

$$y-a(r+p) = -q(x-arp).$$

Subtracting these to eliminate $y$ (since we are interested in the $x$-coordinate of the orthocentre),

$$a(p-q) = -(p-q)x;$$

thus, since $p-q \neq 0$ (the sides of the triangle being distinct), at the point of intersection of the altitudes

$$x+a = 0.$$

This is the equation of the directrix, and so the orthocentre of the triangle lies on it.

An important point to notice in this example is that once the equation of the altitude through $P$ is found, no further work is needed in order to find the equation of the altitude through $Q$. We merely replace $p, q, r$ by $q, r, p$ in the result we have already obtained.

### EXAMPLES III B

In these examples the equation of the parabola is $y^2 = 4ax$.

1. If the normal at $P(at^2, 2at)$ to the parabola intersects the curve again at $Q$ and the position of $P$ is such that $t = 2$, prove that $PQ$ subtends a right angle at the focus $(a, 0)$.

2. A chord of the parabola passes through the focus and makes an angle $\tan^{-1}(\frac{3}{4})$ with the positive direction of the $x$-axis. Find the equations of the chord and of the tangents at its ends. Verify that the line joining the point of intersection of these tangents to the focus is at right angles to the chord.

3. $P$ and $Q$ are two points on the parabola with parameters $p$ and $q$. $O$ is the origin and $OP$ is perpendicular to $OQ$. Show that $pq+4 = 0$ and that the tangents to the curve at $P$ and $Q$ meet on the line $x+4a = 0$.

4. Find the coordinates of the point of intersection $C$ of the tangents to the parabola at the points $A(at_1^2, 2at_1)$ and $B(at_2^2, 2at_2)$. Show that the area of the triangle $ABC$ is $\frac{1}{2}a^2(t_1-t_2)^3$.

5. The tangent to the parabola at the point $P$, with parameter $p$, cuts the axis of the parabola in the point $L$ and any line through $L$ meets the parabola at the points $Q$ and $R$ with parameters $q$ and $r$. Prove that $q, p, r$ are in geometrical progression. Show also that if the tangents at $Q$ and $R$ intersect at $M$, then $MP$ is perpendicular to the axis of the parabola.

6. Find the finite value of $m$ for which the line $y = mx+2a$ through the point $(0, 2a)$ touches the parabola, and find the coordinates of the point of contact $Q$. If the line through $Q$ and the focus $(a, 0)$ meets the parabola again at $R$, find the coordinates of the point of intersection of the tangents at $Q$ and $R$.

7. A line parallel to the $y$-axis meets the parabola at points $P, Q$. If $PQ$ is of length $8a$, find the angle between the lines $OP$, $OQ$, where $O$ is the origin. If the tangent at $P$ meets the $y$-axis at $R$, find the distance $OR$.

8. The normal to the parabola at $P(at^2, 2at)$ meets the $x$-axis in $G$, the tangent meets the $y$-axis in $Z$ and the parallel through $Z$ to the $x$-axis meets the normal at $P$ in $Q$. Prove that if $S(a, 0)$ is the focus, then $SZQG$ is a parallelogram of area $a^2t(1+t^2)$.

9. The normal to the parabola at a point $P(at^2, 2at)$ makes an angle $\theta$ with the $x$-axis and meets the parabola again at $Q$. Show that

$$PQ = 4a \sec\theta \operatorname{cosec}^2\theta.$$

10. The straight line through the vertex of the parabola perpendicular to the tangent at $P(4a, 4a)$ meets this tangent at $Q$. Find the equation of the other tangent from $Q$.

11. $P(ap^2, 2ap)$ and $Q(aq^2, 2aq)$ are two points on the parabola and $p > q$. The tangent at $P$ and the line through $Q$ parallel to the $x$-axis meet at $R$; the tangent at $Q$ and the line through $P$ parallel to the $x$-axis meet at $S$. Prove that $PQRS$ is a parallelogram of area $2a^2(p-q)^3$.

12. Prove that the line $x-2y+4a = 0$ touches the parabola, and find the coordinates of $P$, the point of contact. If the line $x-2y+2a = 0$ meets the parabola in $Q$, $R$ and $M$ is the mid-point of $QR$, prove that $PM$ is parallel to the axis of $x$, and that this axis and the line through $M$ perpendicular to it meet on the normal at $P$ to the parabola.

**8 Geometrical properties.** As well as exemplifying the use of a parameter, the parabola also possesses an interesting chain of properties of a geometrical nature, which can be simply established without using algebra after the first step has been taken. This initial stage can also be obtained by purely geometrical argument, though the proof is rather more cumbersome. Methods of 'geometrical conics', which obtain results without any explicit recourse to algebra, are now largely of historical interest only; in solving a problem we should look for an economical solution, using either algebraic or synthetic reasoning, or probably a judicious mixture of the two.

We refer to the familiar diagram (Fig. 23) with some further points labelled. $P$ is any point on the parabola, vertex $O$, focus $S$.

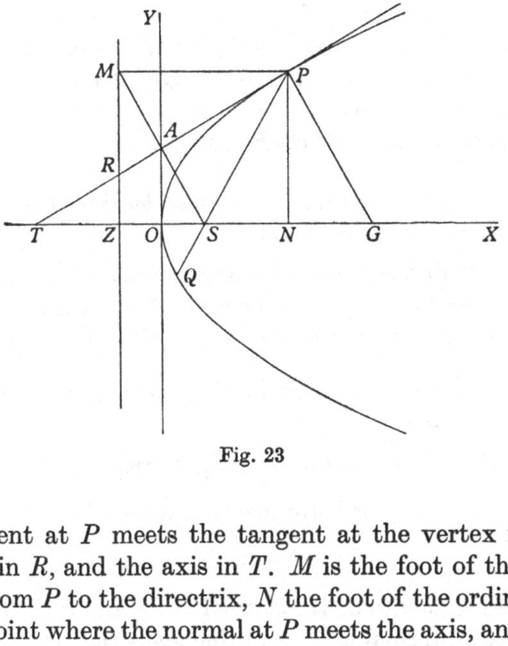


Fig. 23

The tangent at $P$ meets the tangent at the vertex in $A$, the directrix in $R$, and the axis in $T$. $M$ is the foot of the perpendicular from $P$ to the directrix, $N$ the foot of the ordinate at $P$; $G$ is the point where the normal at $P$ meets the axis, and the axis meets the directrix in $Z$. In the diagram $S$, $A$ and $M$ appear collinear, as indeed they are; that is one of the results to be proved.

As a start to the chain of results we find the position of $T$. If $ON$, $OA$ are taken as axes of coordinates in the usual way, and $P$ is the point $(ap^2, 2ap)$, the equation of the tangent at $P$ is

$$x - py + ap^2 = 0.$$

This meets the axis, $y = 0$, where $x = -ap^2$. Thus $OT = ON$.

The properties now follow, with in each case an indication of the proof.

(1) *SPMT is a rhombus.*

For $OT = ON$, and $OS = OZ$ (since $O$ is on the parabola);

so $$ST = ZN = MP$$
$$= SP \text{ (since } P \text{ is on the parabola).}$$

Thus $ST$ is equal and parallel to $PM$, and $ST = SP$.

(2) *The foot of the perpendicular from the focus to a tangent lies on the tangent at the vertex.*

For $OT = ON$, and so $TA = AP$ (since $OA$ is parallel to $PN$). $A$ is therefore the mid-point of the diagonal $PT$, and so lies on and bisects the other diagonal $SM$. The diagonals of a rhombus are perpendicular, so angle $SAP$ is a right angle.

(3) *The circumcircle of a triangle formed by three tangents to the parabola passes through the focus.*

For the feet of the perpendiculars from $S$ to the three sides are collinear (on the tangent at the vertex), and the result follows from the converse of the theorem of the Simson line [G 6].

(4) *SP and the line through P parallel to the axis make equal angles with the tangent at P.*

For if $MP$, $TP$ are produced to $M_1$, $T_1$ respectively,

$$\angle TPS = \angle TPM \text{ (property of a rhombus)}$$
$$= \angle T_1PM_1 \text{ (vertically opposite angle),}$$

which is the result stated. It follows that $SP$ and $PM_1$ are equally inclined to the normal at $P$, and this gives the 'reflecting property', that rays from a point source of light at $S$ will be reflected at a parabola into a parallel beam. This principle is used when making car headlamps and searchlights. Conversely, a parallel beam of light is focused by a parabolic surface at the focus of the parabola (hence the name).

(5) *The portion of the tangent between the point of contact and the directrix subtends a right angle at the focus.*

For $\angle SPR = \angle MPR$ and so the triangles $SPR$, $MPR$ are congruent. It follows that $\angle RSP$ is a right angle.

(6) *The points of contact of tangents from a point on the directrix are the ends of a focal chord* (that is, a chord through the focus).

For if the other tangent from $R$ meets the parabola in $Q$, the angles $RSP$ and $RSQ$ are both right angles; and so $PSQ$ is a straight line.

(7) *The tangents at ends of a focal chord intersect at right angles on the directrix.*

The tangents at $P$ and $Q$ meet on the directrix, since if they meet the directrix in $R$, $R'$, $\angle RSP$ and $\angle R'SQ$ are both right angles, so that $RS$ and $R'S$ must coincide. Since the triangles $SPR$, $MPR$ are congruent, the angles $SRP$, $MRP$ are equal. Similarly, if $M'$ is the foot of the perpendicular from $Q$ to the directrix, the angles $SRQ$, $M'RQ$ are equal, and it follows that $\angle PRQ$ is a right angle.

(8) $SA^2 = SO.SP$.

For the triangles $OAS$, $APS$ are similar. This gives an equation $p^2 = ar$ between $p$, the length of the perpendicular from $S$ to the tangent at $P$, and $r$, the length of $SP$. This is called the $(p, r)$ equation of the parabola, and is occasionally referred to in books on mechanics and calculus.

(9) *$SMPG$ is a parallelogram.*

For $PG$ is parallel to $SM$, both being perpendicular to $PT$.

(10) *The subnormal (the distance along the axis under the normal) of a parabola is a constant length.*

For $SG = ZN$, and so $NG = ZS = 2a$.

(11) *If $T$ is any point of the plane and the polar line of $T$ meets the directrix in $D$, then $\angle TSD$ is a right angle* (Fig. 24).

This property is most easily established by algebra. Let $T$ be the point $(h, k)$; its polar line is $ky = 2a(x+h)$ (§5). This meets the directrix $x = -a$ in $D(-a, 2a(h-a)/k)$. The gradient of $SD$ is

$$\frac{2a(h-a)}{-2a.k},$$

and the gradient of $ST$ is

$$\frac{k}{h-a}.$$

The product of these gradients is $-1$, so the lines are perpendicular.

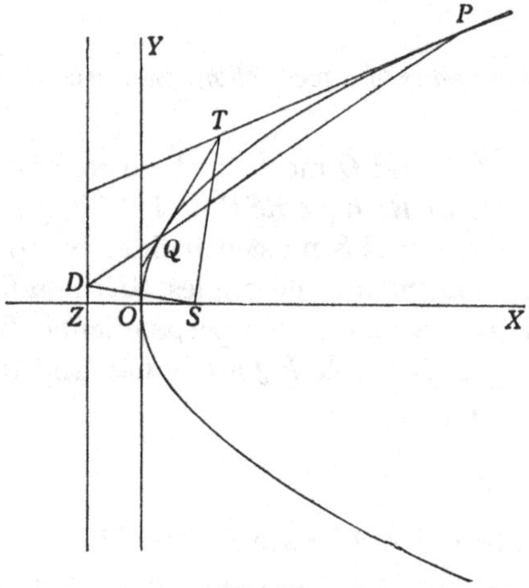


Fig. 24

(12) *If $TP$, $TQ$ are tangents to the parabola, the angles $TSP$, $TSQ$ are equal.*

For if the feet of the perpendiculars from $P$ and $Q$ to the directrix are $M$ and $M'$,

$$\frac{SP}{SQ} = \frac{PM}{QM'} = \frac{DP}{DQ},$$

and so $DS$ bisects $\angle PSQ$ externally. Since $\angle TSD$ is a right angle, $TS$ is the internal bisector of $\angle PSQ$.

An alternative proof of this result is similar to one given on p. 187 for the ellipse.

## EXAMPLES III C

1. With the notation of Fig. 23, prove the following results:

(i) $OA^2 = OS.ON$.

(ii) $PA.AR = SO.SP$.

(iii) $TA = AN$.

(iv) The line through $A$ parallel to the axis bisects $SP$.

(v) The tangent at the vertex touches the circle on $SP$ as diameter.
(vi) $PG = 2.AS$.
(vii) $PG^2 = 2.SL.SP$ (where $L$ is an extremity of the latus rectum).
(viii) If the tangent at $P$ meets $SL$ at $K$ then $SK = SR$.
(ix) If $OP$ meets the directrix in $V$ then $SV$ is parallel to $PT$.
(x) If the foot of the perpendicular from $G$ to $SP$ is $F$, $PF = SL$.

2. $PSQ$ is a focal chord of a parabola whose vertex is $O$ and focus $S$. $M$ is the foot of the perpendicular from $P$ to the directrix, and $N$ is the foot of the perpendicular from $P$ to the axis. Prove the following results:

(i) The tangent at $Q$ is parallel to $SM$.
(ii) The normals at $P$ and $Q$ intersect at right angles.
(iii) If the tangents at $P$ and $Q$ meet at $R$ and $RQ$ meets the latus rectum at $U$ then $RU = SM$.
(iv) The circle on $PQ$ as diameter touches the directrix.
(v) The line through $R$ parallel to the axis bisects $PQ$.
(vi) If $OP$ meets the directrix in $V$ then $VQ$ is parallel to the axis.
(vii) $\angle MSV$ is a right angle.
(viii) If the line through $Q$ parallel to the axis meets $PN$ at $H$ then $\angle PGH$ is a right angle, where $G$ is the point where the normal at $P$ meets the axis.
(ix) $SP.SQ = PQ.OS$.
(x) Circles drawn through $S$ to touch the parabola at $P$ and $Q$ are orthogonal (that is, they cut at right angles).
(xi) If the line through $R$ and the mid-point of $PQ$ meets the parabola in $C$, then $RC = CM$.
(xii) $PQ = 4.SC$.

3. $PSP'$ is a focal chord of a parabola with focus $S$ and vertex $O$. The tangents at $P$ and $P'$ meet in a point $R$, the tangents at $P$ and $O$ in a point $A$. The tangent at $P$ meets the axis of the parabola at $T$ and the latus rectum at $K$. Prove that $A$ is the mid-point of $RK$ and of $TP$. If the normal at $P$ meets the axis at $G$ and the tangent at $P'$ meets the latus rectum at $K'$, prove that $PG = RK'$.

4. Two parabolas have a common focus $S$ and axes in opposite directions. Prove that they intersect at right angles. Prove also that the distance between their vertices is equal to the distance of either point of intersection from $S$.

5. Tangents at two points $P$ and $Q$ on a parabola with focus $S$ meet at $T$. Prove the following results:

(i) If the tangents at $P$, $Q$ meet the tangent at the vertex in $V$, $W$ then $STVW$ is a cyclic quadrilateral.
(ii) $ST$ makes with the tangent at $Q$ the same angle as the axis makes with the tangent at $P$.
(iii) The triangles $SPT$, $STQ$ are similar.
(iv) $ST^2 = SP.SQ$.

(v) If the tangents at $P$ and $Q$ meet the axis in $L$ and $M$, then $ST$ is the tangent at $T$ to the circle $TLM$.

(vi) If from a point $E$ on the axis tangents are drawn meeting any other tangent at $U$ and $V$, then $SU = SV$.

(vii) If the feet of the perpendiculars from $P$, $Q$ to the directrix are $M$, $M'$, circles centres $P$ and $Q$ pass through $S$ and touch the directrix at $M$ and $M'$.

(viii) The line through $S$ perpendicular to $PQ$ meets the directrix in $K$, the mid-point of $MM'$.

(ix) The line through $K$ parallel to the axis bisects $PQ$ in $V$.

(x) $T$ lies on $KV$, and $TV$ meets the parabola in the mid-point of $TV$.

6. $PSP'$, $QSQ'$, $RSR'$ are three focal chords of a parabola. $QR$ meets the line through $P'$ parallel to the axis in $A$, $RP$ meets the line through $Q'$ parallel to the axis in $B$, and $PQ$ meets the line through $R'$ parallel to the axis in $C$. Show that $A$, $B$ and $C$ are on a straight line through $S$.

7. $TP$, $TQ$ are tangents to a parabola vertex $O$, and $OP$, $OQ$ and $OT$ meet the directrix in $P'$, $Q'$ and $T'$ respectively. Show that $P'T' = T'Q'$.

8. The tangents at $P$ and $Q$ meet at $T$, and $ST$ meets the parabola at $A$ and $B$. Show that the tangents at $P$ and $Q$ are equally inclined to the tangents at $A$ and $B$.

**9. Problems about loci.** A problem with which coordinate geometry is particularly well suited to deal is that of finding a locus. The problem is very often as follows: $P$ is a point which can vary on some curve $C$, and another point, $R$, say, is constructed from $P$ and any fixed points and lines which may be given. As a result of the variation of $P$, $R$ will also vary, and the problem is to find this locus of $R$.

The curve $C$ which is given may be any curve; we will illustrate by taking it as the parabola $y^2 = 4ax$. The variable point $P$ on the parabola we take as $(ap^2, 2ap)$, and we notice that when $P$ varies the parameter $p$ will vary also. The problem is then tackled in three stages:

(i) Find the coordinates $(X, Y)$ of $R$, in terms of $p$ and any fixed numbers (usually at least the $a$ involved in $y^2 = 4ax$);

(ii) Eliminate $p$ between the expressions for $X$ and $Y$, thus finding a relation which is satisfied by the coordinates of $R$ for all values of $p$, and so for all positions of $P$;

(iii) Replace $X$ by $x$, $Y$ by $y$ in this equation; the resulting equation is that of the locus of $R$.

This in outline is the method adopted, though in particular circumstances modifications are needed. Frequently $R$ depends on two points $P$ and $Q$ on the parabola, and so on the parameters $p$ and $q$; these are connected by some equation which comes from some condition on $P$ and $Q$, for example, $PQ$ passing through a fixed point. Again, in some more involved examples, stage (i) of this method need not be carried out in full; Illustration 6 shows an instance of this. Another modification arises, when the varying of $R$ results from some variation other than $P$ moving on the parabola. For example, $R$ might depend on a variable point on a fixed line, or (as in Examples III D, 25) on the parabola which itself varies. In each case the important thing is to decide what causes the variation of the point $(X, Y)$ whose locus is required.

***Illustration 4.*** *$P$ is any point on the parabola $y^2 = 4ax$. The tangent at $P$ meets the $y$-axis at $Q$ and $R$ is the mid-point of $PQ$. Prove that the locus of $R$ is a parabola.*

The equation of the tangent at $P(ap^2, 2ap)$ is

$$x - py + ap^2 = 0$$

and it meets the $y$-axis, $x = 0$, in $Q(0, ap)$. $R$ is the mid-point of $PQ$, and so if $R$ is $(X, Y)$,

$$X = \tfrac{1}{2}ap^2, \quad Y = \tfrac{1}{2}(2ap + ap) = \tfrac{3}{2}ap.$$

To eliminate $p$ we write these equations as

$$p^2 = 2X/a, \quad p = 2Y/3a,$$

and so

$$(2Y/3a)^2 = (2X/a).$$

This is the equation between $X$ and $Y$, which does not involve the varying $p$.

Now replace $X$ by $x$ and $Y$ by $y$, and simplify; the equation of the locus of $R$ is thus

$$2y^2 = 9ax.$$

This, we notice, is the equation of a parabola which has the same axis and vertex as the one from which we started; its focus is at $(9a/8, 0)$.

***Illustration 5.*** *The chord $PQ$ of the parabola $y^2 = 4ax$ passes through the foot of the directrix. Find the locus of the intersection of the normals at $P$ and $Q$.*

Here there are two variable points, $P$ and $Q$, connected by the condition that $PQ$ passes through $(-a, 0)$. If $P$ and $Q$ have parameters $p$ and $q$, the equation of the chord $PQ$ is

$$x - \tfrac{1}{2}(p+q)y + apq = 0.$$

This passes through $(-a, 0)$, and so

$$pq = 1.$$

(We could now write $q = 1/p$ all through, but it is tidier to keep both $p$ and $q$ in the working, and remember the connection between them.)

We now find $R(X, Y)$. The equations of normals at $P$ and $Q$ are

$$px + y = ap^3 + 2ap$$

and
$$qx + y = aq^3 + 2aq.$$

Where these meet,

$$X = a(p^2 + pq + q^2 + 2), \quad Y = -apq(p+q).$$

To eliminate $p$ and $q$, we remember that $pq = 1$, so that

$$X = a(p^2 + 2pq + q^2 + 2 - pq)$$
$$= a(p+q)^2 + a$$

and
$$Y = -a(p+q).$$

It follows that
$$Y^2 = a^2(p+q)^2 = a(X-a).$$

The locus of $R$ is therefore given by the equation

$$y^2 = a(x-a);$$

this is the equation of another parabola, whose vertex is at $(a, 0)$, the focus of the given parabola.

***Illustration 6.*** *$H$ is a fixed point on the axis of a parabola whose focus is $S$. A variable chord $PQ$ passes through $H$, the tangents at $P$ and $Q$ meet in $T$, and $R$ is the foot of the perpendicular from $T$ to $PQ$. Find the locus of $R$.*

Let the parabola be $y^2 = 4ax$, $P$ and $Q$ the points with parameters $p$ and $q$, and $H$ the point $(h, 0)$. The equation of $PQ$ is

$$x - \tfrac{1}{2}(p+q)\,y + apq = 0$$

and, since this passes through $(h, 0)$,

$$apq + h = 0.$$

(To keep the symmetry we continue to work with both $p$ and $q$, noting this relation for future use.)

The tangents at $P$ and $Q$ meet in $T\{apq, a(p+q)\}$. The equation of the line through $T$ perpendicular to $PQ$ is therefore

$$(p+q)\,x + 2y = (p+q)\,.\,apq + 2a(p+q).$$

Thus if $R$ is $(X, Y)$, $X$ and $Y$ satisfy this equation, and the equation of $PQ$, so that

$$2X-(p+q)\,Y-2h = 0$$

and
$$(p+q)\,X+2Y = -h(p+q)+2a(p+q),$$

where we have now used the relation $apq = -h$.

To find $X$ and $Y$ separately from these equations would be rather tedious; as our ultimate aim is an equation in $X$ and $Y$, not involving $p$ or $q$, it is quicker to notice that

$$(p+q)\,Y = 2X-2h$$

and
$$(p+q)\,(X+h-2a) = -2Y,$$

so
$$\frac{X+h-2a}{Y} = \frac{-Y}{X-h}.$$

This gives
$$X^2-h^2-2aX+2ah+Y^2 = 0,$$

so the equation of the locus of $R$ is

$$x^2+y^2-2ax+2ah-h^2 = 0,$$

or
$$(x-a)^2+y^2 = (a-h)^2.$$

Remembering Illustration 2 of chapter I, we identify this as a circle, whose centre is $S$ and whose radius is $SH$.

***Illustration* 7.** *A tangent to the parabola $y^2 = 4ax$ meets the parabola $y^2 = 4bx$ at the points $P$ and $Q$, and the tangents at $P$ and $Q$ meet at $R$. Find the locus of $R$.*

In this example there are two parabolas; we shall take the general point on $y^2 = 4ax$ as $(ap^2, 2ap)$, as usual, and on $y^2 = 4bx$ the general point will be $(bt^2, 2bt)$, with parameter $t$. The equation of the tangent to $y^2 = 4ax$ at the point $p$ is

$$x-py+ap^2 = 0.$$

This meets $y^2 = 4bx$ in points whose parameters are the roots of the equation

$$bt^2-2bpt+ap^2 = 0,$$

where this equation is thought of as a quadratic in $t$, the value of $p$ for the moment being fixed. If the roots are $q$ and $r$,

$$q+r = 2p, \quad qr = ap^2/b.$$

The tangents to $y^2 = 4bx$ at $q$ and $r$ meet in the point $\{bqr, b(q+r)\}$, (adapting the result of §5). If then $R$ is the point $(X, Y)$,

$$X = bqr = ap^2$$

and
$$Y = b(q+r) = 2bp.$$

Eliminating $p$, $$(Y/2b)^2 = X/a,$$

and so the equation of the locus of $R$ is

$$ay^2 = 4b^2x,$$

another parabola, with the same vertex and axis as the two given ones, but with a different focus (unless $a = b$).

## EXAMPLES III D

In these examples, the equation of the parabola is $y^2 = 4ax$.

1. $P$ is any point on the parabola, focus $S$, and $Q$ is a point such that $PQ$ is parallel to the $x$-axis and $QS$ makes an angle 45° with the positive $x$-axis. Find the locus of the mid-point of $PQ$, as $P$ varies on the parabola.

2. If the tangent to the paràbola at $P$ meets the $x$-axis at $T$, and $S$ is the focus, find the coordinates of the orthocentre $H$ of the triangle $SPT$ and show that as $P$ varies the equation of the locus of $H$ is the curve $ay^2 = x(a-x)^2$.

3. If the normal to the parabola at $P$ meets the axis of the parabola at $G$ and $GP$ is produced, beyond $P$, to $Q$ so that $P$ is the mid-point of $GQ$, show that the equation of the locus of $Q$ is $y^2 = 16a(x+2a)$.

4. Find the coordinates of $N$, the foot of the perpendicular drawn from the origin to the tangent to the parabola at a general point $P$. Show that, as $P$ varies, the locus of $N$ is the curve $x(x^2+y^2)+ay^2 = 0$.

5. If the tangent and normal at a point $P$ on the parabola meet the line $y = 0$ in $T$ and $G$, find the equation of the locus of the centroid of the triangle $PGT$.

6. $P$ is any point on the parabola, and $O$ is the origin; $Q$ is the foot of the perpendicular from $P$ to the $y$-axis, $R$ is the foot of the perpendicular from $Q$ to $OP$, and $QR$ produced meets the $x$-axis at $K$. Prove that $K$ is a fixed point, and find its coordinates. Prove also that the locus of $R$ is a circle, and find its centre.

7. $PQ$ is a chord of the parabola which is normal at $P$. Find the locus of the point $R$ on the chord such that $PR:RQ = 1:3$.

8. If a tangent to the parabola meets the axis in $T$ and the $y$-axis in $Q$, and the rectangle $TOQR$ is completed ($O$ being the origin), find the locus of $R$.

9. A tangent to the parabola $y^2 = 4bx$ meets the parabola $y^2 = 4ax$ in $P$ and $Q$. Prove that the equation of the locus of the mid-point of $PQ$ is $y^2(2a-b) = 4a^2x$.

10. Find the coordinates of the point of intersection of the tangents to the parabola at the points $P(ap^2, 2ap)$ and $Q(aq^2, 2aq)$. Show that the locus of this point is a straight line whenever $pq$ is constant, and that when $pq+1 = 0$ this line is the directrix.

11. $P$ and $Q$ are variable points on the parabola and the tangents at $P$ and $Q$ meet in $T$. If $PQ$ passes through the fixed point $(c, 0)$, prove that $T$ lies on the line $x+c = 0$. Prove also that the centroid of the triangle $PQT$ lies on the curve $y^2 = a(3x-c)$.

12. Find the coordinates of $R$, the point of intersection of the tangents to the parabola at $P$ and $Q$, the points with parameters $p$ and $q$. Show that, if $PQ$ touches the parabola $y^2 = 2ax$, the point $R$ lies on the parabola $y^2 = 8ax$.

13. Chords $PQ$ of the parabola are drawn parallel to the line

$$lx+my = 1.$$

Find the locus of the point of intersection of tangents at $P$ and $Q$.

14. From a fixed point $H$ on a parabola chords $PH$, $QH$ are drawn at right angles. Find the locus of the point of intersection of tangents at $P$ and $Q$.

15. Show that the locus of the point of intersection of two tangents to a parabola at points whose ordinates are in a constant ratio is a parabola.

16. The two tangents from a point $R$ to the parabola have gradients $m_1$ and $m_2$. Find the locus of $R$ (i) when $m_1+m_2$ is constant, (ii) when $m_1^2+m_2^2$ is constant.

17. Prove that, if the difference of the ordinates of two points on a parabola is constant, then the locus of the point of intersection of the tangents at these points is an equal parabola.

18. The chord $PQ$ of a parabola passes through the focus, and the normals at $P$ and $Q$ meet in $R$. If $P$ and $Q$ have parameters $p$ and $q$ show that $pq+1 = 0$. Find the coordinates of $R$, and hence find its locus.

19. If two perpendicular tangents meet at $T$ and the corresponding normals meet at $N$, show that, for all such pairs of tangents, $TN$ is parallel to the axis of the parabola, and find the locus of $N$.

20. Find the locus of the point of intersection of the normals at two points on a parabola when the chord joining them subtends a right angle at the vertex.

21. A variable chord $PQ$ of the parabola passes through the fixed point $(b, c)$; show that the locus of the meet of the normals at $P$ and $Q$ is given by the equation $2(cx+2ay-2ac-bc)^2 = (4ab-c^2)(2bx+cy-4ab-2b^2)$.

22. The line $y = m(x+a)$ meets the parabola in the two points $P$ and $Q$. Find the coordinates of the mid-point of $PQ$ in terms of $a$ and $m$, and show that the locus of this point is the parabola $y^2 = 2a(x+a)$.

23. The chord $PQ$ of the parabola cuts the $x$-axis in $R$, and the tangent at $P$ cuts the $y$-axis in $S$. If $P$ and $Q$ are variable points on the parabola, and if $RS$ passes through the fixed point $(2a, -a)$ prove that the equation of the locus of the mid-point of $RS$ is $2xy+ax-2ay = 0$.

24. $P$ and $Q$ are two points on a parabola whose focus is $S$. The tangents at $P$ and $Q$ meet at $T$, and the normals at these points meet at $N$; $R$ is the mid-point of $TN$. Prove the angle $TSR$ is a right angle. If the chord $PQ$ passes through the focus of the parabola, prove the locus of $R$ is a parabola with its axis coinciding with that of the first parabola, and vertex $S$.

25. $H(h, k)$ is a fixed point not on the axis, and $R$ is the foot of the perpendicular from $H$ to its polar line (see definition, p. 70) with respect to the parabola $y^2 = 4ax$. Prove that, as $a$ varies, the locus of $R$ is a circle through $H$.

**10*. Locus of mid-points of parallel chords.** We now establish an important locus property of the parabola:

If chords of a parabola are drawn parallel to a fixed direction, their mid-points lie on a line parallel to the axis.

This line is called a *diameter* of the parabola. A special case of this result, when the chords are all parallel to the directrix, is obvious. In general, it follows from Examples III C, 5 (ix). We give two other proofs, as the methods used are interesting.

***Method 1.*** Suppose the chords all have gradient $m \neq 0$. The gradient of a chord joining $P(ap^2, 2ap)$ and $Q(aq^2, 2aq)$ is

$$\frac{2ap-2aq}{ap^2-aq^2} = \frac{2}{(p+q)}, \quad \text{since } p-q \neq 0.$$

Thus, for any chord $PQ$,

$$2/(p+q) = m$$

and so

$$p+q = 2/m.$$

The mid-point of $PQ$ is $\{\tfrac{1}{2}(ap^2+aq^2), \tfrac{1}{2}(2aq+2aq)\}$, and this we see lies on the line $y = 2a/m$, which is a fixed line parallel to the axis.

***Method 2.*** This method is perhaps harder to follow in the first place, but it is included as it has wide applications.

Let $(h, k)$ be the mid-point of a chord with gradient $m$. Then if $\tan\alpha = m$, a variable point on the chord is $(h+r\cos\alpha, k+r\sin\alpha)$, where $r$ is the variable parameter on the line (see p. 32). This point lies on the parabola when

$$(k+r\sin a)^2 = 4a(h+r\cos\alpha)$$

or $$r^2\sin^2\alpha+2r(k\sin\alpha-2a\cos\alpha)+k^2-4ah = 0.$$

Since $(h, k)$ is the mid-point of the chord, the roots of this equation in $r$ must be equal and opposite, and the condition for this [A 4] is that the coefficient of $r$ must be zero. Thus

$$k\sin\alpha-2a\cos\alpha = 0$$

and so $$\tan\alpha = m = 2a/k.$$

It follows that the locus of the mid-point $(h, k)$ is

$$y = 2a/m.$$

***Illustration 8.*** *Find the locus of the mid-points of chords which subtend a right angle at the vertex of the parabola.*

Choose the axes in the usual way, so that the vertex is $(0, 0)$ and the equation of the parabola is $y^2 = 4ax$. Suppose that a typical chord joins the points $P$ and $Q$ with parameters $p$ and $q$. The gradient of $OP$ is

$$2ap/ap^2 = 2/p \quad \text{since} \quad p \neq 0.$$

Similarly, the gradient of $OQ$ is $2/q$. Since $OP$ and $OQ$ are perpendicular

$$(2/p)(2/q) = -1,$$

$$pq = -4.$$

Now if $(X, Y)$ is the mid-point of the chord,

$$2X = ap^2+aq^2, \quad 2Y = 2ap+2aq.$$

To find the equation of the locus of $(X, Y)$, we eliminate $p$ and $q$, using the fact that $pq = -4$.

$$Y^2 = a^2(p+q)^2 = a^2(p^2+q^2)+2a^2pq$$
$$= a.2X+2a^2.-4.$$

The equation of the locus of the point $(X, Y)$ is thus

$$y^2 = 2a(x-4a).$$

From the form of this equation we see that the locus is a parabola with vertex at $(4a, 0)$ and latus rectum $2a$.

***Illustration 9.*** *Find the locus of the mid-points of chords of contact of tangents to the parabola $y^2 = 4ax$ drawn from points on the directrix.*

Let a general point of the directrix be $(-a, \lambda)$ (where $\lambda$ is the number which will later be allowed to vary). Then the equation of the chord of contact of tangents drawn to the parabola from this point (§5) is

$$\lambda y = 2a(x-a).$$

If the mid-point of a chord is the point $(X, Y)$, the gradient of the chord (see method 2, above) is $2a/Y$. The gradient of the given chord is $2a/\lambda$, and so, if its mid-point is $(X, Y)$,

$$2a/\lambda = 2a/Y$$

and so

$$\lambda = Y.$$

(This tells us, incidentally, that the tangents at the ends of a chord intersect on the diameter which bisects the chord.) We must also use the fact that the mid-point $(X, Y)$ lies on the chord; so

$$Y\lambda = 2a(X-a).$$

Eliminating $\lambda$, we find that $(X, Y)$ satisfies

$$Y^2 = 2a(X-a),$$

so that the equation of the locus of the mid-point is

$$y^2 = 2a(x-a).$$

**11*. Conormal points.** We have already shown (§5) that from a point not on the curve there are either two or no tangents to a parabola. We now investigate the corresponding result for normals.

The equation of the normal to a parabola at a point $(at^2, 2at)$ is

$$tx + y = at^3 + 2at.$$

If this passes through a fixed point $(h, k)$,

$$th + k = at^3 + 2at,$$

that is

$$at^3 + t(2a-h) - k = 0.$$

This is a cubic equation in $t$; we deduce that there are at most three normals to a parabola from a point. The feet of these normals are called *conormal* points on the curve.

If the parameters of three conormal points are $p$, $q$ and $r$, then these are the roots of the cubic equation in $t$. The result on the sum and products of the roots [A 6] gives

$$p+q+r = 0,$$

$$qr+rp+pq = \frac{2a-h}{a},$$

$$pqr = k/a,$$

so we have the result that, if the normals at the points with parameters $p$, $q$ and $r$ are concurrent, then

$$p+q+r = 0.$$

Conversely, if $p+q+r = 0$, then the normals at the points $p$, $q$ and $r$ are concurrent, for if the normals at $p$ and $q$ meet in $(h, k)$, then there is a third normal through this point, and if the parameter of its point of contact is $r'$, then $p+q+r' = 0$. It follows that $r = r'$.

Results about conormal points can usually be derived from the three equations found above. Eliminating $r$, which is equal to $-(p+q)$, we find that the point of intersection of the normals at $p$ and $q$ is given by

$$pq+(p+q)r = pq-(p+q)^2 = 2-h/a,$$

and

$$pq \,.\, -(p+q) = k/a;$$

that is, the normals intersect in the point

$$\{a(p^2+pq+q^2+2), \; -apq(p+q)\}.$$

***Illustration 10.*** *A variable chord PQ of the parabola $y^2 = 4ax$ is parallel to the line $y = x$. If P and Q are the points $(ap^2, 2ap)$ and $(aq^2, 2aq)$, show that $p+q = 2$, and prove that the locus of the point of intersection of the normals to the parabola at P and Q is a straight line.*

The gradient of $PQ$ is

$$\frac{2ap-2aq}{ap^2-aq^2} = \frac{2}{p+q}, \quad \text{since} \quad p-q \neq 0.$$

It is given that $PQ$ is parallel to $y = x$; that is, it has gradient 1. Thus

$$2/(p+q) = 1,$$

so

$$p+q = 2, \quad \text{as required.}$$

The equation of the normal to the parabola at the general point $(at^2, 2at)$ is

$$tx + y = at^3 + 2at.$$

This passes through the point $(h, k)$ if

$$at^3 + t(2a - h) - k = 0.$$

If the roots of this equation are $p$, $q$ and $r$

$$p + q + r = 0, \quad (1)$$

$$qr + rp + pq = 2 - h/a, \quad (2)$$

$$pqr = k/a. \quad (3)$$

Now $p + q = 2$, so that from (1), $r = -2$. Then from (2)

$$pq + r(p+q) = pq - 4 = 2 - h/a,$$

so that

$$pq = 6 - h/a.$$

Substituting in (3)

$$(6 - h/a) . -2 = k/a.$$

It follows that the equation of the locus of $(h, k)$ is

$$y = 2x - 12a;$$

this represents a straight line, as required.

## EXAMPLES III E

1. Find the locus of the mid-points of chords of the parabola $y^2 = 4ax$ if the chords: (i) all pass through the focus; (ii) are of given length $d$; (iii) contain the point of intersection of the axis and the directrix; (iv) are each normal to the curve at one of their intersections; (v) all pass through a fixed point $(\lambda, \mu)$.

2. The mid-point of a variable chord $PQ$ of the parabola $y^2 = 4ax$ lies on the fixed line $y = ka$, where $k$ is a constant. Show that the locus of points of intersection of normals at $P$ and $Q$ is a straight line. What is the relation of this line to the parabola?

3. If $OP$, $OQ$ are a variable pair of perpendicular chords through the vertex $O$ of the parabola $y^2 = 4ax$, prove that the chord $PQ$ cuts the axis in a fixed point. Find the equation of the locus of the point of intersection of the normals at $P$ and $Q$ to the parabola.

4. A cubic equation $x^3 + px + q = 0$ has two of its roots equal if and only if $27q^2 + 4p^3 = 0$. Find the equation of the locus of points such that two of the three normals from them to the parabola coincide. Is it possible for all three normals to coincide?

5. Find in each case the equation of the locus of the point of intersection of normals to a parabola $y^2 = 4ax$ at $P$ and $Q$, when (i) $PQ$ is perpendicular to the tangent at $P$; (ii) the normals at $P$ and $Q$ are perpendicular; (iii) $PQ$ passes through a fixed point on the directrix; (iv) $PQ$ is a focal chord.

6. Prove that, in general, three normals can be drawn to the parabola $y^2 = 4ax$ to pass through a given point $P(h, k)$. If two of these three normals are perpendicular, prove that $P$ lies on the parabola

$$y^2 = a(x-3a),$$

and that the length of the chord joining the feet of the perpendicular normals is $h+a$. Prove also that the length of the third normal is

$$3\sqrt{\{a(h-2a)\}}.$$

7. The normals at three points $A, B, C$ of a parabola $y^2 = 4ax$ meet in a point $P$. If the focal distances $SA$, $SB$, $SC$ are in arithmetic progression, find the locus of $P$.

8. If the normals at two points on a parabola meet on the same parabola, show that the chord joining the two points meets the axis in a fixed point.

9. The normals at the point $P$, $Q$ and $R$ to a parabola meet in the point $T$. Prove that $SP+SQ+SR+SO = 2TM$, where $S$ is the focus of the parabola, $O$ is the vertex and $M$ is the foot of the perpendicular from $T$ to the tangent at the vertex.

10. The tangents at the ends of a normal chord of a parabola meet in $P$. Find the locus of $P$.

11. The normals at the points $P$, $Q$ and $R$ of the parabola $y^2 = 4ax$ meet in the point $(\lambda, \mu)$. Find the coordinates of the orthocentre and of the centroid of the triangle $PQR$.

12. The normals at the points $Q$ and $R$ to the parabola $y^2 = 4ax$ meet on the parabola in the point $P$. Find the equation of the locus of (i) the orthocentre, and (ii) the centroid, of the triangle $PQR$.

## MISCELLANEOUS EXAMPLES III

1. The tangent and normal at $P(at^2, 2at)$ to the parabola $y^2 = 4ax$ meet the $x$-axis in $T$ and $G$ respectively. Prove that $P$, $T$ and $G$ are equidistant from the focus. Hence prove that the tangent at $P$ to the parabola is inclined to the tangent at $P$ to the circle through $P$, $T$ and $G$ at an angle $\tan^{-1} t$.

2. If $P_1$, $P_2$, $P_3$, $P_4$ are distinct points on a parabola and if the chords $P_1P_2$, $P_3P_4$ both pass through the focus, prove that the chords $P_1P_3$, $P_2P_4$ meet on (or are parallel to) the directrix.

3. The tangent at the point $P(at^2, 2at)$ to the parabola $y^2 = 4ax$ meets the directrix at $Q$. $O$ is the vertex of the parabola, and $S$ the focus. Prove

that $\angle PSQ$ is a right angle, and obtain the positive value of $t$ for which $P, Q, O$ and $S$ are cyclic. Prove that in this case the areas of the triangles $POQ$, $PSQ$ are in the ratio 2:3.

4. Prove that, if $p^2 > 8$, two chords can be drawn through the point $(ap^2, 2ap)$ which are normal to the parabola $y^2 = 4ax$ at their second points of intersection, and that the line joining these points of intersection meets the axis of the parabola in a fixed point, independent of $p$.

5. A chord $PQ$ of a parabola is normal at $P$, and subtends a right angle at the vertex. If $S$ is the focus, prove that $SQ = 3SP$.

6. Prove that the line $lx+my+n = 0$ touches the parabola $y^2 = 4ax$ if $ln = am^2$. Hence, or otherwise, show that the equation of the common tangent to the parabolas $y^2 = 4x$ and $x^2 = 32y$ is $x+2y+4 = 0$. Determine the point of contact of this tangent with each parabola.

7. Prove that the line $y = mx+a/m$ touches the parabola $y^2 = 4ax$. Hence, or otherwise, find the equation of the common tangent to the parabolas $y^2 = 4ax$ and $2x^2 = ay$. Find also the coordinates of the points of contact.

8. If the variable chord $PQ$ of the parabola $y^2 = 4ax$ subtends a right angle at the vertex, prove that it cuts the axis at a fixed point. Prove also that the tangents at $P$ and $Q$ meet on a fixed line.

9. Find the equation of the locus of a point which moves so that its distance from a fixed point is equal to its distance from a fixed line. Show that the point of intersection of any two perpendicular tangents of this curve lies on the fixed line, and that the foot of the perpendicular from the fixed point to any tangent lies on a fixed tangent, which is parallel to the fixed line. Show further that the locus of mid-points of chords through the point of contact of this fixed tangent is another curve similar to the first, and for this second curve find the corresponding fixed point and line.

10. Obtain the equation of the locus of a point $P$ which moves so that its distance from a fixed point $A$ is equal to its distance from a fixed line $l$. Show that the angles which the tangent at a general point $P$ makes with $l$ and with $PA$ are complementary. Show also that the locus of the mid-points of $PA$ is a similar curve, and find the fixed point and line for this curve.

11. The tangents to a parabola at the points $P$ and $Q$ meet at $T$. Show that if the orthocentre of the triangle $PTQ$ lies on the curve, then either the orthocentre is at the vertex, or the chord $PQ$ is a normal to the parabola.

12. The tangent to a parabola at a point $P$ meets the axis at $T$ and $S$ is the focus. Prove that $ST = SP$. Find the equations of the two parabolas which have the line $y = 0$ as axis, the point $(a, 0)$ as focus, and pass through $Q(aq^2, 2aq)$. Prove that the two parabolas cut at right angles.

13. Find the equation of the parabola whose focus is the origin and directrix $x = -2a$. Two parabolas $S_1$ and $S_2$ have a common focus, and their directrices are perpendicular. Show that, if the directrix of $S_2$ meets $S_1$ in $A$, and the directrix of $S_1$ meets $S_2$ in $B$, then $AB$ is the common tangent of $S_1$ and $S_2$. If, in particular, the common focus is equidistant from the two directrices what is the angle which the common tangent makes with the axis of $S_1$?

14. The tangent at a point $P$ of a parabola meets the axis in $T$, and the normal meets the axis in $G$. Prove that the centre of the circle $TPG$ is the focus of the parabola.

15. $P$ is a point of a parabola whose focus is $S$. The perpendicular at $S$ to $SP$ meets the normal at $P$ in $Q$; $NP$ and $MQ$ are the ordinates of $P$ and $Q$; $D$ is the point of intersection of the axis and the directrix. Prove that $DM = 2DN$.

16. Prove that the equation of the locus of the point of intersection of tangents to the parabola $y^2 = 4ax$ which intercept a constant length $c$ on the directrix is $(y^2-4ax)(x+a)^2 = c^2x^2$.

17. If the normal at the point $P$ on a parabola meets the curve again in $Q$ and the tangents at $P$ and $Q$ intersect in $R$, show that the line through $R$ parallel to the axis meets the parabola in $P'$, where $PP'$ is a focal chord.

18. $PQ$ is a chord of the parabola $y^2 = 4ax$ which is normal at $P$. If $M$ is the mid-point of $PQ$ and $N$ is the mid-point of $PM$, show that, as $P$ varies on the parabola, the locus of $N$ is the parabola $y^2 = a(x-3a)$.

19. A line through $Q(-2a, 0)$, cuts the parabola $y^2 = 4ax$ at $R_1$, $R_2$, and the tangents at these points meet at $P$. If $A$ is the vertex, prove that $AP$ is inclined to the $y$-axis at the same angle that $QR_1R_2$ is inclined to the $x$-axis. Prove also that the normals at $R_1$, $R_2$ meet on the curve.

20. If the tangent at a point $P$ to the parabola $y^2 = 4ax$ meets the $x$-axis at $T$ and the $y$-axis at $K$ and if the normal meets the $x$-axis at $G$, find the coordinates of $T$, $K$ and $G$. If $S$ is the focus find the equations of $PS$ and $GK$ and prove that these lines meet at a point of trisection of each line.

21. The normal at $P$ to the parabola $y^2 = 4ax$ meets the curve again in $Q$. $K$ is the mid-point of $PQ$. $KR$ is drawn parallel to the axis to meet the curve in $R$. Prove that $RK = RP$, and that the area of the triangle $RPQ$ is $2a^2 \sec^3\theta \operatorname{cosec}^3\theta$, where $\theta$ is the angle $PKR$.

22. Prove that three normals cannot be drawn from a point $(h, 0)$ on the axis of the parabola $y^2 = 4ax$ to the curve unless $h > 2a$. Find the area of the triangle whose vertices are the feet of the three normals from the point $(3a, 0)$.

23. The points of contact of the tangents drawn from a point $T$ to a parabola with focus $S$ are $P$ and $Q$. $R$ is the mid-point of $PQ$. Prove that $TP.TQ = 2TS.TR$.

24. $G$ is the centroid of a triangle inscribed in a parabola, and $K$ is the centroid of the triangle formed by the tangents to the parabola at the vertices of the first triangle. Show that $GK$ is parallel to the axis of the parabola and that the point of trisection of $GK$ which is nearer $K$ lies on the parabola.

---

25. The points $P$, $Q$ and $R$ on the parabola $x = at^2$, $y = 2at$ have parameters $p$, $q$ and $r$ respectively. Obtain the coordinates of (i) the centroid, (ii) the orthocentre of the triangle formed by the tangents at $P$, $Q$ and $R$. Deduce the coordinates of the circumcentre of the triangle.

26. If $P$ and $Q$ are the points with parameters $p$ and $q$ on the parabola $y^2 = 4ax$, find the slopes of $PQ$ and of the tangents at $P$ and $Q$. If these tangents meet at $T$, and the angles $TPQ$ and $TQP$ are $\theta$ and $\phi$, prove that $\cot\theta - \cot\phi = 2\cot\alpha$, where $\alpha$ is one of the angles which $PQ$ makes with the axis of $x$.

27. Find the coordinates of the point of intersection of the normals to the parabola $y^2 = 4ax$ at the points where the curve meets the line $lx+my+n = 0$. Find the locus of this point of intersection when the line is a variable focal chord.

28. The point $N$ is taken on the axis of a parabola so that the focus $S$ is the mid-point of the segment joining $N$ to the vertex $A$. The line through $N$ perpendicular to the axis cuts the parabola at $P$. The tangent at $P$ meets the tangent at the vertex in $U$ and the normal at $P$ meets the axis in $G$. Prove that $UG$ is perpendicular to $SP$, and that $UG$ touches the circle on $AN$ as diameter.

29. The parameters of the points $P$, $Q$ on the parabola $y^2 = 4ax$ are $p, q$ respectively. Show that, if $PQ$ passes through the point $(-2a, 0)$, then $pq = 2$, and the normals at $P$ and $Q$ to the parabola meet at a point $R$ on the parabola. If $O$ is the origin, show, by considering the gradients of the sides of the quadrilateral $OPQR$, that the circumcircle of the triangle $PQR$ passes through $O$.

30. The tangents at $P$, $Q$ on the parabola $y^2 = 4ax$ meet in $T$ and the normals meet in $N$. If $T$ lies on the line $y = kx$, find the equation of the locus of $N$.

31. $B(b, 0)$ and $C(c, 0)$ are fixed points on the axis of the parabola $y^2 = 4ax$. $P(at^2, 2at)$ is a variable point on the parabola, and the lines $PB$, $PC$ meet the parabola again at $Q$, $R$. Find the equation of $QR$ and prove that, in general, this line touches a parabola with the same vertex and axis as the given parabola. What is the exceptional case? Prove that in the general case, if the coordinates of $P$ are $(x, y)$ and if $QR$ touches the second parabola at $(x', y')$, then $xx'$ and $yy'$ are both constant.

32. Two parabolas have the same focus, but their axes are at right angles. Show that the locus of intersections of perpendicular tangents, one to each parabola, is a straight line which touches both parabolas.

33. The tangent at any point $P$ on the parabola $y^2 = 4ax$ meets the parabola $y^2 = 4ax + b^2$ at $L$ and $M$. Prove that $P$ is the mid-point of $LM$ The second tangent from $L$ to the inner parabola meets the outer parabola again at $N$. Prove that the area of the triangle $LMN$ is independent of the position of $P$ on the inner parabola.

34. The tangent at $P$ to the parabola $y^2 = 4ax$ meets the parabola $y^2 = 4bx$ (where $b > a > 0$) in the points $Q$ and $R$, the one nearer the common vertex of the parabolas being $Q$. Prove that the ratio $QP/PR$ is equal to

$$\frac{b^{\frac{1}{2}} - (b-a)^{\frac{1}{2}}}{b^{\frac{1}{2}} + (b-a)^{\frac{1}{2}}}.$$

35. The tangents at two distinct points $P$, $Q$ of the parabola $y^2 = 4ax$ meet at the point $T$. If $T$ lies on the curve $y^2 = -4a^3/(x+2a)$, prove that the orthocentre of the triangle $TPQ$ lies on the parabola. Prove also that this triangle can never be equilateral.

36. Prove that the tangents to the parabola $y^2 = 4ax$ at the points where it is met by the line $lx + my + n = 0$ intersect at an angle

$$\tan^{-1}\left(\frac{2\sqrt{(a^2m^2 - anl)}}{al+n}\right).$$

If these tangents are perpendicular, prove that the line passes through the focus of the parabola.

37. Prove that if the tangent at a point $P$ of a parabola cuts the directrix at $Z$, then $PZ$ subtends a right angle at the focus $S$. The tangent at $P$ meets the axis at $T$ and the normal at $P$ meets the axis at $G$. $M$ is the foot of the perpendicular from $P$ to the directrix. Without using coordinates, but making use of the above theorem, show that

$$\text{(i)}\ \angle SPT = \angle TPM; \quad \text{(ii)}\ SP = ST = SG.$$

38. Prove that the chords of the parabola $y^2 = 4ax$ which subtend a right angle at the fixed point $P(a\alpha^2, 2a\alpha)$ pass through a fixed point $Q$, and that the length of $PQ$ is $4a\sqrt{(1+\alpha^2)}$.

39. A system of chords is drawn parallel to the tangent at the point $P(at^2, 2at)$ on the parabola $y^2 = 4ax$. Prove that these chords are bisected by the line $l$ through $P$ parallel to the axis of the parabola. If the perpendicular from the point $Q(h, k)$ to the tangent at $P$ meets the line $l$ in $R$, find the equation of the locus of $R$ as $t$ varies. What is this locus if $Q$ is the focus of the parabola?

40. The tangents at the points $P(ap^2, 2ap)$, $Q(aq^2, 2aq)$ to the parabola $y^2 = 4ax$ meet in $T$, and the perpendiculars from $P$ to $TQ$ and from $Q$ to $TP$ meet in $R$. Find the coordinates of $R$, and prove that $TR$ meets the axis of the parabola where $x = a(2+pq)$. If the chord $PQ$ passes through the focus, show that the tangents intersect at right angles on the directrix, and $TP$, $TQ$ each subtend a right angle at the focus.

41. $Q$ is the point of intersection of the tangents to a parabola at $P$ and $R$. If $F$, $G$ and $H$ are the feet of the perpendiculars from the points $P$, $Q$ and $R$ respectively on to any tangent to the parabola, prove that $FP.HR = GQ^2$.

42. The curves $p$ and $q$ are arcs of parabolas. The equation of $p$ is $y = +2\sqrt{(ax)}$ and that of $q$ is $y = -2\sqrt{(-ax)}$, $a$ being a positive constant. The tangent to $p$ at the point $(at^2, 2at)$ intersects $q$ at the point $Q$, and the tangent to $q$ at the point $Q$ intersects $p$ at the point $(at'^2, 2at')$. Prove that $t' = (\sqrt{2}+1)^2 t$.

43. A tangent to the parabola $y^2 + n^2ax = 0$, where $n$ is positive, intersects the parabola $y^2 = 4ax$ at $P$ and $Q$. Prove that the locus of the mid-point of $PQ$ is another parabola and that, if the latus rectum of this parabola is $16a/17$, then $n = 3$.

44. Two triangles are formed respectively by the tangents and normals to the parabola at any three of its points; prove that the line joining their orthocentres is parallel to the axis of the parabola.

45. $ABC$ is a triangle whose sides touch the parabola; if $AB = AC$ prove that the line joining $A$ to the point of contact of $BC$ with the parabola passes through the focus.

46. Find the locus of the mid-point of the chord $QR$ of the parabola $y^2 = 4ax$ when the tangents at $Q$ and $R$ are perpendicular, and find also the locus of the point of intersection of the normals to the parabola at $Q$ and $R$.

47. Find the equation of the tangent to the parabola $x^2 = 4by$ at the point $(2bs, bs^2)$, and to the parabola $y^2 = 4ax$ at the point $(at^2, 2at)$. If these tangents are at right angles, show that the locus of their point of intersection is the curve $(x^2+y^2)(ax+by)+(bx-ay)^2 = 0$.

48. If two points of the axis of a parabola are equidistant from the focus, show that the difference of the squares of their distances from any tangent is independent of its position.

49. $PQR$ is a triangle inscribed in the parabola $y^2 = 4ax$ such that the polar line of each vertex with respect to the parabola $x^2 = 4by$ is the opposite side. Show that there is one such triangle for each position of $P$ on the first parabola, and find the locus of the centroids of the triangles as $P$ varies.

50. A particle is projected with a velocity $V$ from a point $O$, and reaches a distance $r$ up a plane inclined at an angle $\beta$ to the horizontal. For various angles of projection, the greatest possible value of $r$ is

$$V^2/g(1+\sin\beta).$$

Taking axes through $O$ horizontally and vertically, show that the region accessible to a particle projected from $O$ is bounded by a parabola with focus $O$, and find the position of its directrix.

## METHODS OF DRAWING A PARABOLA

It is instructive to produce accurate diagrams of any curve studied, and we give four methods for a parabola; three of these give tangents to the curve, which is thus obtained as the *envelope* of the lines. In each case the reader should verify that the construction given does in fact yield a parabola.

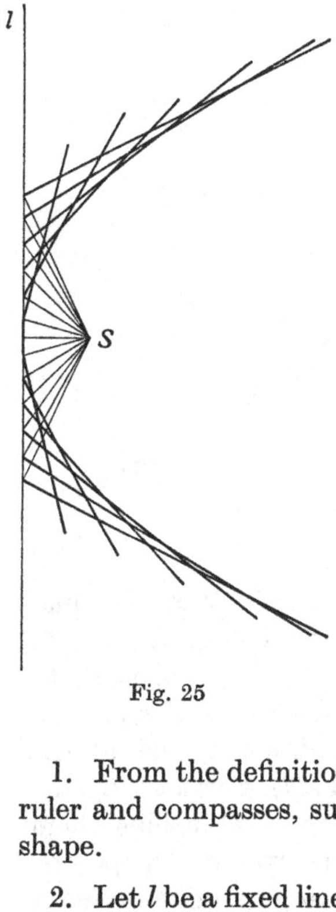


Fig. 25

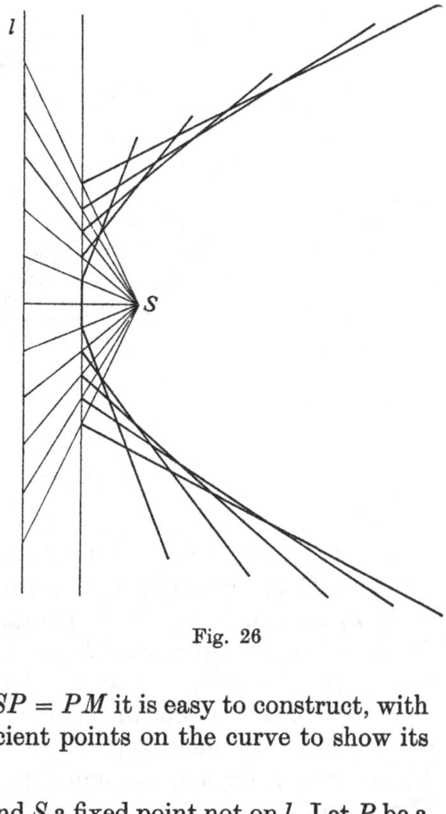


Fig. 26

1. From the definition $SP = PM$ it is easy to construct, with ruler and compasses, sufficient points on the curve to show its shape.

2. Let $l$ be a fixed line and $S$ a fixed point not on $l$. Let $P$ be a variable point on $l$ and draw $PQ$ perpendicular to $SP$. The lines $PQ$ all touch a parabola, which has $S$ as focus and $l$ as tangent at the vertex (Fig. 25).

3. A modification of the last construction gives the next, which is best made on transparent paper (greaseproof, for example). Draw a line $l$ and mark a fixed point $S$ not on $l$. Fold the paper so

that $S$ is on $l$ and mark the crease in the paper clearly. Continue to do this, letting the position of $S$ on $l$ vary, and the different creases on the paper will touch a parabola with $S$ as focus and $l$ as directrix (Fig. 26).

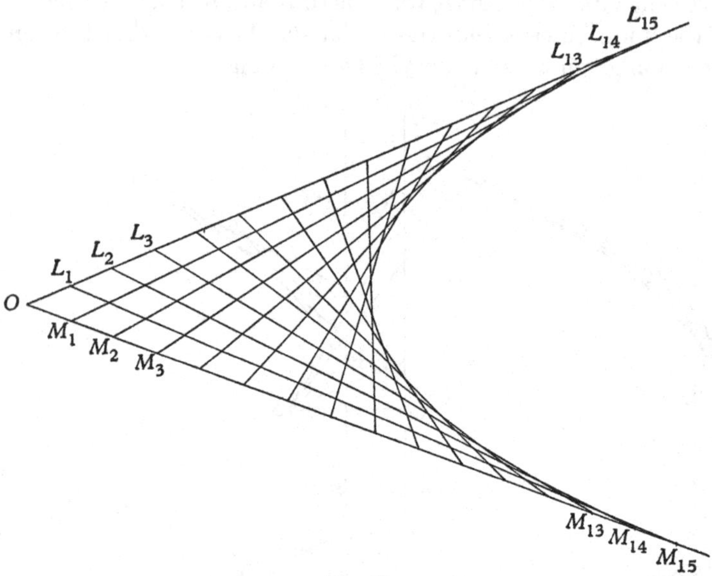


Fig. 27

4. Let $l$ and $m$ be any two lines meeting in $O$. Mark off equal distances ($\frac{1}{4}$ in. or $\frac{1}{2}$ cm., say) on each line, calling the points $O, L_1, L_2, \ldots$, and $O, M_1, M_2, \ldots$, in order from $O$. Now join $L_{15}$ to $M_1$, $L_{14}$ to $M_2$, and in general $L_i$ to $M_{16-i}$ for each $i$. The lines so drawn touch a parabola, which has $L_8M_8$ as tangent at the vertex and $l$ and $m$ among its tangents (Fig. 27). Further tangents can be drawn by marking the points beyond $O$, calling them $L_{-1}$, $L_{-2}$ and so on, and continuing to join corresponding points. The method can obviously be modified to give more or fewer tangents, taking some other number instead of 16.

# IV

# CURVES DEFINED PARAMETRICALLY

In this chapter we shall apply some of the methods illustrated in the last chapter to a variety of curves. These curves will all be defined parametrically; that is, the coordinates of a point on the curve will be given in terms of a single variable, usually denoted by $t$. The parabola, where $x = at^2$, $y = 2at$, is one such curve, and this chapter will throughout follow closely the pattern of the last. The emphasis is on the *methods* adopted, though many of the results obtained are also of interest in themselves.

## A. THE RECTANGULAR HYPERBOLA

**1. Definition and shape.** The rectangular hyperbola we shall consider is defined by the equations

$$x = ct, \quad y = c/t.$$

The reason for the name will appear later, when it will be found that this curve is a special case of a more general one. The equation of the curve, found by eliminating $t$, is $xy = c^2$.

The first step is to find out enough about the curve to sketch it. We shall suppose that $c$ is positive (a negative value of $c$ would give the same curve, but the regions where $t$ is positive and negative would be interchanged). We notice the following facts:

(i) If $t$ is positive, both $x$ and $y$ are positive, and if $t$ is negative, so are both $x$ and $y$. The curve therefore lies entirely in the first and third quadrants. Moreover, if $(x', y')$ lies on the curve, so does $(-x', -y')$; that is, the curve is symmetrical about the origin. For this reason, the origin is called the *centre* of the curve, and a chord through it is called a *diameter*.

(ii) If $x = h \neq 0$, then $t = h/c$ and $y = c^2/h$. Thus a line parallel to the $y$-axis meets the curve in just one point. Similarly, a line $y = k$ parallel to the $x$-axis meets the curve in just one point $(c^2/k, k)$, where $k \neq 0$.

(iii) The lines $x = 0$ and $y = 0$ are specially related to the curve, for they are the only lines among the network parallel to the coordinate axes which do not meet the curve.

The situation becomes clearer when we look at the way a point $(ct, c/t)$ varies on the curve as $t$ varies. When $t$ is large and negative, $ct$ is large and negative, and $c/t$ is negative but nearly zero. It can in fact take a value as near as we please to zero, by taking $-t$ sufficiently large. The point on the curve will be in the region labelled $A$ in the diagram (Fig. 28). As $t$ increases to $-1$, the point moves to $B(-c, -c)$, and as $t$ continues to approach zero from below, the point moves to a position in the region

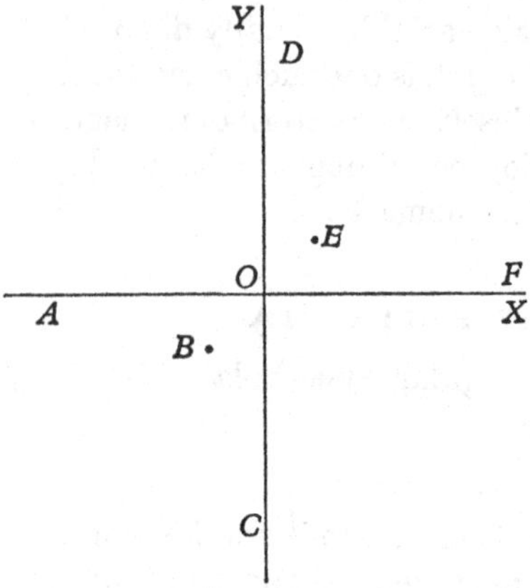


Fig. 28

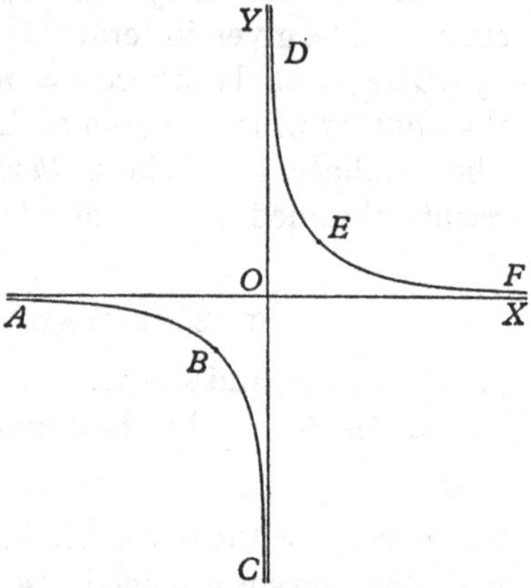


Fig. 29

marked $C$, where $x$ is small and negative, and $y$ is large and negative. There is no point on the curve corresponding to $t = 0$, but when $t$ is just greater than zero, the point is in the region $D$. As $t$ increases the point moves to $E(c, c)$ when $t = 1$, and finally to $F$ when $t$ is large and positive.

The curve may now be sketched (Fig. 29).

**2. Equations of chord, tangent and normal.** A line

$$lx + my + n = 0$$

meets the curve $x = ct$, $y = c/t$ where

$$lct + mc/t + n = 0,$$

that is

$$lct^2 + nt + mc = 0.$$

This is a quadratic equation in $t$, with therefore at most two roots, so that a general line meets the curve in at most two points. If the parameters of the two points are $p$ and $q$, the formula [A 5] for the sum and product of the roots of a quadratic gives

$$p+q = -n/lc, \quad pq = m/l.$$

The equation of the chord joining the points with parameters $p$ and $q$ is therefore (substituting for $m/l$ and $n/l$ in the equation of the line)

$$\mathbf{x+pqy-c(p+q)=0.}$$

If the line is a tangent, $p$ and $q$ coincide, and the equation of the tangent at the point $(cp, c/p)$ is

$$\mathbf{x+p^2y-2cp=0.}$$

This line has gradient $-1/p^2$, and so the line perpendicular to it has gradient $p^2$. The equation of the normal to the curve at the point $(cp, c/p)$ is thus

$$y-c/p = p^2(x-cp),$$

that is

$$\mathbf{p^3x-py=c(p^4-1).}$$

## EXAMPLES IV A

In these examples the curve, unless otherwise stated, is the rectangular hyperbola whose equation is $xy = c^2$, with general point $P(cp, c/p)$.

1. Find the equation of the chord $PQ$, by writing down the equation of the line joining the points $P(cp, c/p)$ and $Q(cq, c/q)$.

2. Find the equation of the tangent at $P$, by differentiating $xy = c^2$ to find the gradient of the curve at $P$.

3. A general line meets the curve in two points (p. 100). A line parallel to an axis meets the curve in just one point (p. 99). Reconcile these two statements.

4. $P$ and $Q$ are the points with parameters 2 and $-\frac{1}{2}$ on the curve. Find the equations of $PQ$, the tangents at $P$ and $Q$, and the normals at $P$ and $Q$.

5. Find the equation of the normal to $xy = 4$ at the point where $y = 1$. Where does this line meet the curve again?

6. A point $(x, y)$ lies on the rectangular hyperbola $xy = 4$. Find the value of $h$ so that $y > 200$ for all $x < h$.

7. Find the coordinates of the points in which the line $x-6y-c=0$ meets the curve. Find the equations of the tangents at these points, and the coordinates of the point where they meet.

8. The tangent to the curve at $P$ meets the coordinate axes in $L$ and $M$ and the normal at $P$ meets the line $y = x$ in $N$. Prove that

$$PL = PM = PN.$$

9. $PQ$ is a chord of the curve which meets the coordinate axes in $A$ and $B$, and $M$ is the mid-point of $PQ$. If $O$ is the origin, show that

$$OM = MA = MB.$$

10. The normal at $P$ meets the curve again in $G$. Find the coordinates of $G$, and if $O$ is the origin prove that $OP^3 = c^2 . PG$.

### 3. Asymptotes.

At an elementary stage it seems to be customary to define an *asymptote* as 'a tangent to the curve at infinity'. This gives the idea of a line which the curve continually approaches, and to which it gets as close as we please, but which it never meets. We saw in §1 that the lines $x = 0$, $y = 0$ have this property, so we say that these two lines are the asymptotes of the rectangular hyperbola $xy = c^2$.

Some students may not be satisfied with this rather inexact idea of a line touching a curve at a point which cannot be reached or drawn. They may wonder why all curves which extend to infinity do not have asymptotes (or, if they do, why all mention of the asymptotes of a parabola has been suppressed!). The account which follows is included in order to give the theory of the asymptote a respectable mathematical basis. It may be found hard, and should be omitted by those who are for the moment satisfied with the definition already given, or by those whose idea of limits is not yet adequate for them to be able to follow the rather sophisticated reasoning.

We are all familiar with the idea that a locus may extend 'to infinity'. For example, the line is such a locus, whereas the circle is not. When using coordinates it is possible to be quite precise about this. If $P(x, y)$ is a point on a curve, and if $P$ moves along the curve so that one or other (or both) of $x$ and $y$ tend to $+\infty$ or to $-\infty$, then $P$ is said to tend to infinity, and we write this as $P \to \infty$.

In the case of curves defined parametrically, the variation of $P$ along the curve depends on the variation of the parameter. The curves already studied provide examples of what may happen.

(*a*) The *line* is given (p. 32) by

$$x = h + r\cos\alpha, \quad y = k + r\sin\alpha,$$

where $r$ is a parameter. Consider the case where $\alpha$ is an acute angle, so that $\cos\alpha$ and $\sin\alpha$ are both positive. As $r \to +\infty$, $x \to +\infty$ and $y \to +\infty$. (This expresses precisely the fact that $x$ and $y$ can be made as large as we please by taking values of $r$ large enough.) Similarly, as $r \to -\infty$, $x \to -\infty$ and $y \to -\infty$.

(*b*) The *parabola* may be taken in the form

$$x = at^2, \quad y = 2at.$$

Then, as $t \to +\infty$, $x \to +\infty$ and $y \to +\infty$; also as $t \to -\infty$, $x \to +\infty$ and $y \to -\infty$.

(*c*) The *rectangular hyperbola* $x = ct$, $y = c/t$ also extends to infinity. As $t \to +\infty$, $x \to +\infty$ and $y \to 0$. Here we have an example of a curve for which only one of the coordinates, in this case $x$, tends to infinity, but we still say that as $t \to +\infty$, $P \to \infty$, and as $t \to -\infty$, $P \to \infty$. These are the regions marked $F$ and $A$ in the diagram (Fig. 29). There are other points at infinity on the curve. As $t \to +0$, $x \to +0$ and $y \to +\infty$ (region $D$), and as $t \to -0$, $x \to -0$ and $y \to -\infty$ (region $C$).

(*d*) As an example of a curve which has no points at infinity, consider the locus given by

$$x = \cos t, \quad y = \sin t.$$

Since $\cos t$ and $\sin t$ both lie between $-1$ and $+1$, for all values of $t$, this curve lies entirely inside the square whose sides are $x = \pm 1$, $y = \pm 1$. (It is in fact a circle, since for all values of $t$,

$$x^2 + y^2 = \cos^2 t + \sin^2 t = 1$$

so that the point is always a constant distance 1 from the origin.)

We return to curves which extend to infinity. We can now define an *asymptote* of such a curve. If the equation of the tangent at $P$ can be written so that as $P \to \infty$ the equation tends to a finite form $ax + by + c = 0$ with *not both of a and b zero*, then this line $ax + by + c = 0$ is called an asymptote of the curve.

The rectangular hyperbola gives an example of this. As $t \to \pm 0$, $P \to \infty$, as we have already seen. The equation of the tangent at the point $(ct, c/t)$ is

$$x + t^2 y = 2ct,$$

and as $t \to 0$, this tends to

$$x = 0,$$

which is therefore an asymptote. Also, $P \to \infty$ as $t \to \pm\infty$. To see what happens to the tangent, we write the equation in the form

$$(1/t^2)\,x + y = (2c/t).$$

As $t \to \pm\infty$, $1/t \to 0$ and $1/t^2 \to 0$, so that this equation tends to

$$y = 0,$$

and this is the equation of the other asymptote.

It is now clear why curves may extend to infinity without having asymptotes. The parabola is such a curve, for $P(at^2, 2at) \to \infty$ as $t \to \infty$, and the tangent

$$x - ty + at^2 = 0,$$

which may be written

$$(1/t^2)\,x - (1/t)\,y + a = 0,$$

then tends to a form in which both the coefficients of $x$ and $y$ are zero. Thus according to the definition the parabola has no asymptote.

**4. Properties of tangents.** Several properties of tangents follow from the equation

$$x + t^2 y - 2ct = 0.$$

The point $(h, k)$ lies on the tangent if

$$h + t^2 k - 2ct = 0,$$

and so, as this is a quadratic equation in $t$, there are at most two tangents from a point to the curve. If the roots of the quadratic are $p$ and $q$, we deduce, [A 5], that

$$p+q = 2c/k, \quad pq = h/k.$$

It follows that $h = 2cpq/(p+q)$ and $k = 2c/(p+q)$, so that the tangents to the curve at the points $P(cp, c/p)$ and $Q(cq, c/q)$ meet at

$$\{\mathbf{2cpq/(p+q)}, \quad \mathbf{2c/(p+q)}\}.$$

If the point $(h, k)$ is given, the equation of the chord of contact of tangents drawn to the curve from $(h, k)$ can be deduced from the same equations; for the chord

$$x+pqy-c(p+q) = 0$$

joining $P$ and $Q$ is $\quad x+(h/k)\,y-c(2c/k) = 0,$

or $$\mathbf{kx+hy = 2c^2}.$$

This is also called the *polar line* of $(h, k)$ with respect to the rectangular hyperbola $xy = c^2$.

Lastly, the line $lx+my+n = 0$ meets the curve where

$$lct^2+nt+mc = 0,$$

and so the line is a tangent if and only if this equation has coincident roots in $t$, and

$$\mathbf{n^2 = 4c^2lm}.$$

**5. Illustrations.**

***Illustration 1.*** *Find the locus of the foot of the perpendicular drawn from the origin to the tangent to the curve $xy = c^2$ at the point $(ct, c/t)$.*

The equation of the tangent at $(ct, c/t)$ to the curve $xy = c^2$ is

$$x+t^2y = 2ct.$$

If $R(X, Y)$ is the foot of the perpendicular to this tangent from the origin, the gradient of $OR$ is $t^2$, and so $Y/X = t^2$. Also $R$ lies on the tangent, so

$$X+t^2Y = 2ct.$$

To find the locus of $(X, Y)$ we eliminate $t$ between these two equations; (notice that it is not necessary to find the separate values of $X$ and $Y$). Squaring the second equation gives

$$X^2+2XYt^2+Y^2t^4 = 4c^2t^2,$$

and substituting $t^2 = Y/X$ we have

$$X^4+2X^2Y^2+Y^4 = 4c^2XY,$$

so that the equation of the locus of $(X, Y)$ is

$$(x^2+y^2)^2 = 4c^2xy.$$

***Illustration 2.*** *The points of the rectangular hyperbola are written in the parametric form $(ct, c/t)$. Prove that if the normals at the points whose parameters are $p$, $q$, $r$, $s$ are concurrent, then*

$$qr+rp+pq+ps+qs+rs = 0$$

*and $pqrs = -1$. Deduce that four conormal points on the curve are such that any one of them is the orthocentre of the triangle formed by the other three.*

The equation of the normal at the point with parameter $t$ is

$$t^3x-ty = c(t^4-1).$$

If this passes through the point $(h, k)$,

$$t^3h-tk = c(t^4-1),$$

or

$$ct^4-ht^3+kt-c = 0.$$

If the roots of this quartic equation in $t$ are $p$, $q$, $r$, $s$, the usual formulae [A 7] for the symmetric functions of the roots give

$$p+q+r+s = h/c, \tag{1}$$

$$qr+rp+pq+ps+qs+rs = 0, \tag{2}$$

$$pqr+qrs+rps+pqs = -k/c, \tag{3}$$

$$pqrs = -c/c = -1. \tag{4}$$

Of these equations, (2) and (4) give the required conditions.

Let the points be $P$, $Q$, $R$, $S$, and consider the triangle $PQR$. The gradient of the chord $QR$ is $-1/qr$ and the gradient of $PS$ is $-1/ps$. Since $pqrs = -1$, the product of these two gradients is $-1$, so that $QR$ is perpendicular to $PS$. Now the relation $pqrs = -1$ is symmetrical, so it follows that also $RP$ is perpendicular to $QS$ and $PQ$ is perpendicular to $RS$. Thus $S$ is the orthocentre of the triangle $PQR$. Again from the symmetry of the given relation we deduce that each of the points is the orthocentre of the triangle formed by the other three.

The method used in this example should be compared with that of Chapter III, §11; see also Examples IV B, 14.

***Illustration 3.*** *Find the locus of the mid-points of chords of the curve $xy = c^2$ drawn parallel to a diameter $y = mx$ through the centre.*

Let $PQ$ be any chord of the set, joining points with parameters $p$ and $q$. The gradient of $PQ$ is $-1/pq$, and so $m = -1/pq$. If $(X, Y)$ is the mid-point of $PQ$

$$X = \tfrac{1}{2}c(p+q)$$

and
$$Y = \tfrac{1}{2}c\left(\frac{1}{p}+\frac{1}{q}\right) = \frac{1}{2}\frac{c}{pq}(p+q).$$

Thus
$$Y = \frac{1}{pq}X = -mX,$$

and so the equation of the locus of $(X, Y)$ is

$$y + mx = 0.$$

This is the equation of a line through the origin, another diameter of the curve. It is said to be *conjugate* to the diameter $y = mx$ with which we began. The relation between conjugate diameters with gradients $m$ and $m'$ is

$$m + m' = 0;$$

we notice the symmetry of this relation.

## EXAMPLES IV B

Unless the contrary is stated, the curve in these examples is the rectangular hyperbola $xy = c^2$, with general point $P(cp, c/p)$.

1. $PN$ is the perpendicular to an asymptote from a point $P$ on a rectangular hyperbola. Prove that the locus of the mid-point of $PN$ is another rectangular hyperbola with the same asymptotes.

2. A tangent to the hyperbola meets the asymptotes in $A$ and $B$, and $O$ is the point of intersection of the asymptotes. Prove that the area of the triangle $AOB$ is independent of the position of the tangent.

3. The normal at $P$ meets the hyperbola again at $R$, the point with parameter $r$. Prove that $p^3r = -1$. The tangent at $P$ meets the $y$-axis at $Q$. Find the area of the triangle $PQR$ in terms of $p$.

4. Prove that there is just one line which is normal to the rectangular hyperbola at both its points of intersection.

5. Find the point of intersection of the tangents to the rectangular hyperbola at the points where it meets the line $3x - y + 2c = 0$.

6. The tangents to $xy = c^2$ at $P$ and $Q$ meet in $R$, and $PQ$ touches the curve $4xy = c^2$. Find the locus of $R$.

7. What condition is satisfied by $l$ and $m$, if $lx+my = 1$ is a normal to the curve $xy = c^2$?

8. The tangent at $P$ meets the asymptotes at $L$ and $M$; $O$ is the centre of the hyperbola and $POQ$ is a diameter. The line $MQ$ meets the other asymptote in $T$. Find the areas of the triangles $MOL$ and $QOT$.

9. The tangent to the rectangular hyperbola at any point meets the asymptotes in $A$ and $A'$, and the normal at the same point cuts them in $B$ and $B'$. If $AB'$ and $A'B$ meet in $C$, find the locus of $C$.

10. Find the finite values of $m$ and $c$ such that the line $y = mx+c$ touches both the hyperbola $4xy = 1$ and the parabola $y^2 = 4x$. Find also the distance between the points of contact.

11. Two parallel tangents to the rectangular hyperbola are met by a third tangent in $P$, $Q$. If $O$ is the origin, prove that $OP$, $OQ$ are equally and oppositely inclined to the axes.

12. Prove that the locus of the intersection of the tangents at the points whose parameters are $t$ and $t^{-1}$ is the line $x-y = 0$.

13. Find the condition for the point $(h+r\cos\alpha,\ k+r\sin\alpha)$ to lie on the curve $xy = c^2$. Hence apply the technique of Method 2 (III, §10) to find the locus of mid-points of chords with a fixed gradient $m$.

14. The parameters $p$, $q$, $r$, $s$ of the points $P$, $Q$, $R$, $S$ are such that $pqrs = -1$ and $qr+rp+pq+ps+qs+rs = 0$, and the normals to the curve at $P$ and $Q$ meet in $A$. If the parameters of the feet of the other two normals through $A$ are $r'$ and $s'$, show that $rs = r's'$ and $r+s = r'+s'$. Deduce that the conditions $pqrs = -1$ and $qr+rp+pq+ps+qs+rs = 0$ are both necessary and sufficient conditions for the points $P$, $Q$, $R$, $S$ to be conormal on the rectangular hyperbola.

15. If the normals at the points $(c, c)$ and $(-2c, -\frac{1}{2}c)$ intersect in the point $K$, find the feet of the other two normals which can be drawn through $K$ to the hyperbola.

## B.* CUBIC CURVES

Both the curves we have studied so far have one basic property in common; in each case a line meets the curve in at most *two* points. This property is shared by all curves of the family called conics, whose equations are quadratics in $x$ and $y$. Before turning to the other members of this family, we shall look about more widely, and apply the methods available to curves which meet a general line in three points; these are called *cubic* curves. There is no reason in theory why we should not look at algebraic curves of any order, but in practice the complications grow as the order (that is, the number of intersections of the curve with a general line) of the curve increases.

In what follows, no attempt has been made to be systematic, or to deal with every type of cubic curve; for further details the reader should consult a more advanced book (for example, E. J. F. Primrose, *Plane Algebraic Curves*, Macmillan, 1955). The curves considered here are chosen for their geometrical interest, and in particular because each exhibits a feature which is not possessed by the conic.

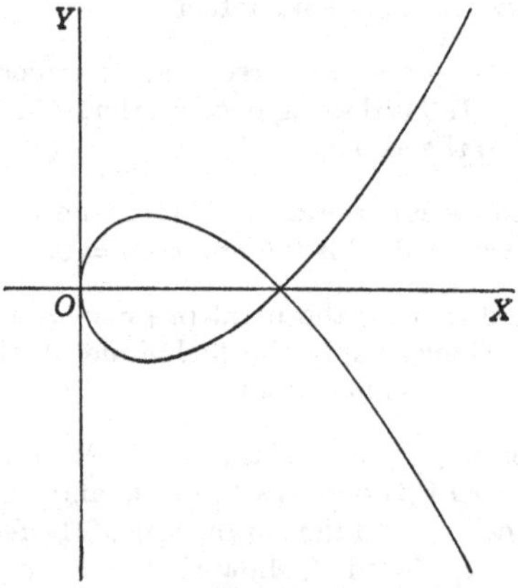

Fig. 30

**6. A curve with a double point.** Consider the locus of the point

$$x = t^2, \quad y = t(t^2 - 1).$$

As a help to sketching the curve we notice the following facts:

(i) The curve is symmetrical about the $x$-axis.

(ii) For all values of $t$, $x \geqslant 0$.

(iii) $x = 0$ when $t = 0$; and $y = 0$ when $t = 0$, $+1$ or $-1$.

(iv) When $t$ is large and positive, so are $x$ and $y$.

The curve is sketched in Fig. 30.

The fact to notice about this curve is that $t = 1$ gives the point $(1, 0)$ and $t = -1$ gives the same point. The curve in fact crosses itself at $(1, 0)$; we say that this is a *double* point, with two *branches* passing through it.

A general line
$$lx+my+n = 0$$
meets the curve where
$$lt^2+mt^3-mt+n = 0.$$

This is a cubic equation in $t$; it follows that a line meets the curve in at most three points. If the parameters of these points are $p$, $q$ and $r$, then [A 6]
$$p+q+r = -l/m,$$
$$qr+rp+pq = -m/m = -1,$$
$$pqr = -n/m.$$

This gives the *condition for collinearity*: if points with parameters $p, q, r$ are collinear, then
$$qr+rp+pq = -1.$$

Conversely, if $qr+rp+pq = -1$, the points $P$, $Q$, $R$ with parameters $p$, $q$, $r$ are collinear. For if the line $PQ$ meets the curve again in $R'$, whose parameter is $r'$,
$$qr'+r'p+pq = -1$$
and so
$$r' = -\frac{pq+1}{p+q} = r,$$
and $R'$ is the same point as $R$.

The equation of the line $PQR$ is
$$\frac{l}{m}x+y+\frac{n}{m} = 0,$$
or
$$(p+q+r)x-y+pqr = 0.$$

From the relation
$$qr+rp+pq = -1$$
for collinearity, we deduce three results.

(i) *Equation of the tangent at P.*

If the above line is a tangent at $P$, then $p = q$, say, and so
$$p^2+2pr = -1$$
and
$$r = -(1+p^2)/2p.$$
Then
$$p+q+r = 2p+r = (3p^2-1)/2p$$
and
$$pqr = p^2r = -\tfrac{1}{2}p(1+p^2),$$

and the equation of the tangent is

$$(1-3p^2)x+2py+p^2(1+p^2) = 0.$$

Alternatively, this result could be obtained by differentiating to find the gradient. At the point with parameter $t$

$$\frac{dy}{dx} = \frac{dy}{dt}\bigg/\frac{dx}{dt} = \frac{3t^2-1}{2t},$$

and so the equation of the tangent at $(p^2, p^3-p)$ is

$$(3p^2-1)x-2py = (3p^2-1)p^2-2p(p^3-p),$$

which gives the same result.

(Those who have read and mastered § 3 will be able to show that this curve has no asymptote.)

(ii) *The tangents at the double point.*

When $p = 1$, the equation of the tangent is

$$x-y = 1,$$

and when $t = -1$ the equation is

$$x+y = 1;$$

these are the distinct tangents to the two branches of the curve through the double point.

(iii) *Condition for a double point.*

Had we not noticed at the start that the values $t = \pm 1$ give the double point, we could have used the following method. *Any* line through a double point meets the curve twice there and so (for a cubic curve) in one further point. If $p$ and $q$ give the double point, and $r$ is the parameter of any other point on the curve,

$$1+pq+r(p+q) = 0,$$

and this is true for *any* value of $r$. It follows that

$$1+pq = 0 \quad \text{and} \quad p+q = 0$$

and so $p = 1$, $q = -1$ (or vice versa).

**7. A curve with a cusp, $x = at^2$, $y = at^3$ $(a > 0)$.**

The first step is to sketch the curve, and as aids to doing so, we note the following facts:

(i) The curve is symmetrical about the $x$-axis.

(ii) For all $t$, $x \geqslant 0$.

(iii) When $t = 0$, $x = y = 0$.
(iv) When $t$ is large, so are $x$ and $y$.
(v) For small values of $t$, $y/x = t$
is also small, and the curve touches the $x$-axis at the origin.

The curve is sketched in Fig. 31. Its equation, found by eliminating $t$, is $ay^2 = x^3$. This curve is called a semi-cubical parabola.

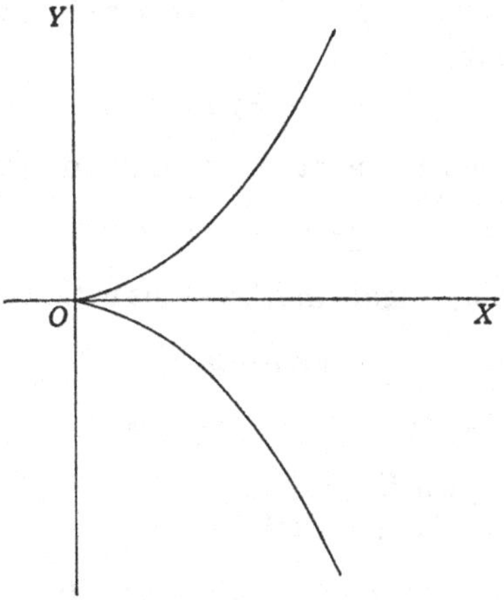


Fig. 31

The obvious new feature of this curve is the shape at the origin $O$. A general line through $O$, $y = mx$ (where $m \neq 0$), meets the curve where

$$at^3 = mat^2,$$

of which the roots are $\quad t = 0, 0, m.$

Thus a general line through $O$ meets the curve in two coincident points at $O$, and in one further point. So far the point has the same property as the double point, just discussed; but here there is only one tangent, not two. When $m = 0$, all three roots are zero, and the line $y = 0$ meets the curve in $O$ repeated three times. A point of this sort is called a *cusp* (from the Latin *cuspis*, a point), and the tangent at $O$ is called the *cuspidal tangent*.

A general line

$$lx + my + n = 0$$

meets the curve where

$$lat^2 + mat^3 + n = 0.$$

This is a cubic in $t$, and so the line meets the curve in at most three points. If the parameters of these points are $p$, $q$, $r$, then [A 6],

$$p+q+r = -l/m,$$
$$qr+rp+pq = 0,$$
$$pqr = -n/am.$$

The *condition for collinearity* follows; if points with parameters $p$, $q$, $r$ are collinear, then

$$qr+rp+pq = 0.$$

Conversely, if $p$, $q$, $r$ are parameters of points $P$, $Q$, $R$ such that

$$qr+rp+pq = 0,$$

let $PQ$ meet the curve again in $R'$, a point with parameter $r'$. Then

$$qr'+r'p+pq = 0$$

and so $$r' = -pq/(p+q) = r,$$

and $R'$ coincides with $R$.

The equation of the line $PQR$ is

$$(p+q+r)x-y+apqr = 0.$$

To find the tangent at a general point $P$, with parameter $p$ ($\neq 0$), we put $p = q$. Then

$$p^2+2pr = 0,$$
$$r = -\tfrac{1}{2}p,$$

and the equation of the tangent is

$$(p+p-\tfrac{1}{2}p)x-y+ap.p.-\tfrac{1}{2}p = 0,$$

or $$3px-2y = ap^3.$$

As usual, differentiation provides an alternative method for this.

***Illustration 4.*** *The tangent to the curve $ay^2 = x^3$ at a point $P(ap^2, ap^3)$ meets it again in $P'$. If $P$, $Q$, $R$ are collinear, show that the same is true of $P'$, $Q'$, $R'$.*

The condition, found above, for $P$, $Q$, $R$ to be collinear is

$$qr+rp+pq = 0.$$

The tangent at $P$ meets the curve again in $P'$, so (proved above) $p' = -\frac{1}{2}p$. Similarly $q' = -\frac{1}{2}q$ and $r' = -\frac{1}{2}r$. Therefore

$$q'r' + r'p' + p'q' = \tfrac{1}{4}(qr + rp + pq) = 0,$$

and so $P'$, $Q'$ and $R'$ are collinear.

This interesting property is true for all cubic curves.

**8. A curve with an inflexion.** Consider the curve $y = x^3$, given parametrically by

$$x = t, \quad y = t^3.$$

This is evidently a curve which lies symmetrically in the first and third quadrants, and passes through the origin $O$ when $t = 0$.

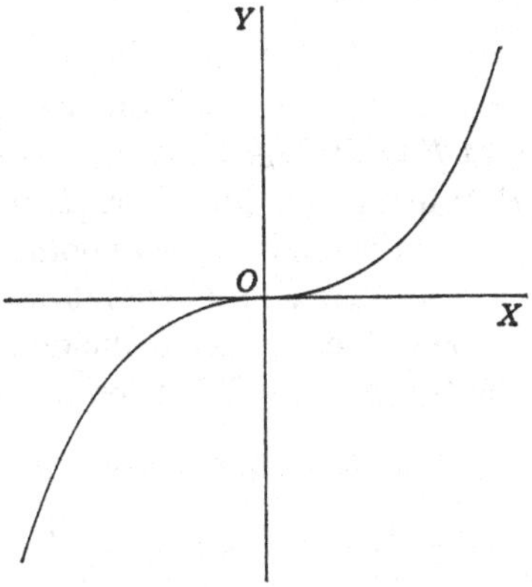


Fig. 32

Near $O$, $y/x = t^2$ and is small when $t$ is small, so the curve touches $OX$ at $O$; when $t$ is large, so are $x$ and $y$. The curve is sketched in Fig. 32.

Employing the usual technique, we find where the curve cuts a general line $lx + my + n = 0$. Points of intersection are given by roots of

$$lt + mt^3 + n = 0;$$

this is a cubic in $t$, so that the roots $p$, $q$, $r$ satisfy

$$p + q + r = 0,$$
$$qr + rp + pq = l/m,$$
$$pqr = -n/m.$$

It follows that the condition for $P$, $Q$ and $R$ to be collinear points on the curve is that their parameters $p$, $q$ and $r$ satisfy

$$p+q+r = 0.$$

(The proof that when this condition is satisfied the points are collinear follows the same pattern as those of the similar results in the last two sections, and is left to the reader.) The equation of $PQR$ is

$$(qr+rp+pq)\,x+y-pqr = 0.$$

If $PQR$ is the tangent at $P$, then $P$ and $Q$ coincide, and $p = q$. The tangent at $P$ therefore meets the curve again in the point $R$ whose parameter $r$ satisfies

$$2p+r = 0, \quad r = -2p.$$

Substituting these values of $q$ and $r$, we find the equation of the tangent as

$$3p^2x-y-2p^3 = 0.$$

Looking more closely at the situation, we notice that when $r \neq p$, the tangent at $P$ meets the curve again in a point $R$ which is different from $P$. When $p = 0$, however, $R$ and $P$ coincide, and the tangent at $O$ meets the curve in *three* points coinciding at $O$. A point like this, which is not a double point or cusp but at which the tangent meets the curve in three coincident points, is called an *inflexion* and the tangent is called the *inflexional tangent*.

***Illustration 5.*** *Investigate the curve given by*

$$x = \frac{2(t^3+t)}{1+3t^2}, \quad y = \frac{2(t^3-t)}{1+3t^2},$$

*showing that it has three inflexions which are collinear. Find the equation of the asymptote, and sketch the curve.*

The parameters of the points which lie on the curve and on the line

$$lx+my+n = 0$$

satisfy

$$2l(t^3+t)+2m(t^3-t)+n(1+3t^2) = 0,$$

or

$$2t^3(l+m)+3nt^2+2t(l-m)+n = 0.$$

If the roots of this cubic are $p$, $q$, $r$,

$$p+q+r = -\frac{3n}{2(l+m)},$$

$$qr+rp+pq = \frac{l-m}{l+m},$$

$$pqr = -\frac{n}{2(l+m)}.$$

Thus, if points $P$, $Q$, $R$ with parameters $p$, $q$, $r$ are collinear

$$p+q+r = 3pqr.$$

Conversely, if $p+q+r = 3pqr$, and if $PQ$ meets the curve again in $R'$,

$$p+q+r' = 3pqr'$$

and so $r = r'$; that is, $P$, $Q$, $R$ are collinear.

When $p = q$,

$$2p+r = 3p^2r,$$

and so

$$r = \frac{2p}{3p^2-1}.$$

Then

$$pqr = \frac{2p^3}{3p^2-1}$$

and

$$qr+rp+pq = \frac{3p^2(p^2+1)}{3p^2-1}.$$

Thus

$$\frac{l-m}{l+m} = \frac{3p^2(p^2+1)}{3p^2-1},$$

so

$$\frac{l+m}{3p^2-1} = \frac{l-m}{3p^2(p^2+1)} = \frac{2l}{3p^4+6p^2-1} = \frac{2m}{-(3p^4+1)} = \frac{2n}{-8p^3}$$

and the equation of the tangent at the point $p$ is

$$(3p^4+6p^2-1)\,x-(3p^4+1)\,y = 8p^3.$$

To find any inflexions of the curve, we put $p = q = r$ in the collinearity condition. This gives

$$3p = 3p^3; \quad p = 0,\ +1,\ -1.$$

There are thus three inflexions, whose coordinates are

$$(0,0), \quad (1,0), \quad (-1,0),$$

lying on $y = 0$. The inflexional tangents at these points are

$$x+y = 0, \quad 2x-y-2 = 0, \quad 2x-y+2 = 0.$$

The curve extends to infinity as $t \to \infty$. Writing the equation of the tangent as

$$\left(3+\frac{6}{p^2}-\frac{1}{p^4}\right)x-\left(3+\frac{1}{p^4}\right)y = \frac{8}{p},$$

we see that as $p \to \infty$, $1/p \to 0$, and the tangent tends to the finite form

$$x - y = 0.$$

As a further aid to sketching we notice that changing $t$ to $-t$ has the effect of changing $x$ to $-x$ and $y$ to $-y$, so the curve is symmetrical about the origin. The sketch is given in Fig. 33.

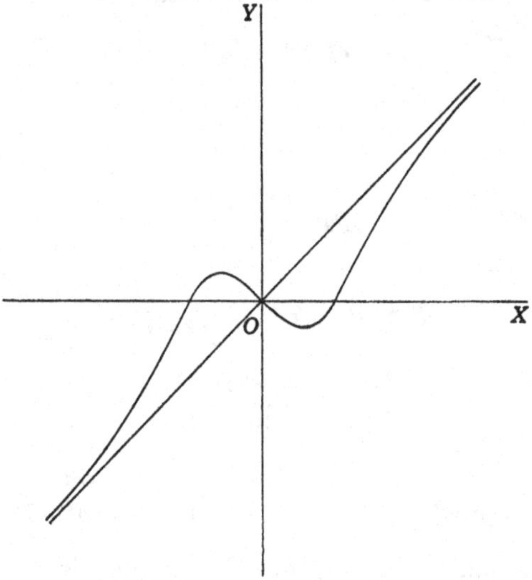


Fig. 33

## EXAMPLES IV C

In every example, illustrate by sketching the curve roughly.

1. A curve is given by the parametric equations

$$x = a^2 - t^2, \quad y = t(a^2 - t^2),$$

where $a$ is a cons*ant. Prove that a general line meets this curve in three points, and the three points of the curve corresponding to parameters $p$, $q$ and $r$ are collinear if and only if $qr + rp + pq = -a^2$. Deduce that the tangent at the point $p$ (where $p \neq 0$) meets the curve again in the point with parameter $-(a^2 + p^2)/2p$, and hence, or otherwise, find the equation of the tangent at $p$.

2. The tangent and normal to the curve $y^2 = x^3$ at the point $(t^2, t^3)$ meet the $x$-axis in $P$ and $Q$ respectively. Prove that $6PQ = t^2(9t^2 + 4)$. Find also the coordinates of the point where the tangent meets the curve again.

3. Prove that the equation of the tangent at the point $P(4at^2, 8at^3)$ of the curve $ay^2 = x^3$ is $y = 3tx - 4at^3$. The tangent meets the curve again in the point $Q$ and the $y$-axis in the point $R$. Show that $Q$ is the point $(at^2, -at^3)$ and that $PQ = 3QR$.

4. Draw a rough sketch of the curve given by the equations

$$x = t^2, \quad y = t^3.$$

Obtain the equations of the tangent and normal to the curve at the point where $t = 1$ and show that this tangent meets the curve again at the point where $t = -\frac{1}{2}$.

5. Find the equation of the tangent at the point $(2t^3, 3t^2)$ on the curve $4y^3 = 27x^2$. Prove that the equation of the locus of the intersections of perpendicular tangents to the curve is $y = x^2 + 1$.

6. If the tangent at the point $P(at^2, at^3)$ on the curve $ay^2 = x^3$ meets the curve again at $Q$, find the coordinates of $Q$. If $N$ is the foot of the perpendicular from $P$ to the $x$-axis, $R$ the point where the tangent at $P$ cuts the $y$-axis and $O$ the origin, prove that $OQ$ and $RN$ are equally inclined to the $x$-axis.

7. The tangent at a point $P(p, p^3)$ on the curve $y = x^3$ cuts the curve again at $Q$. Show that the locus of the mid-point of $PQ$ is the curve $y = 28x^3$. If the tangent at $Q$ to the first curve meets this locus at $R$, prove that $PR$ is parallel to the $y$-axis.

8. Repeat Illustration 4 for the curve $y = x^3$ which was discussed in § 8.

9. Sketch the curve whose general point is $(a/t, at^2)$. If the tangent at $P$ to this curve meets the curve again at $Q$ and the asymptotes at $A$ and $B$, prove that $AP:PB:BQ$ is a constant, independent of the position of $P$ on the curve.

10. Show that the curve

$$x = t^2/(t+1), \quad y = t^3/(t+1)$$

has the line $x+y = 1$ as an asymptote and a cusp at the origin. Sketch the curve.

**9. Other interesting curves.** For further entertainment, details of a few curves, some of historical interest and some which arise in other branches of mathematics, are given here. The more obvious properties are noted, but proofs are omitted, and each curve could provide the basis for a number of examples. A rough sketch is given in each case; in every case the reader should verify the salient features of the curve which have given the guides for the sketch.

The important things to look for, when sketching a curve, are:

(i) symmetry;

(ii) where the curve cuts the axes;

(iii) any restriction of the curve to lie in only one part of the plane;

(iv) regions in which $t$ is large, small, positive, or in some special range.

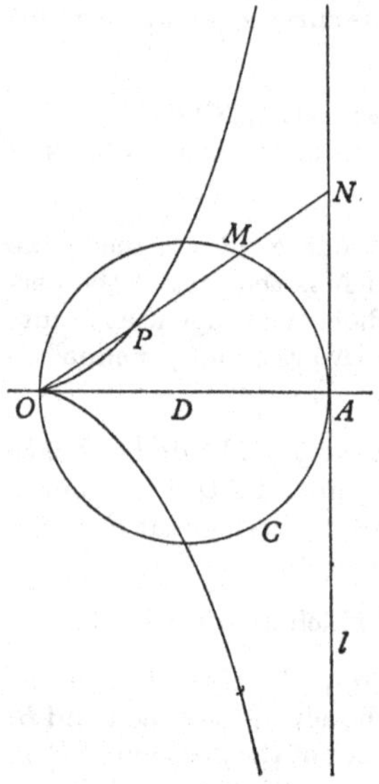


Fig. 34

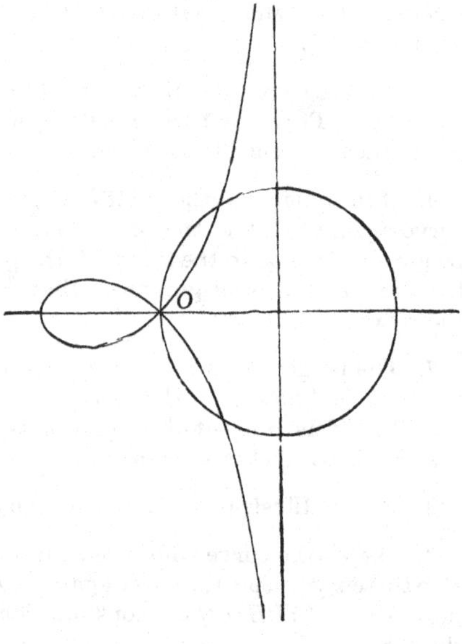


Fig. 35

After that, the gradient at obvious points, such as the origin if the curve goes through it, should be found, and if the knowledge of the calculus is adequate, the positions of maxima and minima will help. Any double points, cusps or inflexions should be traced, and also the position of any asymptotes. With these guides a reasonable sketch of any curve should be possible.

I. *The Cissoid of Diocles,*

$$x = \frac{2at^2}{1+t^2}, \quad y = \frac{2at^3}{1+t^2}.$$

This curve arises as a locus connected with a circle $C$ of radius $a$ and a line $l$ touching it at a point $A$ diametrically opposite $O$

(Fig. 34). A variable line through $O$ meets the circle again in $M$ and the line $l$ in $N$, and $P$ is the point on the line such that $OP = MN$. The locus of $P$, as the line through $O$ varies, is the Cissoid of Diocles.

Taking $OA$ as the $x$-axis and $\angle POA$ as $\theta$, $MN = 2a \tan\theta \sin\theta$ and so the coordinates of $P$ are $(2a\sin^2\theta, 2a\sin^2\theta\tan\theta)$. Putting $\tan\theta = t$, we get the parametric form of the curve as given above. The equation of the curve is

$$y^2(2a-x) = x^3.$$

The curve has a cusp at $O$, it is symmetrical about $OA$, and has the line $l$ as an asymptote (these facts are clear from the geometry of the construction). The collinearity condition for points with parameters $p$, $q$, $r$ is

$$qr+rp+pq = 0,$$

and the equation of the tangent at $P$ is

$$(p^3+3p)x-2y = 2ap^3.$$

Through a general point $(h, k)$ of the plane there are three tangents to the cissoid. If $P$ and $Q$ are points on the curve such that $PQ$ subtends a right angle at the origin, the tangents at $P$ and $Q$ meet in a point which, as $P$ and $Q$ vary, lies on a fixed circle on $(\frac{1}{2}a, 0)$ $(2a, 0)$ as diameter. The mid-point of $PQ$ then lies on the line $x = a$.

This cissoid was discovered by Diocles in the second century B.C., when he was discussing the problem of finding two mean proportionals between two given lengths (that is, a series $a$, $ax$, $ax^2$, $b$ in geometric progression, where $a$ and $b$ are the two given lengths). This is equivalent to finding the cube root of $a/b$. Suppose the centre of the circle is $D(a, 0)$, and $B$ is $(a, b)$; let $AB$ cut the cissoid in $Q$, so that $aq^3 = b$. Then if $OQ$ meets $DB$ in $S$, $DS = aq$, and this is one of the required mean proportionals.

A more modern application of the cissoid is that it is used in the design of planing hulls.

Other, more general, cissoids arise if $l$ is replaced by another line $x = 2(a+b)$. The equation of the curve is then

$$y^2\{x-2(a+b)\} = x^2(2b-x).$$

When $b < 0$, so that the line cuts the circle, the resulting curve has a double point at $O$.

In particular, when $2b = -a$, and the line passes through the centre $D$ of the circle, the curve is called a *Strophoid*. The equation of the curve is then

$$y^2(a-x) = x^2(a+x)$$

and it is given parametrically by

$$x = \frac{a(t^2-1)}{t^2+1}, \quad y = \frac{at(t^2-1)}{t^2+1}.$$

The shape of the curve is shown in Fig. 35. This curve was discovered, about 1670, by Barrow, whose greater claim to mathematical fame is that he taught Isaac Newton.

II. *The Folium of Descartes*,

$$x = \frac{3at}{1+t^3}, \quad y = \frac{3at^2}{1+t^3}.$$

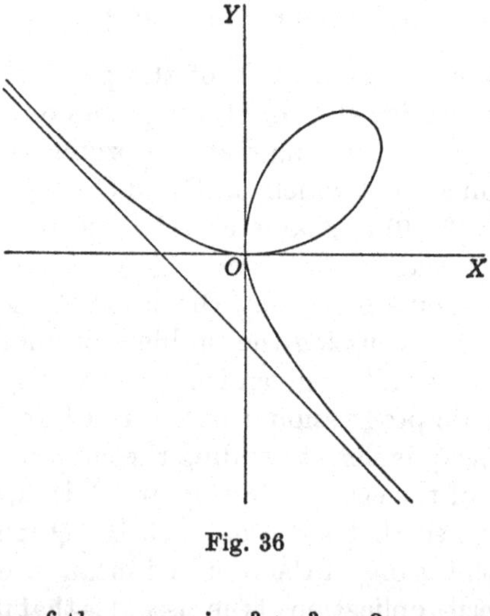


Fig. 36

The equation of the curve is $x^3+y^3 = 3axy$; we see from this that interchanging $x$ and $y$ in the equation leaves it unaltered. Thus if $(x, y)$ lies on the curve, so does $(y, x)$, and it follows that the curve is symmetrical about the line $y = x$ (Fig. 36). Values $t = 0$ and $t = \infty$ give the origin (0, 0), which is therefore a double point of the curve. The collinearity condition for the points with parameters $p, q, r$ is

$$pqr = -1,$$

and the equation of the tangent at $P$ is

$$p(p^3-2)x+(1-2p^3)y+3ap^2 = 0.$$

The curve extends to infinity as $t \to -1$, and the equation of the asymptote is

$$x+y+a = 0.$$

The curve was first discussed by Descartes in 1638.

III. *The Witch of Agnesi*,

$$x = \frac{2a}{1+t^2}, \quad y = 2at.$$

The equation of this curve is $xy^2 = 4a^2(2a-x)$; it is symmetrical about $OX$, and all points of the curve lie between $x = 0$ and $x = 2a$. The construction for a point on the curve is as follows. Take a fixed circle with radius $a$, through the origin $O$ and having $OA$ as a diameter, $A$ being on $OX$ (Fig. 37). Let a chord $OL$ of the circle meet the tangent at $A$ in $M$; $P$ is the point of intersection of the ordinate through $L$ and the line parallel to $OA$ through $M$, and the locus of $P$ is the curve being considered. If $\angle LOA = \theta$, and $P$ is $(x, y)$, $x = 2a\cos^2\theta$, $y = 2a\tan\theta$, and if $\tan\theta = t$ we have the given form of the curve.

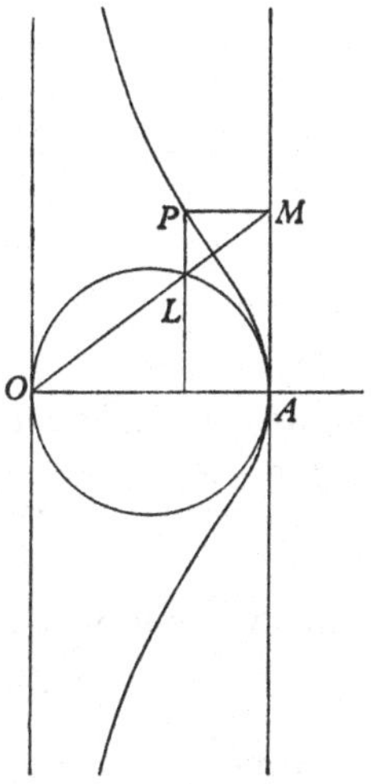

Fig. 37

The curve has inflexions at the points where $x = \frac{3}{2}a$ and the line $x = 0$ is an asymptote. The tangent at the general point with parameter $p$ is

$$(p^2+1)^2x+2py = 2a(3p^2+1).$$

This curve is thought to have been wrongly named by the Italian woman Professor Maria Agnesi (1738), who seems to have confused an old Italian word meaning 'free to move' with another meaning a 'goblin'. This may account for the fact that the shape of the curve comes as something of an anticlimax after one has read the name.

IV. *The Lemniscate of Bernouilli,*

$$x = \frac{2at}{1+t^4}, \quad y = \frac{2at^3}{1+t^4}.$$

The curve lies on the first quadrant when $t$ is positive and symmetrically in the third when $t$ is negative; it passes through the origin, a double point, when $t = 0$ and $t = \infty$. The equation of the curve is

$$(x^2+y^2)^2 = 4a^2xy.$$

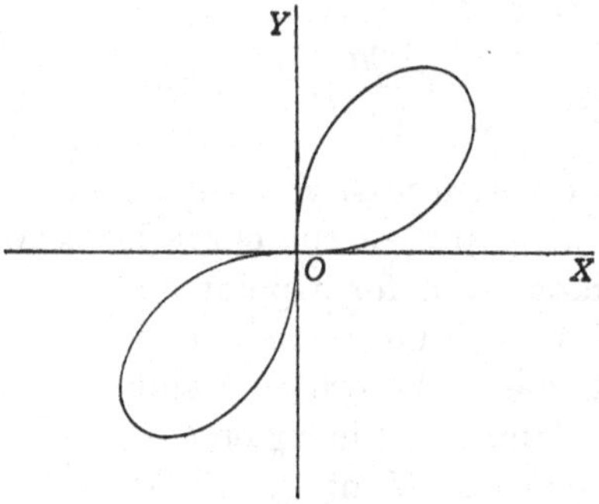


Fig. 38

This curve is a *quartic*, meeting a general line in at most four points; the collinearity conditions for points with parameters $p, q, r, s$ are

$$qr+rp+pq+ps+qs+rs = 0$$

and

$$pqrs = 1.$$

The equation of the tangent at $p$ is

$$p^2(3-p^4)\,x+(3p^4-1)\,y = 4ap^3.$$

Properties of this curve can also be derived from its equation in polar coordinates, and it is mentioned again in chapter IX (p. 268).

V. *Evolute of a parabola.* The *evolute* of a curve $C$ is another curve $C_1$ such that normals to $C$ are tangents to $C_1$; that is, the evolute is *enveloped* by the normals of $C$. This is a general definition; in particular, when $C$ is the parabola $y^2 = 4ax$, its evolute is the curve whose general point is

$$x = 2a+3at^2, \quad y = -2at^3.$$

This is a curve of the type considered in §7, a semicubic parabola with a cusp at $(2a, 0)$, and it also arose in Examples III E, 4. From points on one side of it three normals can be drawn to the parabola, and from points on the other side, only one. It meets the parabola where $x = 8a$, and the tangent to the curve there meets the parabola again on the ordinate through the cusp.

Suggestions of many other curves which are interesting to sketch appear in Chapter I of Cundy and Rollett, *Mathematical Models* (Oxford U.P. 1952). Also well worth consulting is R. C. Yates, *A Handbook on Curves* (Edward, Ann Arbor, 1947). An older treatise, richly illustrated, is P. Frost, *Curve Tracing* (Macmillan, 1872). The recent work by E. H. Lockwood, *A Book of Curves* (Cambridge U.P. 1961) is a gold mine.

## C.* TRIGONOMETRIC CURVES

Many curves are most simply defined in terms of trigonometric functions, $\sin\theta$, $\cos\theta$, $\tan\theta$. Some of these curves could be reduced to the type treated already, by putting $\tan\theta = t$, or $\tan\frac{1}{2}\theta = t$; in the instances to be considered there are advantages in retaining the trigonometric form. In the cases where such a substitution cannot be made, it should be noticed that the idea of the *order* of the curve, defined earlier, does not hold; it is a concept belonging to algebraic curves only.

In the same way, the algebraic method of finding the tangent, which we have used so far, is now no longer possible, and it is necessary to apply [C 1] and find the gradient of the tangent to a curve by differentiation.

Many of the problems about these curves are concerned with arc length and curvature, and these belong more to the realm of the calculus. Details about four curves of special geometrical interest are given here as illustrations.

I. *The Astroid*, $x = a\cos^3\theta,\ \ y = a\sin^3\theta.$

The curve lies symmetrically (Fig. 39) in the four quadrants, within the square bounded by the lines $x = \pm a$, $y = \pm a$. Its equation is

$$x^{\frac{2}{3}} + y^{\frac{2}{3}} = a^{\frac{2}{3}}.$$

The gradient of the tangent at any point is

$$-\frac{3a\sin^2\theta\cos\theta}{3a\cos^2\theta\sin\theta} = -\tan\theta$$

and the equation of the tangent is

$$x \sin\theta + y \cos\theta = a \sin\theta \cos\theta.$$

It follows that the tangent meets the coordinate axes in the points

$$(a\cos\theta, 0) \quad \text{and} \quad (0, a\sin\theta),$$

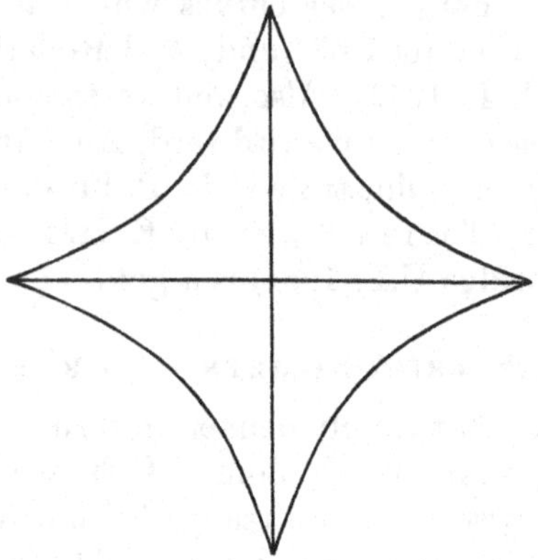

Fig. 39

and so the length of the tangent intercepted between the axes is a constant, $a$. Thus a ladder which is in contact with a wall and a floor and slips down envelopes an astroid, and similarly the fixed bar of the Archimedean Trammel (p. 216) envelopes an astroid. It is thus touched by a family of ellipses. The foot of the perpendicular from the origin to the tangent is

$$V(a\sin^2\theta\cos\theta,\ a\cos^2\theta\sin\theta),$$

and the locus of the mid-point of $PV$ is a circle.

The astroid can be generated by a point $P$ of a circle which rolls without slipping in the inside of another circle whose radius is four times as large. (A curve of this nature is called a *hypocycloid.*)

II. *The Cycloid,*

$$x = a(\theta - \sin\theta), \quad y = a(1-\cos\theta).$$

This is the locus of a fixed point $P$ on a circle of radius $a$ which rolls without slipping on a straight line (Fig. 40). In the form given above, $\theta$ is the angle made by the radius to $P$ with the downward vertical through the centre, and the circle starts initially with $P$ at $O$, the origin. The curve consists of a series of loops.

The equation of the tangent at a general point is

$$x\sin\theta - y(1-\cos\theta) = a(\theta\sin\theta - 2 + 2\cos\theta);$$

at any stage the tangent passes through the highest point of the rolling circle.

This curve arises in a number of mathematical situations; among other things the reflection of the curve (Fig. 40) in the $x$-axis is the path on which a particle travels if the time taken to reach the bottom is independent of the point at which the particle

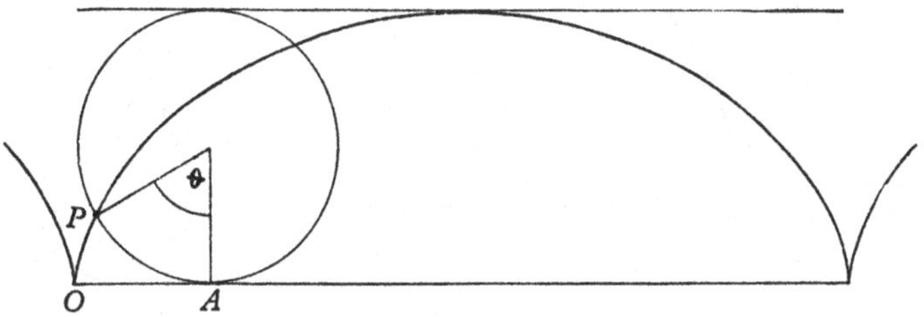


Fig. 40

is released under gravity (see, for example, Pars, *Introduction to Dynamics* (Cambridge U.P., 1953, p. 216)). The curve has been studied by many mathematicians whose names are well known, Descartes, Pascal, Bernouilli among them; Sir Christopher Wren also contributed something to our knowledge of it.

III. *The Cardioid,*

$$x = a(2\cos\theta - \cos 2\theta), \quad y = a(2\sin\theta - \sin 2\theta).$$

This curve is generated by a fixed point $P$ of a circle which rolls outside another circle of equal radius; it is a special epicycloid. It has a cusp at $(a, 0)$ and is symmetrical about the $x$-axis (Fig. 41). If a chord $PQ$ passes through the cusp, $PQ$ is of constant length $4a$, and the tangents at $P$ and $Q$ are perpendicular. The curve is the inverse (p. 160) of a parabola with respect to its focus.

Properties of this curve can also be established from the polar equation (see p. 267).

IV. *The Deltoid,*

$$x = a(2\cos t + \cos 2t), \quad y = a(2\sin t - \sin 2t).$$

This is the locus of a fixed point $P$ on a circle which rolls inside another circle three times the size. It therefore lies entirely inside this circle. The length of a tangent $QR$ intercepted by the curve is a constant, and the normals at $P$, $Q$, $R$ meet in a point on the circumscribing circle. This curve is the envelope of the Simson line [G 6] of a point $P$ as $P$ moves round the circle.

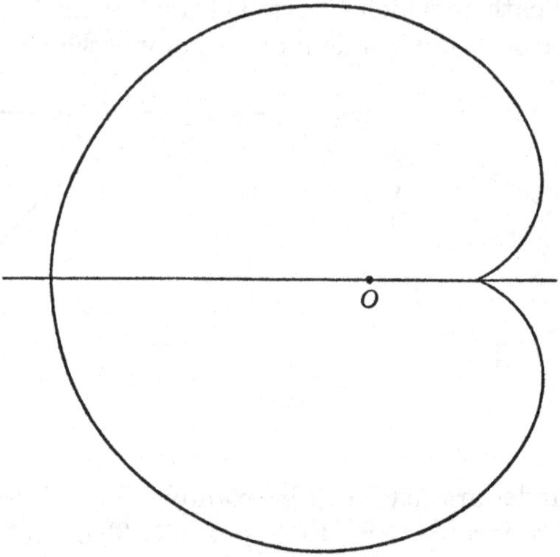


Fig. 41

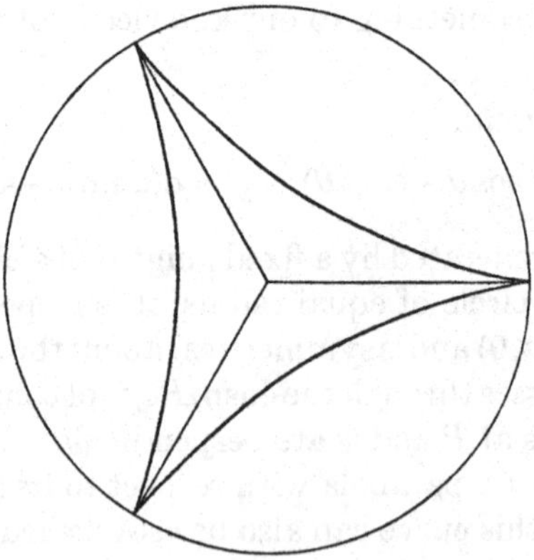

Fig. 42

## MISCELLANEOUS EXAMPLES IV

1. The curves $y^2 = 4ax$ and $xy = c^2$ intersect at right angles. Prove that $c^4 = 32a^4$, and that if the tangent and normal to either curve at the point of intersection meet the $x$-axis at $T$ and $G$, then $TG = 6a$.

2. The points $P$, $Q$, $R$ with parameters $p$, $q$, $r$ on the curve $xy = c^2$ form a triangle right angled at $P$. Prove that the tangent to the hyperbola at $P$ is perpendicular to $QR$.

3. $PQ$ and $PR$ are perpendicular chords of the hyperbola $xy = c^2$. Show that, if $QR$ passes through the foot of the perpendicular from $P$ to the $x$-axis, then the locus of the centroid of the triangle $PQR$ is the curve $y = (2c^2/9x) - (9x^3/8c^2)$.

4. The tangents to a rectangular hyperbola at $P$ and $Q$ meet one of the asymptotes of the hyperbola in $A_1$ and $B_1$ respectively, and meet the other asymptote in $A_2$ and $B_2$ respectively. Prove that the lines $A_1 B_2$ and $A_2 B_1$ are parallel to $PQ$ and equidistant from it.

5. Two points $P$ and $Q$ on the rectangular hyperbola $xy = c^2$ are so related that the tangent at $P$ passes through the foot of the perpendicular from $Q$ to the $x$-axis. Find the locus of the mid-point of $PQ$.

6. Show that the locus of the mid-points of normal chords of the hyperbola $xy = c^2$ is given by the equation $c^2(x^2 - y^2)^2 + 4x^3y^3 = 0$.

7. A chord of the hyperbola $xy = c^2$ passes through the fixed point $(h, k)$. Find the locus of the mid-point of the chord.

8. Find the locus of the mid-point of a chord of the rectangular hyperbola $xy = c^2$ which has a constant length $2a$.

9. $P(ct, c/t)$ lies on the hyperbola $xy = c^2$, and $P'$ is the other end of the diameter through $P$. If $R$ is the intersection of the tangent at $P$ with a line through $P'$ parallel to an asymptote, show that the locus of $R$ is the hyperbola $xy + 3c^2 = 0$.

10. The coordinates of a variable point $P$ are given in terms of a parameter $t$ by the equations $x = 4 - t^2$, $y = 1 + 3t$. Find the value of $t$ for which the tangent to the locus of $P$ is parallel to the $y$-axis.

11. A curve is given by the parametric equations $x = t^2$, $y = 2t - 1$. Obtain the equations of the tangent and normal at the point on the curve where $t = 2$. Prove that this normal intersects the curve again at the point where $t = -3$.

12. Find the equations of the tangent and the normal to the curve $y = x^3 + 1$ at the point $P$ where $x = a$. The $y$-axis cuts the tangent in the point $Q$ and the normal in the point $R$. Show that the length of $QR$ is $3a^3 + 1/(3a)$.

13. The tangent at the point $P(5, 9)$ to the curve with parametric equations $x = 1 + t^2$, $y = 1 - t^3$, meets the curve again at $Q$. Find the value of the parameter corresponding to $Q$ and the coordinates of $Q$.

14. The tangent to the curve with parametric equations

$$x = 2-t^2, \quad y = 3+t^3$$

at the point $P(1, 4)$ meets the curve again at $Q$. Determine the value of the parameter corresponding to $Q$ and calculate the length of the chord $PQ$. Find also the Cartesian equation of the curve.

15. The curves $y = x^3$ and $2y = 19x^2-225$ intersect at $A$, $B$ and $C$. Prove that the tangents at $A$, $B$ and $C$ to the first curve are concurrent at a point $P$ and find its coordinates.

16. A curve is given by the parametric equations $x = t$, $y = t/(1+t^2)$. Find the equation of the tangent at the point $t$, and show that this tangent meets the curve again at the point $2t/(t^2-1)$. Find the values of $t$ corresponding to the points of inflexion of the curve.

17. Find $a$, $b$, $c$, $d$ such that the curve $y = ax^3+bx^2+cx+d$ may pass through the point $(-2, 8)$, touch the $x$-axis at the point $(2, 0)$, and have its tangent parallel to the $x$-axis where $x = -1$ on the curve. Sketch the curve.

18. By putting $y = tx$, find a parametric representation of the curve $3ay^2 = x^2(2x+a)$, and sketch the curve, explaining briefly your procedure. Prove that, in general, any line meets the curve in three points. Find the equation of the normal to the curve at the point $(a/6, -a/9)$ and consider its other points of intersection with the curve, finding their coordinates.

19. A curve whose equation is of the form $y = ax^3+bx+c$ passes through the point $(0, 6)$ and the tangent at $(3/2, 3)$ is parallel to the $x$-axis. Find $a$, $b$, $c$ and the coordinates of the other point where the tangent is parallel to the $x$-axis.

20. Putting $y = tx$, find a parametric representation of the curve $y^2(2a-x) = x^3$, and sketch the curve. $A$ is the origin, $B$ the point $(2a, 0)$, $C$ the mid-point of $AB$. $Q$ is a point on the curve; $BQ$ meets the perpendicular to $AB$ through $C$ in $R$. $AQ$ meets $CR$ in $S$ and $SD$ is drawn perpendicular to $AS$ to meet $AB$ in $D$. Prove that $CB$, $CS$, $CD$, $CR$ are in geometric progression.

21. $O$ is the origin and $P$ the point $(2, 2)$. Find the coordinates of the points $Q$, $R$ on the curve $y = 9x^3-3x^2$ at which the tangents are parallel to $OP$. The normal to the curve at $O$ meets the tangents at $Q$ and $R$ at the points $U$ and $V$ respectively. Prove that $|OU.OV| = 5/243$. Find the perpendicular distance between the tangents $QU$ and $RV$.

22. Show that the point $P(a\cos^4\theta, a\sin^4\theta)$ lies on the curve $x^{\frac{1}{2}}+y^{\frac{1}{2}} = a^{\frac{1}{2}}$ for all values of $\theta$. Draw a rough sketch of the curve. Obtain the equation of the tangent to the curve at $P$ and prove that, if this tangent meets the coordinate axes at $Q$ and $R$, then $OQ+OR = a$ for all positions of $P$ on the curve.

23. Find the equations of the tangent and normal to the curve $x^{\frac{2}{3}}+y^{\frac{2}{3}}=a^{\frac{2}{3}}$ at the point $(a\cos^3\theta, a\sin^3\theta)$. Prove that, if $p, q$, respectively, are the lengths of the perpendiculars drawn from the origin to the tangent and normal, $4p^2+q^2=a^2$.

24. Find the equations of the tangent and normal at the point $t$ of the curve $x=a\cos^2 t\sin t$, $y=a\sin^2 t\cos t$. The normal meets the axis of $y$ in $G$ and the tangent meets the axis of $x$ in $T$. Prove that if $\tan 2t=2$ and $O$ is the origin then $OG=OT$.

25. A point on the curve $x^{\frac{2}{3}}+y^{\frac{2}{3}}=a^{\frac{2}{3}}$ is given parametrically by the equations $x=a\cos^3\theta$, $y=a\sin^3\theta$. Show that the tangent at the point with parameter $\theta$ is also a normal to the curve if $\tan 2\theta=\pm 2$. How many such tangents exist?

26. A curve is given by $x=a(2\cos t+\cos 2t)$, $y=a(2\sin t-\sin 2t)$. Show that the tangent at the point with parameter $t$ has equation $x\sin\frac{1}{2}t+y\cos\frac{1}{2}t=a\sin\frac{3}{2}t$. Hence show that the tangents at the points with parameters $2\theta$, $\pi+2\theta$ meet at right angles on the circle $x^2+y^2=a^2$.

27. A wheel of radius $r$ rolls along the $x$-axis. Let $P$ be that point on the rim of the wheel which was lowest when the centre $C$ crossed the $y$-axis, and let $\phi$ be the angle through which the wheel has turned since. Show that the coordinates of $P$ are $(r\phi-r\sin\phi, r-r\cos\phi)$, and that the direction of motion of $P$ is always parallel to the bisector of the angle between $CP$ and the downward vertical.

28. For the cycloid $x=a(\theta+\sin\theta)$, $y=a(1-\cos\theta)$, determine the equation of the tangent at the point $P$ corresponding to $\theta$ and prove that it passes through the point $(a\theta, 0)$. Sketch the curve. If the tangent at $P$ meets the curve again as the normal at the point $\phi$, prove that $\phi=\pi+\theta$ and that $\tan\frac{1}{2}\theta=2/\pi$.

---

29. A variable chord of a rectangular hyperbola subtends a right angle at a fixed point $A$. Show that the locus of the foot of the perpendicular from $A$ to the chord is a straight line.

30. Each of two perpendicular chords of a rectangular hyperbola subtends a right angle at a point $P$. Prove that the point of intersection of the chords lies on the chord of contact of tangents from $P$.

31. Prove that the coordinates of any point $Q$ on the normal at $P(ct, c/t)$ to the rectangular hyperbola $xy=c^2$ may be expressed in the form $x=c(t+\lambda)$, $y=c(t^{-1}+\lambda t^2)$, where $\lambda$ is a parameter. If the feet of the other three normals from $Q$ to the hyperbola are the points $P_r(ct_r, ct_r^{-1})$, where $r=1, 2, 3$, find the equation whose roots are $t, t_1, t_2, t_3$. Prove that as $Q$ varies the locus of the centroid of the triangle $P_1P_2P_3$ is the diameter of the hyperbola parallel to the normal at $P$.

32. The normals at points $A, B, C, D$ on a rectangular hyperbola meet in a point $P$. Prove that $AB$ is perpendicular to $CD$. If $AB$ and $CD$ meet

the asymptotes in $U$, $U'$ and $V$, $V'$, show that the line joining the mid-points of $UU'$ and $VV'$ divides the line joining $P$ to the centre of the hyperbola in the ratio 3:1.

33. Show that there are two tangents from the point $(x_1, y_1)$ to the curve $xy = c^2$, provided $x_1 y_1 < c^2$, and find the equation of the chord joining their points of contact. Prove that if a variable chord $PQ$ passes through a fixed point $R$, then the tangents at $P$ and $Q$ meet on a fixed straight line $r$. Show that if $R$ lies on the curve $xy + c^2 = 0$, then $r$ touches that curve.

34. The point $(ap^3, ap)$ with parameter $p$, varies on the curve $y^3 = a^2x$. Show that the tangent to the curve at the point $p$ meets the curve again in the point $-2p$, and find the equation of this tangent. A chord is drawn through the point $R$ (not the origin), to meet the curve again in $P$ and $Q$; the tangents at $P$ and $Q$ meet in $L$. Show that the locus of $L$ as the chord varies through $R$ is a parabola, with its axis parallel to the $x$-axis. Show also that, as $R$ varies, the vertex of this parabola lies on a fixed line. Explain why the restriction that $R$ is not the origin is needed.

35. A curve is given in terms of a parameter $t$ by the equations $x = 1/t^2$, $y = t(t+1)$. Prove that (i) the line $lx + my + n = 0$ meets the curve in four points, and that, if $t_r$ $(r = 1, 2, 3, 4)$ are the parameters of these points then $t_1 t_2 t_3 + t_2 t_3 t_4 + t_3 t_4 t_1 + t_4 t_1 t_2 = 0$, and $t_1 + t_2 + t_3 + t_4 + 1 = 0$; (ii) the tangent at the point whose parameter is $t$ $(t \neq 0)$ meets the curve again in points whose parameters $T_1$, $T_2$ are the roots of the quadratic

$$2T^2 + 2(2t+1)T + t(2t+1) = 0;$$

(iii) the curve has one finite point of inflexion.

36. Find the condition that the points $\lambda_1$, $\lambda_2$, $\lambda_3$ of a plane cubic curve $x:y:1 = \lambda:\lambda^2:\lambda^3+1$ be collinear. Find the equation of the tangent to this curve at a general point, and show that four tangents of the curve pass through a general point of the plane.

37. Write down the equation which determines the parameters of the points in which the line $ax + by + c = 0$ meets the curve

$$x:y:1 = t^3 - t : t^3 + t : t^4 - 1,$$

and show that necessary and sufficient conditions that the points $t_1$, $t_2$, $t_3$, $t_4$ of the curve should be collinear are

$$t_1 t_2 t_3 t_4 = -1 \quad \text{and} \quad t_1 t_2 + t_1 t_3 + t_1 t_4 + t_2 t_3 + t_2 t_4 + t_3 t_4 = 0.$$

Hence prove that through a general point of the curve there are four lines which touch the curve elsewhere, and that the four points of contact are collinear.

38. The coordinates of a point on a plane cubic curve are

$$x = t^3 + 3t, \quad y = 3t^2,$$

where $t$ is a parameter; find the condition that the three points with parameters $a$, $b$, $c$ should be collinear, and hence determine the coordinates of the points of inflexion. If the tangents at $A$, $B$ on the curve meet at a point $P$ on the curve, and $AB$ meets the curve at $Q$, prove that $pq+3 = 0$, where $p$, $q$ are the parameters of $P$, $Q$.

39. A plane cubic curve $C$ is defined by $x:y:1 = t^3:t:1$. Prove that the tangent at the point $T$ meets the curve again where $t = -2T$. Show that the three tangents from the point $P(\alpha, \beta)$ touch $C$ at the three points given by $\alpha - 3\beta t^2 + 2t^3 = 0$. The three tangents from $P$ meet the curve again at $P_1, P_2, P_3$. Show that the tangents at $P_1, P_2, P_3$ concur in $Q$. Find the locus of $Q$ when $P$ describes a line.

40. By writing $y = tx$ obtain a parametric representation of the curve $x^3 + y^3 = axy$. Find the condition for the points given by $t = t_1, t_2, t_3$ to be collinear. If the points $A, B, C, A', B', C'$ on the curve have parameters $a, b, c, a', b', c'$ and are such that $A'BC$, $AB'C$, $ABC'$, $A'B'C'$ are straight lines, prove that $a+a' = b+b' = c+c' = 0$. Prove also that the tangents at $A$ and $A'$ meet on the curve, and similarly for $B$, $B'$ and for $C$, $C'$.

41. The Cartesian coordinates $(x, y)$ on a curve are given by

$$x:y:1 = t^3:t^2-3:t-1,$$

where $t$ is a parameter; show that if the points given by $t = a, b, c$ are collinear then $abc-(bc+ca+ab)+3(a+b+c) = 0$, and find the coordinates of the double point on the curve. Prove that $(0, 3)$ is a point of inflexion, and find the equation of the asymptote of the curve.

42. Prove that the equation of the normal at a point of the curve $x = n\cos t - \cos nt$, $y = n\sin t - \sin nt$, where $n$ is an integer greater than unity, is $x\cos\frac{1}{2}(n+1)t + y\sin\frac{1}{2}(n+1)t = (n-1)\cos\frac{1}{2}(n-1)t$. Show that, if $n$ is an even integer, the normals at the points $t$ and $t+\pi$ are perpendicular and intersect on the circle $x^2+y^2 = (n-1)^2$.

43. A circle of radius $a$ has its centre at the origin $O$ and the point $A$ of its circumference lies on the positive $x$-axis. An equal circle rolls on the outside of this fixed circle, the point $P$ on the circumference of the rolling circle coinciding with $A$ in its initial position. If $\theta$ is the angle made with $OA$ by the line of centres, prove that the coordinates of $P$ are

$$x = 2a\cos\theta - a\cos 2\theta, \quad y = 2a\sin\theta - a\sin 2\theta.$$

Prove that the curve traced out by $P$ is the locus of the foot of the perpendicular from $A$ to a variable tangent to the circle with centre $(-a, 0)$ and radius $2a$. Give a sketch of the curve.

44. A curve $C$ is given by $x = a\cos\omega t$, $y = b\sin(\omega t - \alpha)$. Find the Cartesian equation of the curve. Find the equations of the tangents to $C$ at the four points in which $C$ meets the coordinate axes. Show that the diagonals of the parallelogram formed by these tangents meet in the origin.

45. $AB$ is a fixed line of length $2a$ and $ACD$ is a variable triangle in which $\angle ACD = 90°$, $CD = 2a$, $CD$ intersects $AB$ and the point $D$ lies on the perpendicular to $AB$ through $B$. If the mid-point $O$ of $AB$ is taken as origin and the line $AOB$ as $x$-axis, prove that the mid-point $P$ of $CD$ describes the curve given by the parametric equations $x = 2a\sin^2\theta$, $y = 2a\sin^3\theta/\cos\theta$, where $\theta = \angle BOP$. Give a sketch of this curve, and prove that it is the locus of the foot of the perpendicular from the origin to a variable tangent to the parabola $y^2 + 8ax = 0$.

# V

# THE CIRCLE

The circle is a curve whose elementary properties will be familiar to any reader of this book. For example, it is well known that the tangent to a circle is perpendicular to the radius, and that the angle in a semicircle is a right angle. Such results are best established by elementary methods, and no good purpose is served by proving them afresh with the help of coordinates. Indeed the use of coordinates in this range of problems would frequently rather be a hindrance. We shall, therefore, assume all the elementary properties of a circle, and use them whenever it is convenient to do so.

In this chapter we first find the equation of a circle, in several forms, and the equation of a tangent to the circle. In the later paragraphs these results are applied to the investigation of properties of circles, and in particular of systems of circles, which are not so easily accessible without the use of coordinates. Some of these later paragraphs may be found hard, and they may be omitted at a first reading. They do, however, illustrate some of the power of coordinate methods, and are included for that reason as well as for the intrinsic interest of the results.

**1. The equation of a circle.** A circle is the locus of a point $P$ which is a constant distance $r$ (the *radius*) from a fixed point $C$ (the *centre*). If $C$ is the origin $(0, 0)$ and $P$ is the point $(x, y)$, then

$$PC^2 = x^2+y^2$$

and so the equation of the circle is

$$\mathbf{x^2+y^2=r^2.}$$

More generally, if $C$ is the point $(p, q)$, then

$$PC^2 = (x-p)^2+(y-q)^2 = r^2$$

and this reduces to

$$x^2+y^2-2px-2qy+p^2+q^2-r^2 = 0.$$

We notice two things about this equation.

(1) There is a term in $x^2$ and a term in $y^2$, and the coefficients of these are both the same.

(2) There is no term in $xy$.

We now consider the general equation of the second degree

$$ax^2+2hxy+by^2+2gx+2fy+c=0.$$

The selection of the coefficients $a, b, c, f, g, h$ will no doubt seem arbitrary, if not perverse, but there is a simple explanation. In more advanced work great importance attaches to the general homogeneous equation of the second degree in three variables $x$, $y$, $z$; this is

$$ax^2+by^2+cz^2+2fyz+2gzx+2hxy=0.$$

In the work we are doing there are only two variables $x$ and $y$, and the third variable $z$ is replaced by 1. This gives

$$ax^2+by^2+c+2fy+2gx+2hxy=0$$

and when rearranged this is the expression above.

If this general equation is of the form obtained for the circle, then (1) $a=b\neq 0$, and (2) $h=0$. Since $a\neq 0$, we may divide through by it, leaving the terms in $x^2$ and $y^2$ both with coefficient 1. The equation then is

$$\mathbf{x^2+y^2+2gx+2fy+c=0.}$$

This may be written

$$(x+g)^2+(y+f)^2=g^2+f^2-c,$$

and so we see that it represents a circle whose centre is $(-g, -f)$. Further:

(i) If $g^2+f^2-c>0$, the circle has radius $+\sqrt{(g^2+f^2-c)}$.

(ii) If $g^2+f^2-c=0$, the circle has zero radius, and there is then only one point $(-g, -f)$ satisfying the equation

$$(x+g)^2+(y+f)^2=0;$$

the locus is then called a *point-circle*.

(iii) If $g^2+f^2-c<0$, there are no real values of $x$ and $y$ satisfying the equation (since the left-hand side is greater than or equal to zero and the right-hand side is less than zero), and so there are no points on the locus. (Some writers count this as a circle, which they call—when they observe the distinction between real and unreal points—a *virtual* circle).

Since the equation of the general circle involves three coefficients $g, f$ and $c$, which are all at our disposal in any particular problem, it follows that a circle can be made to satisfy *three* independent linear conditions. In particular, through three points which are not collinear there is a unique circle.

The circle divides the plane into two regions; a point may be either outside or inside the circle, if it is not on the circle itself which is the boundary between the regions. The algebraic conditions for this are easily found, for if the point is $P$, and $C$ is the centre of the circle of radius $r$, then $P$ is outside or inside the circle according as

$$PC^2 > r^2 \quad \text{or} \quad PC^2 < r^2.$$

If $P$ is the point $(x_1, y_1)$, $C$ is $(-g, -f)$ and $r^2 = g^2+f^2-c$, these give

$$(x_1+g)^2+(y_1+f)^2 \gtrless g^2+f^2-c,$$

or

$$x_1^2+y_1^2+2gx_1+2fy_1+c \gtrless 0.$$

We can express this result more briefly by using an abridged notation (cf. Chapter II, §8). Write the equation of the circle as $S = 0$, where

$$S \equiv x^2+y^2+2gx+2fy+c.$$

(Here all the terms in the equation are on one side, and the coefficients of $x^2$ and $y^2$ are both 1.) Let $S_{11}$ be the expression obtained by putting $x_1$ for $x$ and $y_1$ for $y$ in $S$; that is†

$$S_{11} \equiv x_1^2+y_1^2+2gx_1+2fy_1+c.$$

We have then proved that $P_1(x_1, y_1)$ is

*outside* the circle if $S_{11} > 0$,

and

*inside* the circle if $S_{11} < 0$.

**2. Parametric form of the circle.** If the origin $O$ is taken as the centre of the circle and $OP$ makes an angle $\theta$ with $OX$, then (Fig. 43) the coordinates of $P$ are $(r\cos\theta, r\sin\theta)$, where $r$ is the radius of the circle. This point evidently lies on $x^2+y^2 = r^2$, for all values of $\theta$. Provided that $0 \leqslant \theta < 2\pi$, $\theta$ serves as a parameter on the curve (see the definition on p. 31).

† It will become clear in a later chapter (p. 250) why this expression is called $S_{11}$ rather than $S_1$.

More generally, if the circle has centre $(p, q)$ and radius $r$, the general point on the curve may be taken as

$$(p+r\cos\theta,\ q+r\sin\theta).$$

The reader should draw a diagram and identify the angle $\theta$ in this case.

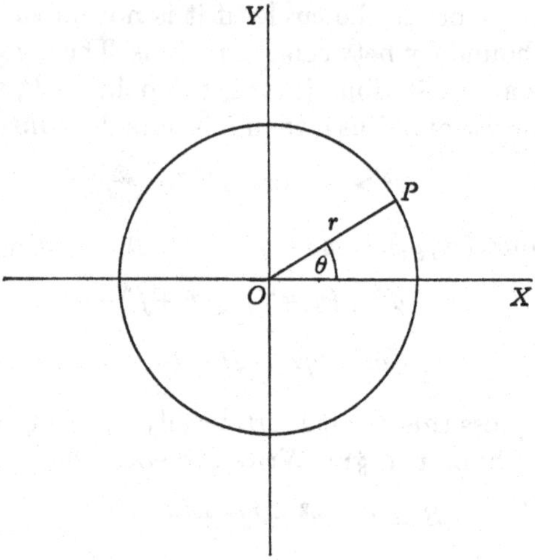


Fig. 43

In either of these cases the trigonometric functions $\cos\theta$, $\sin\theta$ can be replaced by algebraic ones, if we write $\tan\frac{1}{2}\theta = t$. In the first case, for example, the coordinates of $P$ are then

$$\left(r\frac{1-t^2}{1+t^2},\ \frac{2rt}{1+t^2}\right).$$

**3. Tangents to a circle.** The equation of a tangent to a circle is most easily found by using the fact that the tangent is perpendicular to the radius at the point of contact.

Take first the circle $x^2+y^2 = r^2$, with centre $(0, 0)$ and radius $r$. The tangent at $P(r\cos\theta, r\sin\theta)$ is a line through $P$ and perpendicular to $OP$. The gradient of $OP$ is $\tan\theta$, so that the gradient of the tangent is $-1/\tan\theta$, and the equation of the tangent is

$$y-r\sin\theta = -\cot\theta(x-r\cos\theta),$$

or

$$x\cos\theta+y\sin\theta = r(\sin^2\theta+\cos^2\theta),$$

that is $$\mathbf{x\cos\theta+y\sin\theta=r.}$$

(We notice that this is the normal form of the equation of a line (p. 31).)

The same method may be applied in the general case, and we first give a numerical example. Consider the circle

$$x^2+y^2-4x+2y-20 = 0;$$

and suppose we wish to find the equation of the tangent at $P(5, 3)$, a point on the circle. The centre $C$ of the circle is, by inspection, $(2, -1)$, and so the gradient of $CP$ is $\frac{4}{3}$. The gradient of the tangent is then $-\frac{3}{4}$, and its equation is

$$y-3 = -\tfrac{3}{4}(x-5),$$

or $$3x+4y = 27.$$

Again, if the equation of the circle is

$$x^2+y^2+2gx+2fy+c = 0,$$

and we wish to find the equation of the tangent at $P(x_1, y_1)$, we proceed in the same way. The centre of the circle is $C(-g, -f)$, and so the gradient of $CP$ is $(y_1+f)/(x_1+g)$. The gradient of the tangent at $P$ is then $-(x_1+g)/(y_1+f)$ and so the equation of the tangent is

$$y-y_1 = -\frac{(x_1+g)}{(y_1+f)}(x-x_1),$$

or $$xx_1+yy_1+g(x+x_1)+f(y+y_1) = x_1^2+y_1^2+2gx_1+2fy_1.$$

Now $(x_1, y_1)$ lies on the circle and so satisfies its equation, so that

$$x_1^2+y_1^2+2gx_1+2fy_1+c = 0.$$

The equation of the tangent is thus

$$\mathbf{xx_1+yy_1+g(x+x_1)+f(y+y_1)+c=0.}$$

**4. The form of the equation of a tangent.** We have now found the equations of the tangents to three curves which have equations which are quadratic in $x$ and $y$, namely the parabola, rectangular hyperbola and circle. A comparison of these suggests a useful mnemonic. Consider the results:

Parabola $y^2 = 4ax$, tangent at $(x_1, y_1)$ is $yy_1 = 2a(x+x_1)$ (p. 67).

Hyperbola $xy = c^2$, tangent at $(x_1, y_1)$ is $xy_1+yx_1 = 2c^2$ (p. 101).

Circle $x^2+y^2+2gx+2fy+c=0$, tangent at $(x_1, y_1)$ is

$$xx_1+yy_1+g(x+x_1)+f(y+y_1)+c=0.$$

The general rule, suggested (though not yet *proved*) by these results, is that, to form the equation of a tangent at $(x_1, y_1)$, we

(i) in the equation, replace $x^2$ by $xx_1$ and $y^2$ by $yy_1$;

(ii) in the equation, replace $2xy$ by $xy_1+yx_1$, $2x$ by $x+x_1$, and $2y$ by $y+y_1$;

(iii) leave any constant term unaltered.

The result which has here been indicated is later proved (p. 252) for *any* curve whose equation is quadratic. At this stage it is useful to notice that the result makes it easier to recollect and write down at once the equation of the tangent to any curve of this type. (It does *not* apply to curves whose equations are not of quadratic form.)

**5. The equation of the chord of contact.** The reader may have noticed that, in the first two cases in the last paragraph, the equations given are also the equations of the *chords of contact*, or *polar lines*, of a point $(x_1, y_1)$, (see pp. 70, 104). The difference is that the point $(x_1, y_1)$ lies on the curve in the first case and not in the second.

The equation of the chord of contact is easily found for the circle. If $P_2(x_2, y_2)$ and $P_3(x_3, y_3)$ are the points of contact of tangents drawn from $P_1(x_1, y_1)$ to the circle

$$x^2+y^2+2gx+2fy+c=0,$$

then the tangent

$$xx_2+yy_2+g(x+x_2)+f(y+y_2)+c=0$$

goes through $(x_1, y_1)$ and so

$$x_1x_2+y_1y_2+g(x_1+x_2)+f(y_1+y_2)+c=0.$$

By similar reasoning, since $(x_1, y_1)$ lies on the tangent at $(x_3, y_3)$

$$x_1x_3+y_1y_3+g(x_1+x_3)+f(y_1+y_3)+c=0.$$

Thus $(x_2, y_2)$ and $(x_3, y_3)$ both satisfy the equation

$$x_1x+y_1y+g(x_1+x)+f(y_1+y)+c=0.$$

This is the equation of a line through $P_2$ and $P_3$, and so it is the equation of the chord of contact.

Beginners may find this reasoning hard to follow at first. If so, the equation is easily, though less elegantly, found by writing down the equation of the line through $P_2$ perpendicular to the line joining $P_1$ to the centre.

## EXAMPLES V A

1. Write down the coordinates of the centre, and the radius, of the following circles:

(i) $x^2+y^2 = 25$; (ii) $x^2+y^2-4x+2y-11 = 0$;

(iii) $x^2+y^2+3x-y-1 = 0$; (iv) $2x^2+2y^2+x+3y = 0$;

(v) $x^2+y^2-10x+14y+74 = 0$.

2. Give the equations of the circles for which the centre $C$ and radius $r$ are

(i) $C(4, 1)$, $r = 3$; (ii) $C(3, -4)$, $r = 5$;

(iii) $C(-a, b)$, $r = a$; (iv) $C(at^2, 2at)$, $r = a(t^2+1)$.

3. Find the values of $g$ for which the circle $x^2+y^2+2gx-4y+15 = 0$ has radius 5.

4. Find the equation of the circle through the points $(0, 3)$, $(2, -1)$, $(1, 4)$.

5. Find the equation of the nine-points circle of the triangle whose vertices are $(0, 0)$, $(4, 0)$ and $(3, 6)$.

6. Which of the points $(2, 1)$, $(0, -2)$, $(-1, 5)$, $(-7, 5)$, $(-3, 7)$ lie on the circle $x^2+y^2+6x-4y-12 = 0$? Write down the equations of the tangents at the points which do lie on it.

7. The tangents to the circle $x^2+y^2-2x-8y-8 = 0$ at $(-2, 8)$, $(5, 7)$ and $(1, -1)$ form a triangle. What are its vertices, and which vertex is the nearest to the centre of the circle?

8. Find the coordinates of the points of intersection of the line

$$x-2y+10 = 0$$

with the circle, centre the origin, whose radius is 10.

9. Find where the line $3x-4y+14 = 0$ meets the circle

$$x^2+y^2-2x-4y-20 = 0.$$

10. Find the vertices of the triangle formed by the tangents at $A$, $B$ and $C$ to the circumcircle of the triangle whose vertices are $A(2, 3)$, $B(-2, 1)$ and $C(-3, -2)$.

11. Show that the line $y = mx+a\sqrt{(1+m^2)}$ touches the circle

$$x^2+y^2 = a^2,$$

for any value of $m$.

12. A circle touches the $x$-axis and intercepts a fixed length $2a$ on the $y$-axis. Find the locus of its centre.

13. Find the equations of the two circles which touch the $x$-axis and pass through the points $(1, -2)$ and $(3, -4)$. Obtain also the coordinates of their centres and the lengths of their radii.

14. Find the expression for the length of the perpendicular from the point $(x_1, y_1)$ to the line $Ax+By+C = 0$. Hence, or otherwise, determine the two values of $C$ for which the line $3x+4y+C = 0$ is a tangent to the circle $x^2+y^2-6x-2y = 15$.

15. Calculate the coordinates of the centre and the radius of the circle $x^2+y^2-6x-4y = 12$. If this circle cuts the $y$-axis in the points $A$ and $B$, find the length of $AB$. If one end of a diameter of the circle is $(7, -1)$ calculate the coordinates of the other end of the diameter.

16. Find the centre and radius of the circle which passes through the points $(7, 5)$, $(6, -2)$, $(-1, -1)$.

17. The line joining $(5, 0)$ to $(10\cos\theta, 10\sin\theta)$ is divided at $P$ internally in the ratio $2:3$. If $\theta$ varies, show that the locus of $P$ is a circle, and find its centre and radius.

18. Show that the circle through the origin and the points $(0, 6)$ and $(8, 0)$ has equation $x^2+y^2-8x-6y = 0$. A line $y = mx$ cuts the circle in $O$ and $P$, and $Q$ lies on $OP$ such that $OQ:QP = 2:3$; find the equation of the locus of $Q$.

19. Find the equations of the circles which touch the line $4y = 3x$, have their centres on the $y$-axis, and pass through the point $(0, 3)$. Find also the coordinates of the points in which the circles touch the given line.

20. A circle of radius 3 is to be drawn with its centre on the line $y = x-1$ and passing through the points $P(7, 3)$. Show that two such circles are possible and find their equations. Show that one of these circles touches the axis of $x$.

21. Show that the points $A(4{\cdot}9, 6{\cdot}2)$ and $B(3{\cdot}5, 7)$ lie on a circle whose centre is the point $(1, 1)$. Find the equation of the perpendicular bisector of $AB$, and the equation of the line through $A$ which is perpendicular to the tangent at $B$ to the circle.

22. Prove that the circle whose centre is $(3, 5)$ and which touches the $y$-axis is $x^2+y^2-6x-10y+25 = 0$. Find the equation of the other tangent from the origin, and the coordinates of the point of contact.

23. A circle $C$ has centre $(4, 0)$ and radius 2. The line $y = mx$ cuts the circle in the points $P$, $Q$, and $M$ is the mid-point of $PQ$. Find the coordinates of $M$, and its locus as $m$ varies.

24. Find the coordinates of the orthocentre and the circumcentre of the triangle whose vertices are the points $(c, 0)$, $(-c, 0)$ and $(a, b)$. Find also the equation of the circumcircle of the triangle.

25. Find the equations of the two real circles that touch the lines $x = 0$ and $y = 1$ and go through the point $(2, 0)$. Show that the distance between their centres is $4\sqrt{2}$.

**6. Elementary problems on circles.** The use of the geometry of the figure is illustrated by the following techniques, each of which should be noted.

(*a*) *Equation of a circle having $AA'$ as diameter.* Here we use the fact that the angle in a semicircle is a right angle. Suppose $A(h, k)$ and $A'(h', k')$ are the ends of the diameter and $P(x, y)$ is any point of the required circle. The gradient of $AP$ is

$$(y-k)/(x-h),$$

and of $A'P$ is

$$(y-k')/(x-h').$$

Since $AP$ is perpendicular to $A'P$, the product of these gradients is $-1$, and this gives

$$(x-h)(x-h')+(y-k)(y-k')=0$$

as the equation of the required circle.

For example, the equation of the circle on the line joining $(-1, 3)$ and $(2, 5)$ as diameter is

$$(x+1)(x-2)+(y-3)(y-5) = 0,$$

or

$$x^2+y^2-x-8y+13 = 0.$$

As a check of accuracy we notice that the centre of this circle is $(\frac{1}{2}, 4)$ which is the mid-point of the given diameter.

(*b*) *Condition for a line to touch a circle.* The geometrical fact used here is that the tangent is perpendicular to the radius. Consider the circle $x^2+y^2 = r^2$ and the general line $lx+my+n = 0$. If this line touches the circle, the perpendicular distance from the centre $(0, 0)$ to the line is equal to the radius $r$. This gives

$$\pm\frac{n}{\sqrt{(l^2+m^2)}} = r,$$

or

$$r^2(l^2+m^2) = n^2,$$

as the condition for the line to touch the circle.

A similar method will apply for the condition for a general line to touch a general circle.

(*c*) *Length of a tangent from a point to a circle.* Here we use the theorem of Pythagoras. $P(h, k)$ is any point, and we wish to find the length of a tangent from $P$ to the circle

$$x^2+y^2+2gx+2fy+c = 0$$

with centre $C$. If $T$ is the point of contact of the tangent (Fig. 44), $PT^2 = PC^2-CT^2$. Now $C$ is $(-g, -f)$, so

$$PC^2 = (h+g)^2+(k+f)^2,$$

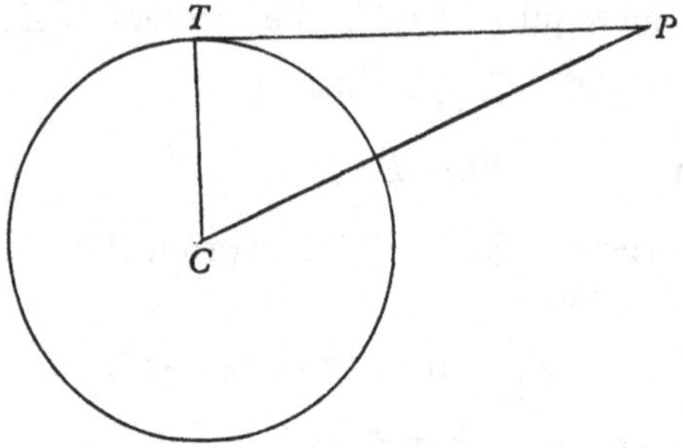


Fig. 44

and $CT$ is the radius of the circle, so $CT^2 = g^2+f^2-c$. Thus

$$PT^2 = (h+g)^2+(k+f)^2-(g^2+f^2-c)$$
$$= h^2+k^2+2gh+2fk+c.$$

This expression is just the left-hand side of the equation of the circle, written with *unit* coefficients of $x$ and $y$, with $h$ and $k$ replacing $x$ and $y$. If $(h, k)$ lies on the circle, the length of the tangent is zero, as we should expect. The expression, which gives the *square* of the length of the tangent from $(h, k)$ to the circle, is called the *power* of the point $(h, k)$ with respect to the circle.

(*d*) *Condition for two circles to cut at right angles.* Suppose the circles are

$$x^2+y^2+2gx+2fy+c = 0$$

and

$$x^2+y^2+2g'x+2f'y+c' = 0,$$

with centres $C$ and $C'$. Then, if $P$ is a point of intersection (Fig. 45), the condition for the tangents at $P$ to be perpendicular is

$$CP^2+C'P^2 = CC'^2,$$

since the tangent to each circle is the radius of the other. Now $CP$ is the radius of the first circle, so $CP^2 = g^2+f^2-c$ and similarly, $C'P^2 = g'^2+f'^2-c'$. $C$ is the centre $(-g, -f)$, and $C'$ is $(-g', -f')$, and so

$$CC'^2 = (g-g')^2+(f-f')^2.$$

Substituting these values, we have

$$g^2+f^2-c+g'^2+f'^2-c' = (g-g')^2+(f-f')^2,$$

or

$$\mathbf{2gg'+2ff' = c+c'}$$

as the condition for the two circles to cut at right angles.

Circles which satisfy this relation are said to cut *orthogonally*.

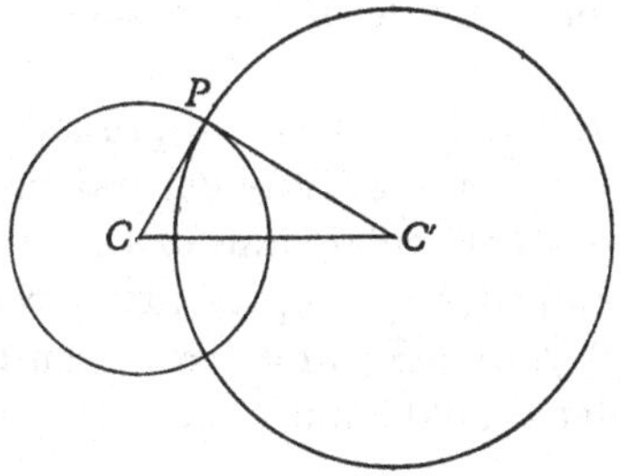


Fig. 45

**7. Circles with a common chord.** Given the equations of two circles which meet in two points, both the equation of the line joining these two points (the common chord), and the equation of any other circle through the points, can be deduced. We use the method of *inspection* (cf. p. 41).

Let the circles be

$$S \equiv x^2+y^2+4x-6y-7 = 0$$

and

$$S' \equiv x^2+y^2-2x+6y-55 = 0,$$

and consider the expression $S-S' = 0$. This is

$$(x^2+y^2+4x-6y-7)-(x^2+y^2-2x+6y-55) = 0,$$

or

$$x-2y+8 = 0.$$

We notice two things about this expression. First, it is the equation of a line (since the terms are linear). Secondly, it is an equation satisfied by the points which lie on both circles. It is therefore the equation of the common chord.

Now consider the expression $S-\lambda S' = 0$, where $\lambda$ is any constant. This is

$$(x^2+y^2+4x-6y-7)-\lambda(x^2+y^2-2x+6y-55) = 0$$

or $$x^2(1-\lambda)+y^2(1-\lambda)+x(4+2\lambda)-y(6+6\lambda)-(7-55\lambda) = 0.$$

Unless $\lambda = 1$ (the case already considered), this is the equation of a circle, since there is no term in $xy$, and the coefficients of $x^2$ and $y^2$ are equal and not zero. It is satisfied by the coordinates of the points where both $S = 0$ and $S' = 0$, and so it represents any circle through the points of intersection of the two circles. The constant $\lambda$ can be chosen to make the circle satisfy one other linear condition; for example, if the circle contains the origin, $\lambda = \frac{7}{55}$.

To sum up, if $S = 0$ and $S' = 0$ represent two circles, the equation $S-\lambda S' = 0$ represents any general circle through their intersections, or (for a particular value of $\lambda$), their common chord.

By similar reasoning, if $S = 0$ represents a circle and $l = 0$ a line meeting it in $P$ and $Q$, then $S-\lambda l = 0$ represents a circle through $P$ and $Q$. The reader should supply the details of the argument.

**8. Illustrations.** In the following worked examples, the advantages of using the geometrical properties of the circle should be noticed.

***Illustration 1.*** *Find the equations of the tangents to the circle* $x^2+y^2-2x-6y+6 = 0$ *which pass through the point* $(-1, 2)$.

Let $lx+my+n = 0$ be the equation of one of the required tangents. There are two conditions to be satisfied. First, the line contains $(-1, 2)$, so

$$-l+2m+n = 0.$$

Secondly, the line touches the circle, so its perpendicular distance from the centre $(1, 3)$ is equal to the radius

$$\sqrt{(1+9-6)} = 2.$$

Thus $$\pm\frac{l+3m+n}{\sqrt{(l^2+m^2)}} = 2,$$

or $$(l+3m+n)^2 = 4(l^2+m^2).$$

Substituting for $n$, $$(2l+m)^2 = 4(l^2+m^2),$$

so that $$4lm - 3m^2 = 0,$$

$$m = 0, \quad \text{or} \quad 4l/3.$$

Then $$n = l - 2m = l \quad \text{or} \quad -5l/3,$$

so that the tangents are $$x + 1 = 0$$

and $$3x + 4y - 5 = 0.$$

Checking accuracy mentally, we note that $(-1, 2)$ does lie on these lines.

This example illustrates the fact that it is often unwise to work with the line in the form $y = mx + c$. It was noted on p. 30 that a line parallel to the $y$-axis cannot be expressed in this form, and if in this example we had begun with the line $y = mx + c$, hoping from the two conditions to find the values of $m$ and $c$, we should have found that $m = -\frac{3}{4}$ and $c = \frac{5}{4}$. We should thus have 'lost' the second tangent $x + 1 = 0$.

***Illustration 2.*** *Show that the circles*

$$x^2 + y^2 = 4 \quad \text{and} \quad x^2 + y^2 + 6x - 8y + 16 = 0.$$

*touch externally* (Fig. 46.)

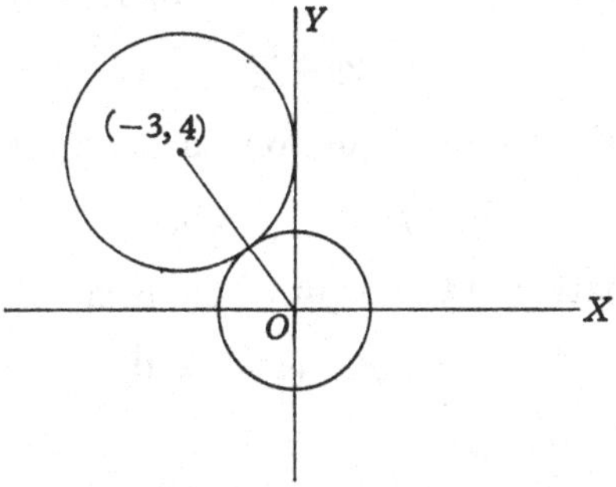


Fig. 46

The centres of the circles are $(0, 0)$ and $(-3, 4)$, and their radii are 2 and 3. The distance between the centres is 5, and as this is the sum of the radii, the circles touch.

***Illustration 3.*** *A variable circle passes through the point* $(-2, 3)$ *and cuts the circle* $x^2 + y^2 = 5$ *in points which are ends of a diameter of the latter circle. Find the locus of the centre of the circle. Find also the equation of the particular circle which also cuts*

$$x^2 + y^2 - 2x + 5y + 1 = 0$$

*orthogonally.*

Let the equation of the circle be $x^2+y^2+2gx+2fy+c=0$. The first condition, that the circle passes through $(-2, 3)$, implies that

$$4+9-4g+6f+c=0.$$

The common chord of the circles is

$$(x^2+y^2+2gx+2fy+c)-(x^2+y^2-5)=0,$$

or

$$2gx+2fy+c+5=0,$$

since this is a linear equation satisfied by the points which lie on both circles. This line passes through the centre $(0, 0)$, so that

$$c+5=0.$$

It follows that

$$-4g+6f+8=0$$

and so the locus of the centre $(-g, -f)$ is the line

$$2x-3y+4=0.$$

The circle which also cuts $x^2+y^2-2x+5y+1=0$ at right angles satisfies

$$2g.-1+2f.\tfrac{5}{2}=c+1,$$

that is

$$-2g+5f=-4.$$

The solution of these two equations in $g$ and $f$ is

$$g=2, \quad f=0,$$

and so the equation of the required circle is

$$x^2+y^2+4x-5=0.$$

## EXAMPLES V B

1. Write down the equations of the circles on the lines $AB$ as diameter, where:

(i) $A$ is $(4, -8)$ and $B$ is $(3, 5)$;
(ii) $A$ is $(2, 1)$ and $B$ is $(-1, 0)$;
(iii) $A$ is $(4, 7)$ and $B$ is $(-6, -6)$;
(iv) $A$ is $(-2, -2)$ and $B$ is $(1, 2)$.

2. $A$ and $B$ are the centres of the circles whose equations are

$$x^2+y^2+4x-2y+4=0 \quad \text{and} \quad x^2+y^2-2x+6y+1=0.$$

Find the equations of the circles whose centres are on $AB$ and which: (i) touch both the circles internally; (ii) touch both the circles externally; (iii) cut both the circles orthogonally.

3. Find the equations of the tangents to the circle
$$x^2+y^2-6x+8y+16 = 0$$
which pass through $(8, -4)$.

4. Show that the six points $(0,0)$, $(6,0)$, $(4,8)$, $(1,2)$, $(5,4)$ and $(-3,0)$ are the vertices of a figure consisting of four lines. (This is called a complete quadrilateral.) Find the equations of the circumcircles of the four triangles formed by sets of three of the four lines in this figure and show that these four circles all pass through the point $(-\frac{3}{2}, \frac{9}{2})$. (This is a general theorem for all such quadrilaterals; it is called Wallace's theorem.)

5. Find the equation of the circle having the points $(3,7)$ and $(9,1)$ as the ends of a diameter. Show that the circle touches the straight line $x+y = 4$ and find the coordinates of the point of contact.

6. A circle has the points $A(1,2)$, $B(5,8)$ as the opposite ends of a diameter. Find the centre of the circle, and the equation of the diameter $CD$ perpendicular to $AB$. Find also the coordinates of the points $C$ and $D$.

7. $A$ is the point $(-1,1)$ and $C$ is the point $(3,3)$. Calculate the equation of the perpendicular bisector of $AC$ and the equation of the circle on $AC$ as diameter. Hence, or otherwise, find the coordinates of $B$ and $D$ in order that $ABCD$ may be a square with a diagonal $AC$.

8. Find the equation of the circle passing through the three points $(1,0)$, $(4,0)$ and $(0,2)$. Show that this circle and the circle $x^2+y^2 = 4$ intersect at right angles.

9. Three circles, each of radius $a$, have their centres at $(0,0)$, $(b,0)$, $(0,c)$. Find the equation of the circle which intersects all three circles at right angles and state the condition between $a$, $b$ and $c$ which must be satisfied if this circle is to be real. If $P$, $Q$ are its points of intersection with the $x$-axis and $O$ is the origin, find the values of $OP+OQ$ and $OP.OQ$.

10. A circle touches the $x$-axis at the point $(5,0)$ and cuts off a chord of length 24 on the negative $y$-axis. Find the equation of the circle, and give its radius and the coordinates of its centre. Find also the equation of the tangent to the circle (other than the $x$-axis) which passes through the origin.

11. $A$ is the point $(2,1)$; $B$ is the point $(3,2)$. Find the equations of the circle on $AB$ as diameter, the circle through $A$, $B$ and the origin, and the two circles that go through $A$ and $B$ and touch the $x$-axis.

12. Show that two circles can be found which go through the two points $(a,0)$ and $(-a,0)$ and touch the line $lx+my+n = 0$. Prove that these two circles cut at right angles if $a^2(2l^2+m^2) = n^2$.

13. $A$ and $B$ are the points $(2,5)$ and $(6,3)$. A circle is described on $AB$ as diameter. Find the equation of the circle and the coordinates of the extremities of the diameter at right angles to $AB$.

14. Find the condition for the line $lx+my+n = 0$ to be a tangent of the circle $x^2+y^2 = a^2$. This circle is the inscribed circle of an equilateral triangle. If one side of the triangle is given by $x+a = 0$, find the equations of the other two sides, and of the circumcircle of the triangle.

15. One diagonal of a square lies along the line $x-2y+2=0$ and one vertex of the square is at the point $(1, 4)$. Prove that the slope of one side of the square is 3 and find the coordinates of the other three vertices. Find also the equation of the circumcircle of the square.

16. Prove that the equation $x^2+y^2-2kx-2ky+k^2=0$ represents a circle and that it touches the axes of coordinates. If the circle also touches the line $3x+4y=6$, find the possible values of $k$.

17. Find the equations of the two circles of radius 5 which pass through the origin and whose centres lie on the line $x+y=1$. Show that the point $(2, 0)$ is inside one of these circles, and find the length of the tangent from this point to the other circle.

18. A variable point $P$ on the circle $x^2+y^2=a^2$ has coordinates $(a\cos\theta, a\sin\theta)$, and $A$ is the fixed point $(c, 0)$. If $Q$ is the point of $AP$ such that $AQ:QP=\lambda:1$, find the coordinates of $Q$, and show that the locus of $Q$ is another circle. Calculate the coordinates of the centre of this circle and the length of its radius. Show that the condition for the two circles to intersect at right angles is $c^2=a^2(1+2\lambda+2\lambda^2)$.

19. The circle $x^2+y^2=a^2$ $(a>0)$ and $x^2+y^2-10x+9=0$ intersect in two distinct points; prove, using a diagram, that $1<a<9$. Prove that, if the length of the common chord is 24/5, then $a=3$ or $\sqrt{73}$. Prove also that, when $a=3$, the circles cut at right angles.

20. Obtain the equations of the two circles passing through the points $(0, 0)$ and $(2, 0)$ and touching the line $y=x+2$. Show that these circles also touch the line $y+x=4$ and find the coordinates of their points of contact.

21. Find the equations of the two circles which pass through the point $(2, 4)$ and touch both axes of coordinates. Find the angle at which the circles intersect, the equation of their common chord and the coordinates of their other point of intersection.

22. Show that the equation $x^2+y^2+2gx+2fy+g^2=0$ represents a circle touching the $x$-axis. Find the equation of a circle touching the $x$-axis at the point $(5, 0)$ and passing through $(7, 4)$. Find also the equation of the other tangent to this circle from the origin.

23. Show that the length of the tangent from the point $(h, k)$ to the circle $x^2+y^2+2gx+2fy+c=0$ is $\sqrt{(h^2+k^2+2gh+2fk+c)}$. Find the locus of the point such that the length of the tangent from it to the circle

$$x^2+y^2+14ay-a^2=0$$

is double the length of the tangent from it to the circle

$$x^2+y^2+2ay-a^2=0.$$

Show that the locus cuts the latter circle orthogonally.

24. The centre of a circle is on the line $y+3x=5$ and the points $(1, -3)$ and $(5, -1)$ lie on the circle. Find by calculation the centre and radius of the circle and the length of the tangent to the circle from the point $(5, 7)$.

25. Find, but do not simplify, the equation of a circle which lies in the first quadrant, touches the $y$-axis at $(0, 12)$ and has radius 9 units. Show that this circle touches externally the circle whose equation is $x^2+y^2 = 36$ and find the coordinates of their point of contact.

**9. Concyclic points on the parabola and on the rectangular hyperbola.**† Suppose we wish to find the points where a circle meets a parabola. We may choose the axes of coordinates in such a way that the equation of *one* of these takes a simple form, and as the circle in the general form is easier to deal with than the parabola in the general form, we take the axis and tangent at the vertex of the parabola as axes, and obtain its equation as $y^2 = 4ax$. The equation of a general circle,‡ referred to the same axes, is

$$x^2+y^2+2gx+2fy+c = 0.$$

Now the general point on the parabola $y^2 = 4ax$ is $(at^2, 2at)$, and this lies on the circle if and only if

$$a^2t^4+4a^2t^2+2gat^2+4fat+c = 0.$$

This is a quartic equation in $t$, so that it has at most four roots. We deduce that a circle meets a parabola in at most four points (as indeed we should expect from considering a diagram). If the parameters of the four points are $p, q, r, s$ then the usual conditions [A 7] give

$$p+q+r+s = 0,$$

$$qr+rp+pq+ps+qs+rs = (4a^2+2ga)/a^2 = 4+2g/a,$$

$$pqr+qrs+rps+pqs = -4fa/a^2 = -4f/a,$$

$$pqrs = c/a^2.$$

The first of these conditions gives the important result that, *if four points on the parabola $y^2 = 4ax$ are concyclic, the sum of their parameters is zero.* Conversely, if the sum of the parameters of four points on the parabola is zero, the points are concyclic. For the circle through three of the points, say those with parameters $p, q, r$, meets the parabola again in the point $s'$ where

$$p+q+r+s' = 0,$$

† If Chapter III has not yet been read, this paragraph must be omitted.

‡ The reader should decide what is wrong with the all too common habit of 'taking' the parabola as $y^2 = 4ax$, and the circle as $x^2+y^2 = r^2$, or, worse still, as $x^2+y^2 = a^2$.

and so $s = s'$; that is, a circle through three of the points also contains the fourth.

The power of the method adopted here is shown by the following corollary. It may be recollected that if $p$, $q$, $r$ are the parameters of conormal points on a parabola (p. 89) then $p+q+r = 0$. Suppose the circle determined by these three points meets the parabola again in the point with parameter $s$. Then $p+q+r+s = 0$, and so $s = 0$. Thus, *any circle through conormal points on a parabola passes through the vertex.*

A similar method applies to the rectangular hyperbola, which we take as $xy = k^2$ (the $c$ which we used in Chapter IV now appearing as the constant term in the equation of a circle). The general point $(kt, k/t)$ lies on the circle

$$x^2+y^2+2gx+2fy+c = 0,$$

if and only if

$$k^2t^2+k^2/t^2+2gkt+2fk/t+c = 0,$$

or

$$k^2t^4+2gkt^3+ct^2+2fkt+k^2 = 0.$$

As before, this is a quartic equation in $t$, and we deduce that a circle and a rectangular hyperbola meet in at most four points.

From among the four conditions on the roots $p$, $q$, $r$ and $s$ of this equation (which the reader should write down) we note

$$pqrs = k^2/k^2 = 1$$

as the one which is independent of $g$, $f$ and $c$. The other conditions give $g$, $f$ and $c$ in terms of $p$, $q$, $r$ and $s$.

The proof of the converse, that if $pqrs = 1$ the points are concyclic, follows in just the same way as before, and is left for the reader to supply.

## EXAMPLES V C

1. Show that, if a circle touches a parabola at two distinct points, its centre lies on the axis of the parabola.

2. Show that if a circle cuts a parabola at its vertex the normals to the parabola at the other three points of intersection with the circle are concurrent.

3. A variable circle touches a parabola at a fixed point $P$ and cuts it again in $H$ and $K$. The normals to the parabola at $H$ and $K$ meet in $R$. Prove that the locus of $R$ is a straight line.

4. The circle through the origin $O$ which touches the parabola $y^2 = 4ax$ at a point $P$ meets it again at $Q$. Find the locus of the point of intersection of $OQ$ with the tangent at $P$.

5. A circle meets a parabola in $P, Q, R, S$. Show that $PQ$ and $RS$ are equally inclined to the axis of the parabola [i.e. have slopes $m, -m$].

6. A circle touches the parabola $y^2 = 4ax$ at $P$ and meets it again in points on the line $lx+my+n = 0$. Show that the parameter of $P$ is $m/l$, and find the coordinates of the centre of the circle.

7. A circle is described on a chord $PQ$ of a parabola as diameter, and meets the parabola again in $R$ and $S$. $PQ$ and $RS$ meet the axis of the parabola in $H$ and $K$. Prove that $HK$ is of constant length.

8. A variable circle passes through the fixed points $A, B$ of a rectangular hyperbola, and meets the hyperbola again at $P$ and $Q$. Show that the direction of $PQ$ is fixed.

9. A circle cuts a rectangular hyperbola in four points $A, B, C, D$. If $AB$ is a diameter of the hyperbola, prove that $CD$ is a diameter of the circle.

10. A circle meets a rectangular hyperbola in $A, B, C, D$. Prove that the angle between $AB$ and one of the asymptotes is equal to the angle between $CD$ and the other asymptote.

11. $A, B, C, D$ are conormal points on a rectangular hyperbola. Show that if the circle through $A, B, C$ meets the curve again in $D'$ then $DD'$ is a diameter of the hyperbola.

12. A rectangular hyperbola is cut by any circle in four points. Prove that the sum of the squares of the distances of these four points from the centre of the hyperbola is equal to the square of the diameter of the circle.

**10.* Radical axis.** We have already considered the relations between two circles, if they are in special positions, touching, or cutting orthogonally. We now consider two general circles

$$S \equiv x^2+y^2+2gx+2fy+c = 0$$

and
$$S' \equiv x^2+y^2+2g'x+2f'y+c' = 0,$$

and we shall use, where convenient, a symbolic notation so that the circles are referred to as $S$ and $S'$ ,with equations $S = 0$ and $S' = 0$.

The problem to be investigated is to find the locus of a point equidistant from the two circles; that is, a point $P$ such that the length of the tangent $PT$ to $S$ is the same as the length of the tangent $PT'$ to $S'$ (Fig. 47). If $P(x_1, y_1)$ is such a point, then (p. 142)

$$PT^2 = x_1^2+y_1^2+2gx_1+2fy_1+c$$

(since the equation of $S$ is expressed with *unit* coefficients of $x^2$ and $y^2$). A similar result holds for $PT'^2$, and since $PT = PT'$

$$x_1^2+y_1^2+2gx_1+2fy_1+c = x_1^2+y_1^2+2g'x_1+2f'y_1+c'.$$

Thus the locus of $P(x_1, y_1)$ is

$$2(g-g')x+2(f-f')y+(c-c') = 0$$

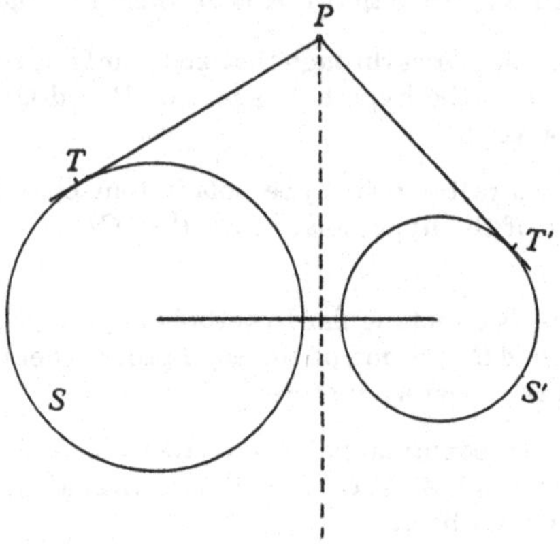


Fig. 47

We notice several things about this result.

(i) The locus of $P$ is a *line*; this is called the *radical axis* of the two circles.

(ii) The radical axis is perpendicular to the line joining the centre of the circles (as can be verified at once by writing down the gradients).

(iii) The equation of the radical axis is obtained by subtracting the left-hand sides of the equations of the two circles; symbolically it is $S-S' = 0$.

Distinct cases arise, depending on whether or not the circles intersect. The radical axis cannot meet one of the circles without meeting the other (for a point of intersection is a zero distance from one circle and therefore also from the other), and conversely the radical axis passes through any common point of the circles.

It follows that:

(i) If the circles do not meet, the radical axis lies outside both of them (Fig. 48).

(ii) If the circles meet, the radical axis is their common chord (Fig. 49).

(iii) If the circles touch, the radical axis is their common tangent (Fig. 50).

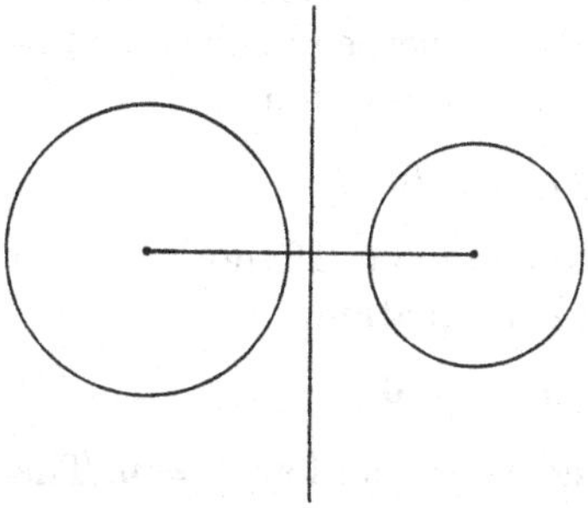

Fig. 48

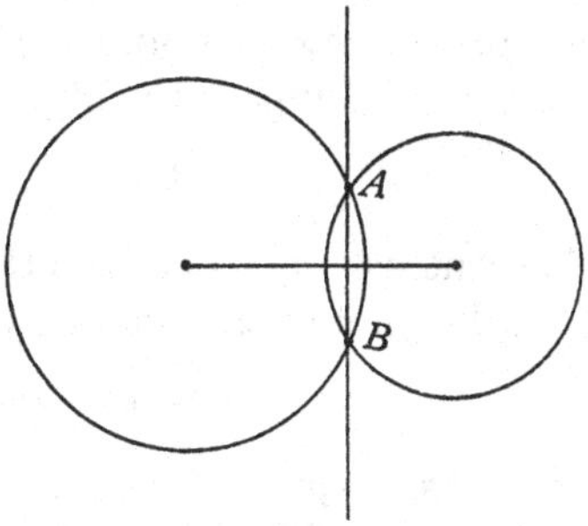


Fig. 49

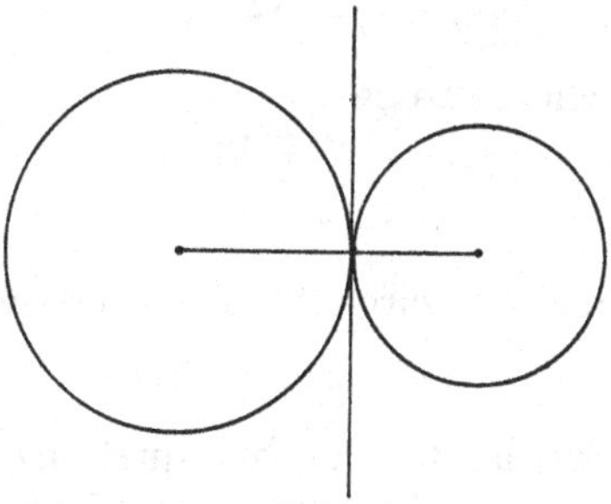

Fig. 50

The reader may notice that in the second case, points which lie inside both circles (this is, between $A$ and $B$ in the diagram) do not strictly speaking belong to the locus, since tangents cannot be drawn from them to the circles. This anomaly may be avoided by extending the definition of the power of a point $P(x_1, y_1)$ (p. 142) to be the expression

$$x_1^2 + y_1^2 + 2gx_1 + 2fy_1 + c,$$

*whether this expression is positive or negative*, and then defining the radical axis as the locus of a point whose powers with respect to the two circles are equal. Alternatively, the distinction between real and imaginary tangents may be (and frequently is) ignored. In any case, we shall regard the radical axis as the *whole* line whose equation was found above.

**11.* Coaxal circles.** A system of circles which all have the same radical axis are said to form a *coaxal system*. Suppose the radical axis, $l$, and one circle, $S$, with centre $C$, are given, and let $C'$ be the centre of another circle of the coaxal system. $CC'$ is perpendicular to $l$, and it follows that $C'$, and in general the centre of each circle of the system, lies on the line through $C$ perpendicular to $l$.

It is convenient to simplify the algebra by choosing the line of centres as the $x$-axis, and the radical axis (perpendicular to it) as the $y$-axis. Suppose the given circle $S$ has equation

$$x^2+y^2+2gx+2fy+c = 0.$$

The centre $(-g, -f)$ lies on the $x$-axis, so $f = 0$. If another circle $S'$, belonging to the coaxal system, has equation

$$x^2+y^2+2g'x+2f'y+c' = 0,$$

its centre $(-g', -f')$ also lies on the $x$-axis, and so $f' = 0$. The radical axis of $S$ and $S'$ is

$$2x(g-g')+(c-c') = 0$$

and this we have chosen to be

$$x = 0;$$

it follows that $$c-c' = 0.$$

Thus the equation of any circle $S'$ of the system is

$$x^2+y^2+2g'x+c = 0.$$

We emphasize the fact that in this equation $g'$ is arbitrary, by writing the equation of the general circle of the coaxal system as

$$S_\lambda \equiv x^2+y^2+2\lambda x+c = 0,$$

where the constant term $c$ is fixed, for the whole system, and the coefficient $\lambda$ varies from one circle to another. In fact, $\lambda$ serves as *a parameter* for the circles of the system; given a circle belonging to the system, a value of $\lambda$ is determined, and conversely any value of $\lambda$ determines just one circle of the system (provided $\lambda$ is restricted to the range where $\lambda^2 \geqslant c$).

A circle $S_\lambda$ meets the radical axis $x = 0$ where $y^2+c = 0$. As we should expect, three cases arise.

(i) If $c > 0$, there are no points of intersection.

(ii) If $c = 0$, every circle of the system touches the radical axis, at the origin $(0, 0)$.

(iii) If $c < 0$, say $c = -k^2$, there are two points of intersection $(0, \pm k)$, and these lie on every circle of the system.

Three cases also arise when we consider point circles (circles with zero radius, see p. 134) of the system. The radius of a general circle is $\sqrt{(\lambda^2 - c)}$, and so point circles arise when $\lambda^2 = c$.

(i) If $c > 0$, there are two point circles, for $\lambda = \pm\sqrt{c}$, and these are the points $(\pm\sqrt{c}, 0)$. These points are called *limiting points* of the system; we see that they are equidistant from the radical axis $x = 0$.

(ii) If $c = 0$, the point circles coincide, at the origin.

(iii) If $c < 0$, there are no point circles.

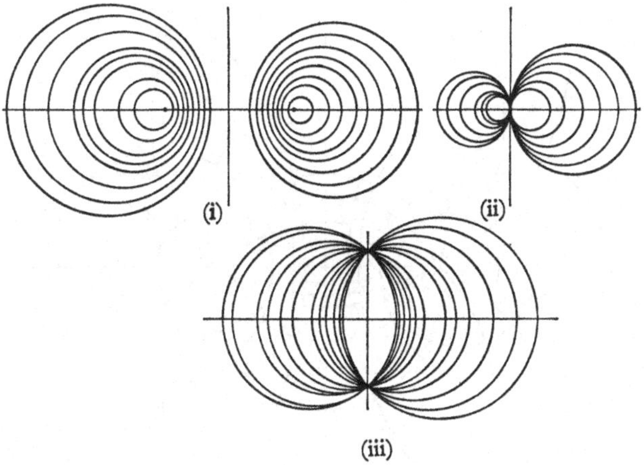

Fig. 51

The three cases are illustrated in Fig. 51. We see that, except in the second case which is a compromise between the other two, there are *either* limiting points, *or* intersections, but not both.

Consider now the first case, in which there are limiting points $(\pm\sqrt{c}, 0)$. A circle

$$x^2 + y^2 + 2g'x + 2f'y + c' = 0$$

passes through these points if

$$c \pm 2g'\sqrt{c} + c' = 0;$$

that is, if

$$g' = 0, \quad c' = -c.$$

The family of circles through these two points is therefore

$$x^2 + y^2 + 2\mu y - c = 0,$$

for any value of the coefficient $\mu$, and we see that this is another coaxal system, consisting of circles whose centres lie on the radical axis of the first system, and with radical axis the line of centres of the first system.

Similarly, if the first system of circles has intersections $P$ and $Q$, there is another system of circles of which $P$ and $Q$ are limiting points.

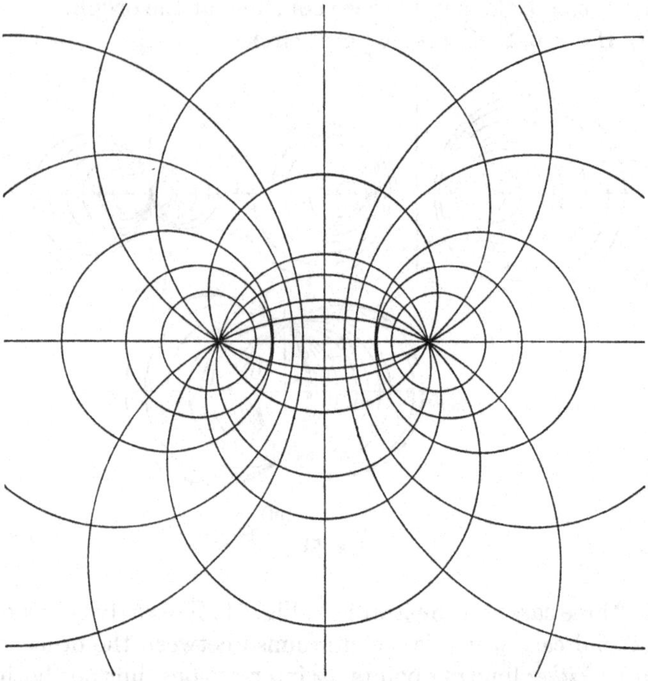

Fig. 52

An important property of these two systems is that any circle of one system cuts any circle of the other system orthogonally, for in this case

$$2(gg'+ff')-(c+c') = 2(\lambda.0+0.\mu)-(c-c)$$

$$= 0,$$

and so the orthogonality condition is satisfied. The two systems of circles are drawn in the diagram (Fig. 52).

This gives an example of an important geometrical configuration, that of two families of loci such that any member of one family cuts any member of the other at right angles. The most simple and familiar example of this is two families of parallel lines, all of one family perpendicular to all of the other; we see this 'orthogonal net' whenever we look at a piece of ordinary graph paper. Another example is of a system of concentric circles, with a system of radii through their common centre. Some readers may have used 'polar' graph paper which exhibits these orthogonal families. The importance of such systems lies in the fact that through a general point of the plane there is just one member of each family; this property is easily checked for each of the three examples mentioned here.

An example of this orthogonal property of curves appears in everyday life in the contour lines and lines of greatest slope, on a map. Again, in an electric or magnetic field lines of force and equipotentials form orthogonal families.

The coaxal system has been determined by the radical axis and one circle of the system. It could equally well be determined by any two circles of the system, one or both of which could be point circles.

We consider finally a general coaxal system, not referred to special axes. If

$$S \equiv x^2+y^2+2gx+2fy+c = 0$$

and

$$S' \equiv x^2+y^2+2g'x+2f'y+c' = 0$$

are any two circles, the equation $S-\lambda S' = 0$ represents a general circle of the coaxal system determined by $S$ and $S'$. For if $(h, k)$ is any point on the radical axis of $S$ and $S'$ let $d$ be the length of the tangent from $(h, k)$ to either circle. Then

$$\begin{aligned} d^2 &= h^2+k^2+2gh+2fk+c \\ &= h^2+k^2+2g'h+2f'k+c'. \end{aligned}$$

Now $S-\lambda S' = 0$ represents a circle, for general $\lambda$, and when expressed with unit coefficient of $x^2$ and $y^2$ its equation is

$$\frac{1}{1-\lambda}(S-\lambda S') = 0.$$

The square of the length of the tangent from $(h, k)$ is then

$$\frac{1}{1-\lambda}(d^2-\lambda d^2) = d^2,$$

and so any point on the radical axis of $S$ and $S'$ is also on the radical axis of $S$ and $S-\lambda S'$, for each $\lambda$.

In the case of intersecting circles this paragraph gives just the same result as §7; indeed, this theory of coaxal systems is just a natural extension of that section.

## EXAMPLES V D

1. Find the radical axes of the following pairs of circles:

(i) $x^2+y^2 = 4$ and $x^2+y^2-4x+3y-17=0$;

(ii) $x^2+y^2+4x-3 = 0$ and $x^2+y^2+6x-8y+7 = 0$;

(iii) $x^2+y^2+2ax+2by+c = 0$ and $x^2+y^2-2ax-2by+c = 0$.

2. $S = 0, S' = 0$ and $S'' = 0$ are three circles. Show that in general the three radical axes of the three pairs of these are concurrent. (This point is called the *radical centre* of the circles.)

3. $PT$ and $PT'$ are the tangents from any point $P$ to two circles $S$ and $S'$. Prove that $PT^2-PT'^2$ is proportional to the distance of $P$ from the radical axis of $S$ and $S'$.

4. Show that the length of the tangent from a point of the radical axis to any circle of the coaxal system $x^2+y^2+2\lambda x+c = 0$ is a constant independent of $\lambda$.

5. Write down the equation of the polar line of $(h, k)$ with respect to the circle $x^2+y^2+2\lambda x+c = 0$. As $\lambda$ varies, show that all the polar lines obtained are concurrent.

6. Show that the locus of a point which is such that the ratio of the lengths of the tangents from it to two fixed circles is constant is a circle, belonging to the coaxal system to which the two given circles belong.

7. Show that the point $P(6, 8)$ lies on the common chord of the circles $x^2+y^2-8y+12 = 0$ and $x^2+y^2-4x-4y+4 = 0$. Calculate the length of the tangent from $P$ to either circle.

8. A system of coaxal circles is defined by one of the limiting points $(-1, 2)$ and the circle $x^2+y^2+18x+4y-35 = 0$. Find the coordinates of the second limiting point. Find also the equation of the other circle of the system which has the same radius as the given circle.

9. Prove that the circles $x^2+y^2 = 1$ and $x^2+y^2-8x-8y+7 = 0$ intersect in real points and show that there are two circles which pass through these points and touch the line $x = 2$. Find the equation of the line from any point of which the lengths of the tangents to all four circles are equal.

10. Find the centres and radii of the circles $x^2+y^2-4x-5 = 0$ and $x^2+y^2+6x-2y+6 = 0$. $P$ is a point $(h, k)$ such that the tangents from $P$ to both circles are equal. Prove that $10h-2k+11 = 0$. Hence show that the locus of $P$ is a line perpendicular to the line joining the centres of the circles.

11. A system of coaxal circles is defined by two of its members, $x^2+y^2-6x-4y+3=0$ and $x^2+y^2+10x+4y-1=0$. Obtain the equations of the radical axis and the line of centres. Find also the equations of the circle of the system which passes through the origin and of the two point circles of the system, and hence or otherwise obtain the coordinates of these two limiting points.

12. Find the point of the line $x+2y=7$ from which tangents of equal length can be drawn to the circles $x^2+y^2=2y$ and $x^2+y^2-4x+3=0$.

13. Circles are drawn on a system of parallel chords of a rectangular hyperbola as diameters. Show that they form a coaxal system, and find the radical axis.

14. Find the coordinates of the limiting points of the system of circles coaxal with the circles $x^2+y^2-6x-6y+4=0$, $x^2+y^2-2x-4y+3=0$, and find also the equations of the circles of this coaxal system which touch the line $x+y-5=0$.

15. Find the radical axis of

$$x^2+y^2+6x+2y+1=0 \quad \text{and} \quad x^2+y^2-6x-2y+1=0.$$

Find the equation of the circle coaxal with these two circles and passing through the point $(1,1)$.

16. Find the limiting points of the coaxal system given by

$$x^2+y^2+2\lambda(x+y-4)-6=0.$$

Find the equations of the two circles through these points which have radius 3.

17. The circle $x^2+y^2+2gx+2fy+c=0$ belongs to a coaxal system, and a limiting point is $(0,-f)$; find the equation of the radical axis, and the coordinates of the other limiting point.

18. Show that as $\lambda$ varies the circles

$$x^2+y^2+2ax+2by+c+2\lambda(ax-by+1)=0$$

form a coaxal system, and find the equations of the radical axis and the line of centres. Find the equations of the circles which are orthogonal to all the circles of the given system.

**12.* Inversion.** The study of correspondences between sets of points is an important feature of more advanced geometry; at this stage we discuss just one example.

Given a circle $S$ whose centre is $O$ and radius $a$ (Fig. 53), two points $P$ and $P'$, collinear with $O$ and on the same side of $O$, are said to be *inverse points* with respect to $S$ if and only if

$$OP.OP' = a^2.$$

The circle $S$ is called the *circle of inversion*, and the radius $a$ is called the *radius of inversion*. Very frequently the actual size of the circle $S$ and so of $a$ is unimportant, affecting only the scale of the diagram; the centre $O$ is always of paramount importance, and is called the *centre of inversion*. The process of changing from a figure consisting of points $P, Q, R, \ldots$, to the one consisting of their inverse points $P', Q', R', \ldots$, is called *inversion*.

It follows at once from the definition that:

(i) $P$ and $P'$ coincide if and only if $P$ lies on the circle $S$;

(ii) carrying out the process of inversion twice leaves all points unchanged;

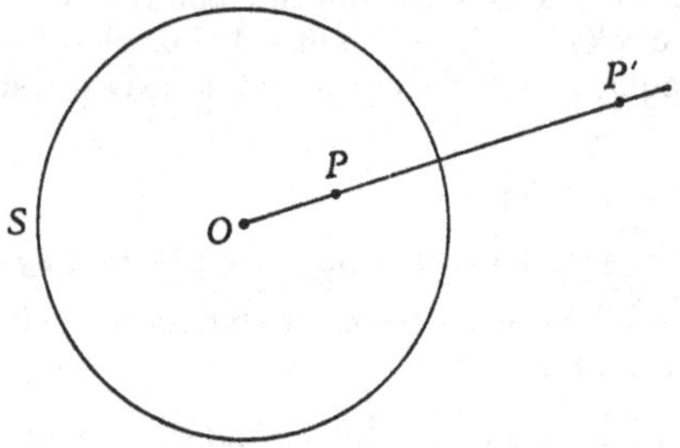


Fig. 53

(iii) inversion replaces any point inside the circle $S$ by a point outside $S$, and vice versa (this property gives rise to the name, inversion);

(iv) inversion replaces a line through $O$ by the same line (though the individual points on the line are interchanged in pairs);

(v) a circle centre $O$ inverts into another circle centre $O$ (with radius magnified or diminished).

A result which is not so obvious is the following:

If curves $C_1$, $C_2$ cut at an angle $\theta$, then their inverse curves cut at the same angle $\theta$. To prove this, we first consider points $P$ and $Q$ on a curve $C$, and their inverse points $P'$ and $Q'$ on $C'$ (Fig. 54). Then

$$OP.OP' = a^2 = OQ.OQ',$$

and so $P, Q, P', Q'$ are concyclic; thus $\angle OQP = \angle OP'Q'$.

Now let $Q \to P$, so that $Q' \to P'$. The chord $PQ$ becomes the tangent at $P$ (Fig. 55), and it follows that if the tangents to $C$ and $C'$ at $P$ and $P'$ make angles $\alpha$ and $\alpha'$ with $OPP'$, then

$$\alpha = \alpha'.$$

Now if $C_1$, $C_2$ are two curves intersecting at $P$ and inverting into $C_1'$, $C_2'$ intersecting at $P'$, and angles $\alpha_1$, $\alpha_2$, $\alpha_1'$, $\alpha_2'$ are defined similarly, then

$$\alpha_1 = \alpha_1' \quad \text{and} \quad \alpha_2 = \alpha_2'.$$

Thus

$$\alpha_1 - \alpha_2 = \alpha_1' - \alpha_2'$$

and these angles are the angles between the tangents to $C_1$, $C_2$ at $P$, and between the tangents to $C_1'$, $C_2'$ at $P'$ (Fig. 56).

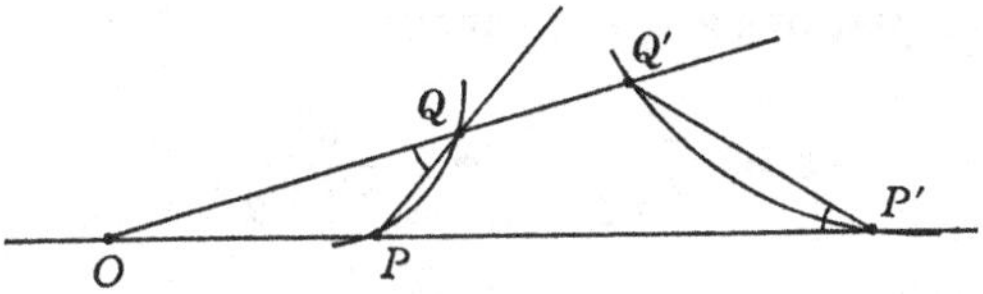


Fig. 54

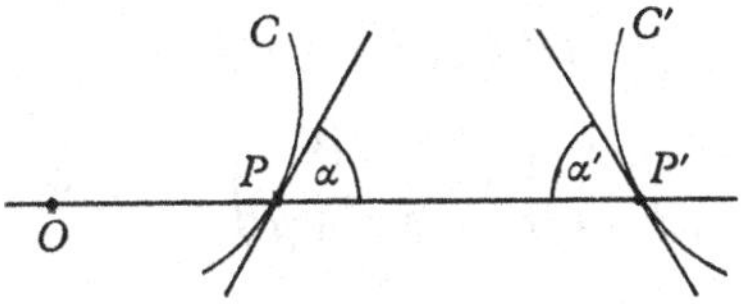


Fig. 55

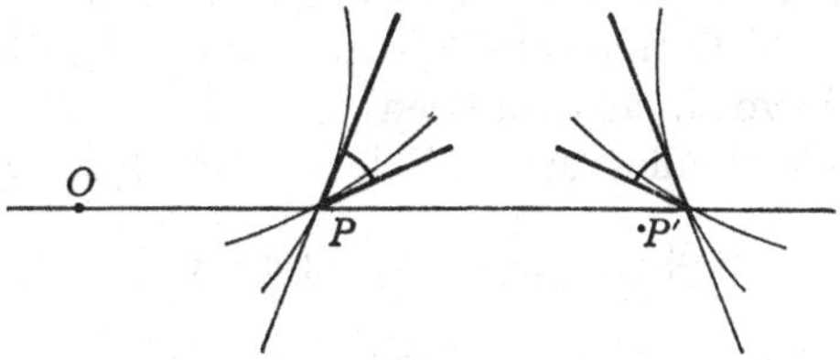


Fig. 56

The important corollary of this result is that orthogonal curves invert into orthogonal curves.

In order to investigate the exact nature of inverse curves, it is convenient to use coordinates. Take $O$ as the origin; let $P$ be $(x, y)$ and let $P'$ be $(x', y')$. Since $O$, $P$, $P'$ are collinear,

$$x' = \lambda x \quad \text{and} \quad y' = \lambda y,$$

for some value of $\lambda(>0)$. Also

$$OP^2 = x^2 + y^2 \quad \text{and} \quad OP'^2 = x'^2 + y'^2$$
$$= \lambda^2(x^2 + y^2),$$

and, since $$OP.OP' = a^2,$$

it follows that $$\lambda^2(x^2+y^2)^2 = a^4,$$

giving $$\lambda = \frac{a^2}{x^2+y^2}.$$

Thus $$x' = \frac{a^2x}{x^2+y^2}, \quad y' = \frac{a^2y}{x^2+y^2}.$$

The reverse equations are, similarly,

$$x = \frac{a^2x'}{x'^2+y'^2}, \quad y = \frac{a^2y'}{x'^2+y'^2}.$$

We apply these equations to find the inverse curves of a line and of a circle. First, suppose $P'$ lies on a line $lx+my+n = 0$. Then

$$lx'+my'+n = 0,$$

and so $$l\left(\frac{a^2x}{x^2+y^2}\right)+m\left(\frac{a^2y}{x^2+y^2}\right)+n = 0,$$

and $P$ lies on $$n(x^2+y^2)+la^2x+ma^2y = 0.$$

This is either, if $n \neq 0$, a circle through $O$, or, if $n = 0$, a line through $O$; indeed in this case it is the same line as the one inverted (as we have already foreseen).

Next, suppose $P'$ lies on a circle $x^2+y^2+2gx+2fy+c = 0$. Then

$$x'^2+y'^2+2gx'+2fy'+c = 0,$$

and so
$$\frac{a^4x^2}{(x^2+y^2)^2}+\frac{a^4y^2}{(x^2+y^2)^2}+2g\frac{a^2x}{x^2+y^2}+2f\frac{a^2y}{x^2+y^2}+c = 0,$$

or $$c(x^2+y^2)+2ga^2x+2fa^2y+a^4 = 0.$$

This is either, if $c \neq 0$, a circle, not through $O$,

or, if $c = 0$, a line, not through $O$.

These results are summarized in the phrase—circles and straight lines invert into circles and straight lines.

By means of the process of inversion, theorems about circles can be transformed into different theorems, mainly about straight lines, and so proved more simply. As an example, consider

Wallace's theorem. This states that, if $ABC$, $AQR$, $PBR$, $PQC$ are four lines, the circumcircles of the four triangles $PQR$, $PBC$, $AQC$, $ABR$, meet in a point $M$. Let us invert with respect to $A$, and let the inverse points of $B$, $C$, $P$, $Q$, $R$ be $B'$, $C'$, $P'$, $Q'$, $R'$ respectively. The four given lines invert into the two lines $AB'C'$ and $AQ'R'$ and the circles $P'B'R'$ and $P'Q'C'$ (fig. 57). Let $B'R'$ and $C'Q'$ meet in $M'$. It follows at once from the diagram that the exterior angle at $M'$ equals $\angle B'P'C'$, and so the points $M'B'P'C'$ are concyclic. Similarly, the points $M'Q'P'R'$ are concyclic. Now invert again, thus regaining the original figure. $B'R'$ is a line not

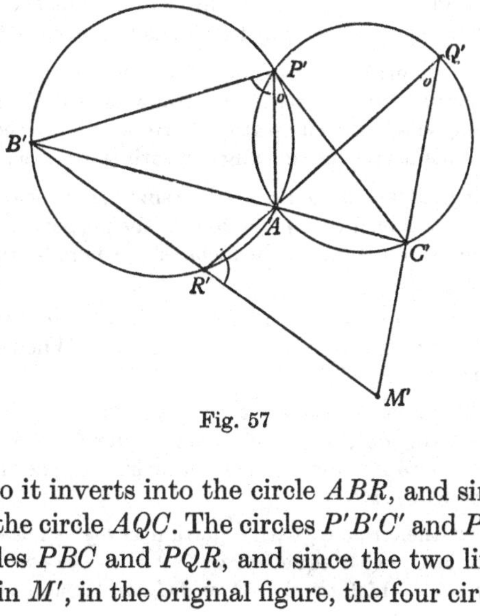


Fig. 57

through $O$, so it inverts into the circle $ABR$, and similarly $Q'C'$ inverts into the circle $AQC$. The circles $P'B'C'$ and $P'Q'R'$ invert into the circles $PBC$ and $PQR$, and since the two lines and two circles meet in $M'$, in the original figure, the four circles all meet in $M$.

## MISCELLANEOUS EXAMPLES V

1. A circle, one of whose diameters lies along the line $4y = x+7$, cuts the axis of $x$ in points where $x = 2, 6$. Find the equation of this circle, and the coordinates of the ends of the above diameter. Show also that the circle does not cut the axis of $y$.

2. A point $P$ moves so that its distance from the line $4x-3y+5 = 0$ is proportional to the square of its distance from the point $(1, 2)$. Prove that the locus of $P$ is a circle. If the circle passes through the origin, find its centre and radius.

3. The circle $x^2+y^2-2x+4y = 0$ touches the circle

$$x^2+y^2+2gx+3y = 0.$$

Find the value of $g$, and give a reason why there is only one solution, illustrating with a rough sketch showing clearly the centres and radii of the circles.

4. Prove that the points $(-3, 4)$, $(1, -4)$, $(3, 7)$ are the vertices of a right-angled triangle, and find the equation of the line joining the mid-point of the hypotenuse to the opposite vertex. Find also in its simplest form the equation of the circumcircle of this triangle, and the equation of the tangent at the point $(-3, 4)$.

5. Find the equation of the circle whose centre is the point $(4, 3)$ and which touches externally the circle $x^2+y^2 = 4$. If $A$ is the point $(4, 3)$ and $O$ is the origin, find the equation of the locus of a point $P$ which moves so that $PA^2-PO^2 = 5$, and show that this locus touches the two circles.

6. $P, Q$ are points on $x^2+y^2 = a^2$ such that the radii to $P, Q$ make angles $\theta, \phi$ with the positive $x$-axis. Find the coordinates of $T$, the point of intersection of the tangents to the circle at $P, Q$. Hence find the equation of the locus of $T$ if $PQ$ makes a constant angle $\alpha$ with the $x$-axis.

7. Show that the locus of $\{a(1+\cos\theta), a\sin\theta\}$ as $\theta$ varies is a circle. If $P$ is a point on the circle, and $N$ is the foot of the perpendicular from $P$ to the $x$-axis, obtain the equation of the locus of $Q$, where $Q$ is a point on $PN$ such that $PQ = \frac{1}{2}QN$.

8. Find the equation of the circle $C$ which touches the axis of $x$ at $(2a, 0)$ and which touches the circle $x^2+y^2-2y = 0$. When $a$ varies, what is the locus of the centre of $C$?

9. $A(3, 9)$, $B(8, 4)$ and $C(0, 0)$ are the three vertices of a triangle. Find the coordinates of the foot, $D$, of the perpendicular from $A$ on to $BC$, and of the point $A'$ where $AD$ meets the circumcircle of the triangle. Show that $A'B$ is parallel to $AC$.

10. Show that the circle $C_1$ with equation $x^2+y^2-18x-10y+86 = 0$ touches the circle $C_2$ with equation $x^2+y^2-6x-4y+8 = 0$. Find the equation of the circle which touches $C_1$ and $C_2$, and inside which $C_1$ and $C_2$ lie.

11. Obtain the condition for the line $y = mx+c$ to touch the circle $x^2+y^2 = a^2$. Find the area of the parallelogram formed by the tangents to the circle $x^2+y^2 = 4$ which have slopes 3 and $-2$.

12. The length of the tangent from the point $P(1, 1)$ to the circle $x^2+y^2-4x-6y+k = 0$ is 2 units. Find the value of $k$. Show that the circle whose centre is $P$, and which touches each axis of coordinates, does not cut the first circle.

13. Find the coordinates of the centre and the radius of the circle $2x^2+2y^2+4x-12y+15 = 0$. Find also the equation of the tangent to the circle which is furthest from the origin. Calculate the length of the chord intercepted by the circle on the line $x+y = 0$.

14. Find the equations of the tangents to the circle

$$x^2+y^2-4x-6y+9=0$$

which are parallel to the line $3x+4y=0$. Find also the coordinates of the points of contact of these tangents and the equation of the diameter which is parallel to them.

15. Find the coordinates of the points $P$ and $Q$, where the line

$$x+2y=11$$

meets the circle $x^2+y^2-4x-4y-2=0$, and calculate the length of the chord $PQ$. Show that the tangents at $P$ and $Q$ are perpendicular and find the area bounded by these two tangents and the minor arc $PQ$.

16. Prove that the circles

$$x^2+y^2-4y-5=0 \quad \text{and} \quad x^2+y^2-6x-12y+29=0$$

intersect each other at right angles. Find the equation of their common chord and that of the tangent to the first circle at each point of intersection.

17. Obtain the equation of the circle touching the line $2y=x-1$ at the point $A(1,0)$ and passing through the point $B(-1,0)$. Find the coordinates of the point $T$ of intersection of the tangents at $A$ and $B$ and the equation of the circle through $A$, $B$ and $T$.

18. Find the equation of the circle which touches the line $2y=x$ at the origin $O$ and passes through the point $A(0,8)$. Find the angle subtended at the centre of the circle by the chord $OA$ and calculate, correct to two decimal places, the areas of the two segments into which the circle is divided by $OA$.

19. Prove that two circles $C_1$ and $C_2$, each of radius 12 and with centres not on the $x$-axis, can be drawn to pass through the origin and to touch the circle $C$ whose equation is $x^2+y^2-40x+384=0$, and find the coordinates of their centres. Prove that the tangent to $C_1$ and $C$ at their point of contact meets the tangent to $C_2$ and $C$ at their point of contact in $(15,0)$.

20. Find the equation of the circle through the origin and the points $P(1,2)$ and $Q(2,1)$. Prove that the tangents to this circle at $Q$ and $P$ meet the tangent at the origin in the points $(\frac{5}{2},-\frac{5}{2})$ and $(-\frac{5}{2},\frac{5}{2})$.

21. Show that the circles

$$x^2+y^2+2x-8y+8=0 \quad \text{and} \quad x^2+y^2+10x-2y+22=0$$

touch each other. Find the coordinates of the point of contact $A$ and the equation of the common tangent at that point. Find the equation of the circle which touches each of these circles at $A$ and passes through the point $(3,7)$.

22. Find the equation of a circle, in the first quadrant, whose centre lies on the circle $x^2+y^2=16$, and which has the axis of $x$ and the line $y=\sqrt{3}x$ as tangents.

23. Find the equations of the two circles which pass through the points $(0, 4)$ and $(8, 20)$ and have the $x$-axis as a common tangent. Show that the line of centres of these two circles meets the $x$-axis at the point $(28, 0)$ and find the equation of their second common tangent.

24. Show that the line $x+3y = 1$ touches the circle

$$x^2+y^2-3x-3y+2 = 0$$

and find the coordinates of the point of contact. Prove that the point $P(3, 2\frac{1}{2})$ lies outside the circle and calculate also the length of the tangent to the circle from $P$.

25. Find the equation of a circle $C_1$ with centre at the point $P(2a, 0)$ which cuts the line $y = x$ in two points $Q$ and $R$ distant $2a$ apart. Prove that the circle $C_2$ passing through the points $P$, $Q$, $R$ is

$$2x^2+2y^2-5ax-3ay+2a^2 = 0.$$

Show also that the tangent to the circle $C_2$ at the point $P$ is parallel to the line $QR$.

26. Tangents are drawn to the circle $x^2+y^2-6x-4y+9 = 0$ from the origin. If $\theta$ is the angle between them, find the value of $\tan\theta$. Find the area of the rhombus formed by these tangents and the tangents parallel to them.

27. If

$$S \equiv x^2+y^2+7x-12y+20 = 0 \quad \text{and} \quad S' = 4x^2+4y^2+3x+2y-20 = 0$$

for what value of $\lambda$ does the equation $S-\lambda S' = 0$ represent (i) a circle through the origin; (ii) a circle through the point $(2, 1)$; (iii) a circle meeting $S$ orthogonally; (iv) the common chord of the circles? Find also the equation of the common chord, and the coordinates of the common points.

28. Prove that the points $(10, 5)$, $(1, 8)$ form a cyclic quadrilateral with the points of intersection of the circles given by

$$x^2+y^2-40x+175 = 0, \quad 3x^2+3y^2-40y-75 = 0.$$

29. A circle has its centre at the point $(1, 2)$ and passes through the point $(0, 3)$. Find its equation and also that of the circle whose centre is at the point $(4, 3)$ and which cuts the first circle at right angles. What is the equation of the circle which passes through the origin and is such that all three circles have a single chord in common?

30. The circle $C_1$ has equation $x^2+y^2-5x+7y+8 = 0$. Find the equation of the circle $C_2$ which passes through $(3\frac{1}{2}, -1\frac{1}{2})$ and cuts $C_1$ where $2x-3y = 6$ cuts $C_1$. Show that a circle $C_3$ orthogonal to $C_1$ and $C_2$ can be drawn with $(6, 2)$ as centre. Find its equation.

31. Find the equation of the circle $C$ which passes through the origin and the points of intersection of the circles $x^2+y^2-4x-8y+16 = 0$ and $x^2+y^2+6x-4y-3 = 0$. Find the equations of the tangent to $C$ at the origin and of the line through the centres of the circles.

32. A point $P$ moves in a plane so that the ratio of its distances from two fixed points $A$, $B$ in the plane is constant. Prove that its locus is a circle. If $PA/PB = \lambda > 1$, and $AB = 2a$, prove that the radius of the circle is $2a\lambda/(\lambda^2 - 1)$, and find the length of the tangent to the circle from the mid-point of $AB$.

33. $A$ is the point $(a, b\sqrt{3})$ and $B$ the point $(a+2b, b\sqrt{3})$, $O$ being the origin. Find the coordinates of $C$, the centre of the circle through $O, A, B$. If $b$ is fixed and $a$ varies, prove that the locus of $C$ is the curve

$$x^2 + 2by\sqrt{3} = 4b^2.$$

34. Find the equations of the common tangents of the circles whose equations are $x^2 + y^2 = 2x$, $x^2 + y^2 + 8x = 0$.

35. The line $4x + 3y = 5$ meets the axes $OX$, $OY$ in the points $A$ and $B$. Find the equation of the escribed circle of the triangle $AOB$ with centre on the internal bisector of the angle $AOB$.

36. The tangents from $P$ to the circle $x^2 + y^2 + 2x + 4y = 4$ touch the circle at $Q$ and $R$ such that the angle $QPR$ has a fixed value $\theta$. Find the locus of the centroid of the triangle $PQR$. What is the locus when $\theta = 60°$?

37. Find the coordinates of the incentre and of the three excentres of the triangle formed by the lines $y = 0$, $3x - 4y = 0$, $4x + 3y = 20$. Prove that the area of the triangle formed by the three excentres is five times that of the triangle formed by the three given lines.

38. The three points $A$, $B$, $C$ on the circle $x^2 + y^2 + 4x - 2y = 20$ have coordinates $(-7, 1)$, $(-5, 5)$, $(1, 5)$ respectively. The points $A'$, $B'$, $C'$ are the opposite ends of the diameters through $A$, $B$, $C$; the lines $AB'$ and $BC'$ meet in $P$, and the lines $A'B$ and $B'C$ meet in $Q$. Find the coordinates of $P$ and $Q$, show that $BQB'P$ is a parallelogram, and calculate its area.

39. Prove that the circles

$$x^2 + y^2 - 6x - 4y + 9 = 0 \quad \text{and} \quad x^2 + y^2 + 2x + 2y - 7 = 0$$

touch each other, and obtain the equation of the common tangent at their point of contact $P$. Find the equations of the two circles which touch this common tangent at $P$ and also touch the straight line $y = 1$.

40. Find the equations of the two circles of radius $\frac{1}{2}$, each of which is orthogonal to both of the circles $x^2 + y^2 + 2x = 0$, $x^2 + y^2 - 2x - 2y = 0$.

41. Show that all circles passing through the two points $(-3, 0)$ and $(7, 0)$ bisect the circumferences of each of the circles

$$x^2 + y^2 + 4x = 5 \quad \text{and} \quad x^2 + y^2 - 8x = 5.$$

42. $P$ is the point $(at^2, 2at)$ on the parabola $y^2 = 4ax$. A triangle is formed by the tangents at $P$, at the vertex $O$ and at the point $Q$ whose parameter is $-t$. Determine the equation of the circumcircle of this triangle and hence prove that it passes through the point $(a, 0)$.

43. If $O$ is the vertex of the parabola $y^2 = 4ax$, show that the normal at $P(at^2, 2at)$ meets the perpendicular bisector of the line $OP$ at the point $(2a+\frac{3}{4}at^2, -\frac{1}{2}at^3)$. Hence obtain the equation of the circle through $O$ which touches the parabola at the point $P$.

44. $P$ and $Q$ are points with parameters $p$ and $q$ on the parabola $y^2 = 4ax$. If the chord $PQ$ subtends a right angle at a point with parameter $T$ on the parabola, show that $(T+p)(T+q)+4 = 0$. If the circle on $PQ$ as diameter meets the parabola again at $H$ and $K$, show that the chords $PQ$, $HK$ are equally inclined to the axis of the parabola.

45. A circle touches the parabola $y^2 = 4ax$ at the points where the line $x = b$ cuts it. Show that the centre of the circle is at the point $(b+2a, 0)$, and find the radius.

46. Prove that the line $x-ty+at^2 = 0$ touches the parabola $y^2 = 4ax$ for any value of $t$. Find the equation of the circumcircle of the triangle formed by the $y$-axis and two other tangents to the parabola, and show that this circle passes through the focus.

47. The curve $x^2 = a^2-by$ cuts the $y$-axis in $P$ and the $x$-axis in $Q$ and $R$. Find the centre and radius of the circle through $P$, $Q$, $R$ and show that its equation may be expressed in the form $x^2 = (a^2-by)(1+y/b)$. Show that the arc $QPR$ of the given curve lies entirely within this circle.

48. Show that the circle on which $(x_1, y_1)$ and $(x_2, y_2)$ are diametrically opposite points is $(x-x_1)(x-x_2)+(y-y_1)(y-y_2) = 0$. A variable circle passes through the fixed point $A(x_1, y_1)$ and touches the $x$-axis. Show that the locus of the other end of the diameter through $A$ is given by

$$(x-x_1)^2 = 4y_1y.$$

49. Prove that the circle which has as diameter the common chord of the two circles $x^2+y^2+2x-5y = 0$ and $x^2+y^2+6x-8y = 1$ touches the axes of coordinates.

50. Prove that the origin, the point $(1, 1)$ and the intersections of the circles $x^2+y^2-2x-3 = 0$ and $x^2+y^2-4x+2y-4 = 0$ are concyclic, giving the equation of the circle on which they lie.

51. Find the equation of the inscribed circle of the triangle formed by $y = 0$, $12x = 5y$ and $3x+4y = 27$.

52. Obtain the locus of a point such that the tangents drawn from it to the circle $4x^2+4y^2 = 25$ are inclined to each other at $60°$. Find the coordinates of the point $P$ such that each of the two circles $4x^2+4y^2 = 25$ and $x^2+y^2-18x-24y+200 = 0$ subtends an angle of $60°$ at $P$.

53. Prove that any circle which passes through the point $(b, 0)$ and cuts the circle $x^2+y^2 = a^2$ orthogonally also passes through the point $(a^2/b, 0)$.

54. Prove that, if the chord joining the points $(au^2, 2au)$ and $(av^2, 2av)$ on the parabola $y^2 = 4ax$ goes through the focus, then $uv = -1$. Prove that the circle drawn on a focal chord as diameter touches the directrix.

55. Show that the circle $x^2+y^2-2x-10y+1=0$ is inscribed in the triangle whose vertices are $(-9,0)$, $(16,0)$, $(0,12)$.

56. Find the equation of the circle circumscribing the triangle whose sides are the lines $3x+4y=24$, $x+3y=8$, $3x-y+6=0$.

---

57. The parameters of the points $A$, $B$ and $C$ on the rectangular hyperbola $x=ct$, $y=c/t$ are $t_1$, $t_2$ and $t_3$ respectively. Prove that the orthocentre $H$ of the triangle $ABC$ also lies on the curve, and express its parameter $t_4$ in terms of $t_1$, $t_2$ and $t_3$. If $AH$ meets $BC$ at $D$, by considering the circle having $AB$ for diameter, or otherwise, prove that

$$HA\,.\,HD = -c^2(t_1-t_4)(t_2-t_4)(t_3-t_4)/t_4.$$

58. $P$, $Q$, $R$, $S$ are four points on the rectangular hyperbola $xy=c^2$ such that $PQ$ is perpendicular to $RS$. Prove that each of the points is the orthocentre of the triangle formed by the other three. If $A$ is the mid-point of the chord joining any two of the points and $B$ is the mid-point of the chord joining the other two, show that the circle on $AB$ as diameter passes through the origin.

59. A point $P(x,y)$ moves so that $OP^2=\lambda y$, where $O$ is the origin and $\lambda$ is a constant. Show that the locus of $P$ is a circle. State the coordinates of the centre of the circle and the length of its radius. Using the same axes, find the locus of a point $Q(x',y')$ which moves so that $QR^2=\lambda' y'$, where $R$ is the fixed point $(\alpha,0)$ and $\lambda'$ is a constant. If the two loci touch each other and $\alpha \neq 0$, prove that $\alpha^2=\lambda\lambda'$.

60. A circle passes through the points $(\alpha,0)$, $(\beta,0)$ and $(0,\gamma)$. Prove (do not merely verify) that its equation is

$$\gamma(x^2+y^2)-\gamma(\alpha+\beta)x-(\gamma+\alpha\beta)y+\alpha\beta\gamma=0.$$

A circle passes through the point $(0,1)$ and cuts off a chord of length 2 from the $x$-axis. Find the equation of the locus of the centre of the circle.

61. Two circles have centres $(a,0)$, $(-a,0)$ and radii $b$, $c$ respectively, where $a>b>c$. Prove that the points of contact of the exterior common tangents lie on the circle $x^2+y^2=a^2+bc$. Find the corresponding result for the points of contact of the interior common tangents.

62. Prove that the coordinates of any point on the circle through the points $A(4,1)$, $B(-4,-3)$, $C(2,5)$ are $(-1+5\cos\theta,\ 1+5\sin\theta)$, where $\theta$ is a parameter. Find the coordinates of the point in which the perpendicular from $A$ on $BC$ meets the circle again.

63. Show that the coordinates of a point $P$ on the circle

$$(x-a)^2+(y-b)^2=r^2$$

may be written $x=a+r\cos\theta$, $y=b+r\sin\theta$, where $\theta$ is the angle which the radius to $P$ makes with the $x$-axis. Prove that the equation of the tangent at $P$ is $(x-a)\cos\theta+(y-b)\sin\theta=r$. Prove also that, if $N$ is the foot of the perpendicular from the origin to this tangent, the coordinates of $N$ satisfy the equation $y\cos\theta - x\sin\theta = 0$; and deduce that the locus of $N$ as $P$ moves round the circle is $\{x(x-a)+y(y-b)\}^2 = r^2(x^2+y^2)$.

64. If $O$ is the centre of the rectangular hyperbola through the four points $A, B, C, D$ whose coordinates are $(t, 1/t)$ where $t = a, b, c, d$, and if the perpendiculars from $O$ to $BC, AD$ meet $AD, BC$ respectively in $P, P'$, prove that, if $abcd+1 \neq 0$, the equation of the circle on $PP'$ as diameter is

$$(x+bcy-b-c)(x+ady-a-d)+(bcx-y)(adx-y) = 0.$$

65. Find the condition that the tangent to the parabola $y^2 = 4ax$ at the point $(at^2, 2at)$ should touch the circle $x^2+y^2+2gx = 0$. Hence (or otherwise) find the radius of the circle inscribed in the triangle formed by the tangents at $(at^2, 2at)$, $(at^2, -2at)$ and the vertex.

66. A circle with centre at the point $(a, 0)$ and radius greater than $a$ meets the parabola $x = at^2$, $y = 2at$ at the points $P$, $Q$. Prove that the tangents to the parabola at $P$ and $Q$ meet on the circle.

67. The tangent at $P(at^2, 2at)$ to the parabola $y^2 = 4ax$ cuts the $y$-axis at $T$. If $S$ is the point $(a, 0)$, prove that $ST$, $PT$ are at right angles. Show also that the locus of the centre of the circle through $P$, $T$ and $S$ is the curve $2ax = a^2+y^2$ and sketch this curve.

68. $Q$ is the foot of the perpendicular from a point $P$ on the polar of $P$ with respect to the parabola $y^2 = 4ax$. Show that, if $P$ does not lie on the line $y = 0$ and if $PQ$ goes through the vertex of the parabola, then the locus of $P$ is the line $x+2a = 0$ and the locus of $Q$ is the circle $x^2+y^2 = 2ax$.

69. A point $Q$ is taken on the line joining $P$ to the origin $O$ such that $OP.OQ = k^2$, a constant. Show that when $P$ moves round a circle about the centre $C$, $Q$ moves, in general, round another circle whose centre is in line with $OC$.

70. The line $lx+my = 1$ meets the circle $x^2+y^2+2gx+c = 0$ in the points $M$ and $N$, and $O$ is the origin. Prove that, if $OM$ and $ON$ are perpendicular, $c(l^2+m^2)+2gl+2 = 0$, and that, when this is the case, the foot of the perpendicular from $O$ on the line $MN$ lies on the circle

$$2(x^2+y^2)+2gx+c = 0.$$

71. A circle is drawn with centre $P$ and radius equal to the length of the tangent from $P$ to the circle $x^2+y^2 = a^2$. If this second circle passes through the point $(p, q)$ prove that it passes also through the point

$$\{a^2p/(p^2+q^2),\quad a^2q/(p^2+q^2)\}.$$

72. A circle of variable radius $r$ moves so that it always touches each of the two fixed circles $(x-1)^2+y^2 = 4$; $(x+1)^2+y^2 = 9$. If both the fixed circles lie inside the variable circle, prove that its centre lies on the curve whose equation is $12x^2-4y^2 = 3$.

73. If the length of the tangent from a point $P$ to the circle $x^2+y^2 = r^2$ is $n$ times the length of the tangent from $P$ to the circle

$$(x-2r)^2+y^2 = 4r^2,$$

prove that the locus of $P$ is a circle. If the radius of this circle is $2r$ show that $n = \sqrt{(3/7)}$.

74. A coaxal system is defined by the circles $x^2+y^2+2ax+2by+c = 0$, $(a^2+b^2 > c)$ and $(a^2+b^2)(x^2+y^2)+2acx+2bcy+c^2 = 0$. Show that the origin is one of the limiting points and find the equation of the orthogonal coaxal system of circles.

75. Show that the equations of any two circles may be put in the form $x^2+y^2+2gx+c = 0$ and $x^2+y^2+2Gx+c = 0$. What are the conditions satisfied by $\alpha$, $\beta$, $\gamma$ if the circle $x^2+y^2+2\alpha x+2\beta y+\gamma = 0$ cuts both these circles orthogonally? Two circles $S_2$ and $S_3$ cut a circle $S_1$ orthogonally. $P$ is any point on $S_1$. Show that the polars of $P$ with respect to $S_2$ and $S_3$ intersect in a point which lies on $S_1$.

76. $OX$ and $OY$ are two perpendicular lines. $P$ is a variable point on $OX$, and the polar of $P$ with respect to a given fixed circle in the plane $OXY$ cuts $OY$ at the point $Q$. Prove that, for different positions of $P$ on $OX$, the circles on $PQ$ as diameter form a coaxal system whose radical axis passes through the centre of the given circle.

77. Prove that the circles with respect to which the given line

$$lx+my+n = 0$$

is the polar of the origin form a coaxal system with radical axis

$$2lx+2my+n = 0.$$

Find the coordinates of the limiting points of this system and the equation of the system of circles which cuts the system orthogonally.

78. Prove that, in general, there exists one and only one circle (real or imaginary) which cuts three given coplanar circles orthogonally. State any special case when this is not true. If $S_r \equiv x^2+y^2-2\alpha_r s-2\beta_r y+\gamma_r = 0$ $(r = 1, 2, 3)$ represent three given circles whose centres are not collinear, prove that the circle which cuts these three circles orthogonally also cuts a circle represented by $S_1+\lambda S_2+\mu S_3 = 0$ orthogonally, where $\lambda$ and $\mu$ are constants. Prove that those circles of the family $S_1+\lambda S_2+\mu S_3 = 0$ which have their centres on a given line are coaxal.

79. Find the system of circles which cuts orthogonally each of the circles $x^2+y^2+2\lambda x+c = 0$, where $\lambda$ is a variable parameter. If the origin is one of the limiting points of a coaxal system of circles of which the circle $x^2+y^2+2ax+2by+c = 0$ is a member, prove that the system of orthogonal circles is given by $a(x^2+y^2)+cx = \lambda\{b(x^2+y^2)+cy\}$.

80. Prove that if $AB$ is a diameter of a circle $S$ which cuts a circle $S'$ orthogonally, the polar of $A$ with respect to $S'$ passes through $B$. Prove that the equation of the circle which cuts the three circles

$$x^2+y^2+2gx+2fy+c=0,$$

$$x^2+y^2+2g'x+2f'y+c'=0,$$

$$x^2+y^2+2g''x+2f''y+c''=0$$

orthogonally is $\begin{vmatrix} x+g & y+f & gx+fy+c \\ x+g' & y+f' & g'x+f'y+c' \\ x+g'' & y+f'' & g''x+f''y+c'' \end{vmatrix} = 0.$

# VI

## THE ELLIPSE AND HYPERBOLA

In Chapter III we discussed the curve which is defined by the property $SP = PM$, where $S$ is a fixed point, and $M$ the foot of the perpendicular from a variable point $P$ on the locus to a fixed line $DD'$. It is natural to ask what modifications are made when one considers the locus of $P$ subject to the condition $SP = e.PM$, where $e$ is any positive constant. The cases $e < 1$ and $e > 1$ have to be considered separately to some extent, and the two resulting curves, the ellipse and the hyperbola, are the subject of this chapter.

The ellipse occurs commonly in everyday life; it is the shape seen when any circle is viewed obliquely, the section of a circular cylinder by any plane not perpendicular to the axis of the cylinder. It is also the general path of any planet, including our own, in its orbit round the sun. The hyperbola is frequently seen when a branch of it is the part of the road illuminated by a car headlamp. Mathematically speaking, it is the section of a cone by a plane which makes an angle with the axis of the cone less than that made by a generator of the cone. The orbits of certain comets are observed to be arcs of hyperbolas; for example, the Arend–Roland comet (seen in 1957) was computed to be moving on a hyperbola whose eccentricity $e$ was 1·0001778.

**1. Definition of the ellipse.** The ellipse is the locus of a point $P$ which moves so that its distance from a fixed point $S$ is a constant multiple, $e$, of its distance from a fixed line $DD'$, where $0 < e < 1$. $S$ is called the *focus*, $DD'$ the *directrix* and $e$ the *eccentricity* of the curve.

It is convenient to take as $x$-axis the perpendicular through $S$ to $DD'$. Suppose this meets $DD'$ in $Z$ (Fig. 58). On the line there will be two points $A$ and $A'$ of the locus (dividing $SZ$ internally and externally in the ratio $e:1$). Take the mid-point of $AA'$ as $O$, the origin, and let $A$ and $A'$ be $(a, 0)$ and $(-a, 0)$, where $a$ is any positive constant. Then, if $S$ is $(c, 0)$

and $Z$ is $(k, 0)$, since $A$ and $A'$ divide $SZ$ internally and externally in the ratio $e:1$,

$$a = \frac{c+ek}{1+e} \quad \text{and} \quad -a = \frac{c-ek}{1-e}.$$

Thus

$$a(1+e) = c+ek,$$
$$-a(1-e) = c-ek,$$

so that

$$c = ae, \quad k = a/e.$$

Thus the focus $S$ is the point $(ae, 0)$ and the directrix $DD'$ is the line $x = a/e$. We notice that $ae < a$ and $a/e > a$; the diagram was drawn to accord with this.

If now $P(x, y)$ is any point of the locus, $SP^2 = (x-ae)^2 + y^2$ and $PM = a/e - x$, where $M$ is the foot of the perpendicular from $P$ to $DD'$. As $SP = e.PM$,

$$(x-ae)^2 + y^2 = e^2(a/e - x)^2,$$

or

$$x^2(1-e^2) + y^2 = a^2(1-e^2).$$

Since $e < 1$, $1-e^2$ is positive, so that we may write

$$a^2(1-e^2) = b^2$$

and choose $b$ to be positive. Thus the equation of the locus of $P$ reduces to the simple form

$$\mathbf{\frac{x^2}{a^2} + \frac{y^2}{b^2} = 1.}$$

In order to sketch the graph represented by this equation, we notice that:

(i) The curve is symmetrical about both axes.

(ii) Both $x^2/a^2$ and $y^2/b^2$ are positive and their sum is 1, so neither can exceed 1; thus $-a < x < a$ and $-b < y < b$, the curve lying entirely inside a rectangle.

(iii) The curve meets the $x$-axis in $A, A'$ $(\pm a, 0)$ and the $y$-axis in $B, B'$ $(0, \pm b)$, and by definition $b < a$.

The curve is shown in Fig. 59. The axes $AA'$ and $BB'$ are called the *principal* axes (*major* and *minor* respectively); they meet in the *centre* of the ellipse. The ordinate through $S$ is called the *latus rectum*. Its length is $2l$ where

$$a^2e^2/a^2 + l^2/b^2 = 1;$$
$$l^2 = b^2(1-e^2) = b^4/a^2$$

so that

$$2l = 2b^2/a.$$

From the symmetry of the equation it follows that there is a second focus $S'(-ae, 0)$ and a second directrix, $x = -a/e$, since the whole figure remains unaltered by reflection in the $y$-axis.

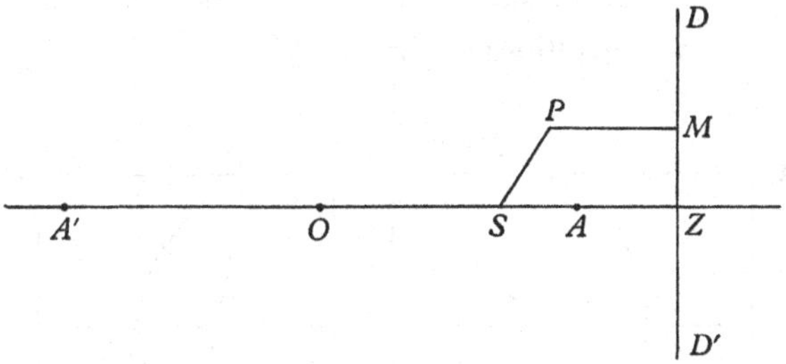


Fig. 58

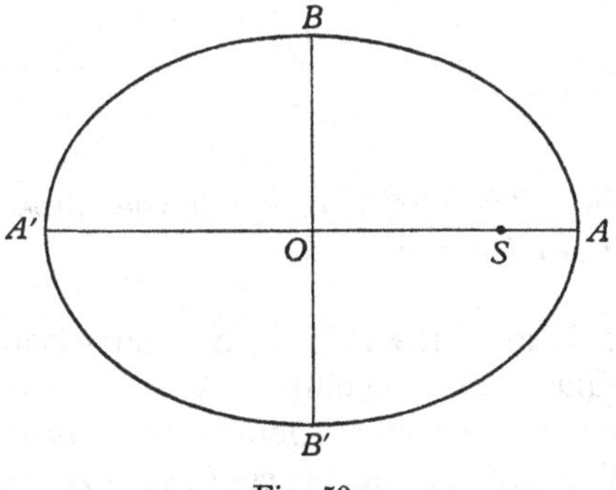


Fig. 59

A very important property follows. If $M'$ is the foot of the perpendicular from any point $P$ of the locus to the second directrix (Fig. 60)

$$SP = e.PM$$

and

$$S'P = e.PM',$$

so that

$$SP + S'P = e(PM + P'M)$$

$$= e.2a/e = 2a.$$

Thus the sum of the focal distances, of $P$ from $S$ and $S'$, is a constant, $2a$, and this fact gives rise to a speedy and accurate method for drawing the curve.

As for the circle (p. 135) we notice that the plane is divided into three regions. The point $P(x_1, y_1)$ is

*outside* the ellipse if $x_1^2/a^2 + y_1^2/b^2 > 1$,

*inside* the ellipse if $x_1^2/a^2 + y_1^2/b^2 < 1$,

and *on* the ellipse if $x_1^2/a^2 + y_1^2/b^2 = 1$.

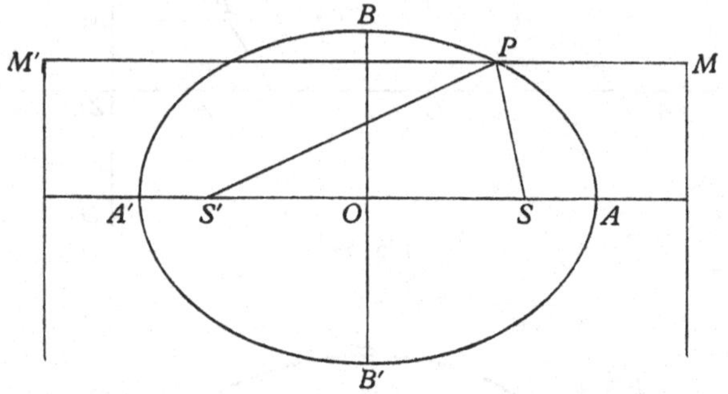

Fig. 60

In the same way, $P$ is outside or inside the ellipse according as $SP + S'P > 2a$ or $SP + S'P < 2a$.

**2. Parametric form of the ellipse.** A comparison between the forms of the equation of the ellipse $x^2/a^2 + y^2/b^2 = 1$ and a concentric circle $x^2 + y^2 = a^2$, whose general point is $(a\cos\theta, a\sin\theta)$, leads us to notice that the point $(a\cos\theta, b\sin\theta)$ lies on the ellipse for all values of the parameter $\theta$. In order to preserve the one-one nature of the relation between a point and its parameter, we restrict $\theta$ to lie in the range $0 \leqslant \theta < 2\pi$.

The angle $\theta$ has a geometrical significance. Suppose $C$ is a circle drawn on the major axis $AA'$ as diameter, and let the ordinate through any point $P$ on the ellipse meet $C$ in $Q$ (in the same quadrant as $P$) and the major axis in $N$ (Fig. 61). If

$$\angle QON = \theta,$$

$$x = \overrightarrow{ON} = a\cos\theta,$$

and so

$$y = b\sin\theta.$$

In fact $\overrightarrow{PN}/\overrightarrow{QN} = (b\sin\theta)/(a\sin\theta) = b/a$, a constant for all positions of $P$. The angle $\theta$ is called the *eccentric angle* and the circle $C$ is called the *auxiliary circle*.

From this trigonometrical form it is easy to obtain a parametric representation in algebraic form. If $\tan \frac{1}{2}\theta = t$, then

$$\cos\theta = (1-t^2)/(1+t^2) \quad \text{and} \quad \sin\theta = 2t/(1+t^2),$$

and so the point
$$\left(a\frac{1-t^2}{1+t^2}, \frac{2bt}{1+t^2}\right)$$

lies on the ellipse for all values of the parameter $t$. The reader should trace the variation in $t$ as $P$ moves on the curve.

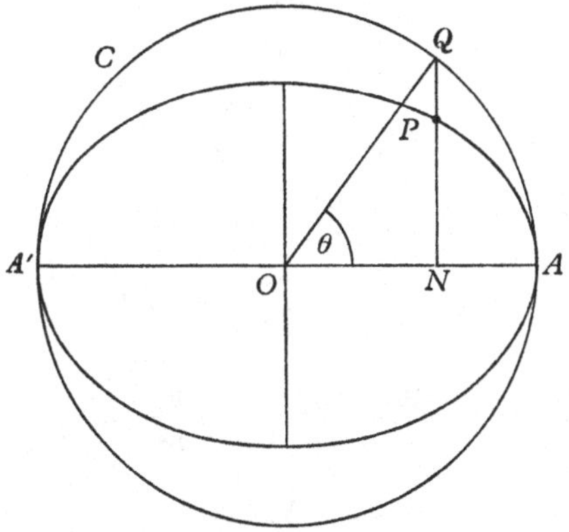


Fig. 61

**3. Equations of chord, tangent and normal.** The method of obtaining these results is neither new nor difficult, though the algebraic details are less easy to keep tidy. We consider the ellipse with general point $(a\cos\theta, b\sin\theta)$. The gradient of the chord joining the points $\theta = \alpha, \beta$ is

$$\frac{b\sin\alpha - b\sin\beta}{a\cos\alpha - a\cos\beta} = \frac{b\cos\frac{1}{2}(\alpha+\beta)\sin\frac{1}{2}(\alpha-\beta)}{-a\sin\frac{1}{2}(\alpha+\beta)\sin\frac{1}{2}(\alpha-\beta)}$$
$$= -(b/a)\cot\tfrac{1}{2}(\alpha+\beta),$$

using [T 7].

The equation of the chord is then

$$y - b\sin\alpha = -(b/a)\cot\tfrac{1}{2}(\alpha+\beta)\,(x - a\cos\alpha),$$

or
$$\frac{x}{a}\cos\tfrac{1}{2}(\alpha+\beta) + \frac{y}{b}\sin\tfrac{1}{2}(\alpha+\beta) = \cos\alpha\cos\tfrac{1}{2}(\alpha+\beta) + \sin\alpha\sin\tfrac{1}{2}(\alpha+\beta),$$

that is $\frac{x}{a}\cos\tfrac{1}{2}(\alpha+\beta)+\frac{y}{b}\sin\tfrac{1}{2}(\alpha+\beta) = \cos\tfrac{1}{2}(\alpha-\beta).$

The equation of the tangent at $\theta = \alpha$ is obtained by letting $\beta \to \alpha$ in this equation, giving

$$\frac{x}{a}\cos\alpha+\frac{y}{b}\sin\alpha = 1.$$

Alternatively, the gradient at a general point is got by differentiating,

$$\frac{2x}{a^2}+\frac{2y}{b^2}\frac{dy}{dx} = 0,$$

so that at $(x_1, y_1)$
$$\frac{dy}{dx} = -\frac{b^2x_1}{a^2y_1}.$$

The equation of the tangent is then

$$y-y_1 = -\frac{b^2x_1}{a^2y_1}(x-x_1),$$

or
$$\frac{xx_1}{a^2}+\frac{yy_1}{b^2} = \frac{x_1^2}{a^2}+\frac{y_1^2}{b^2},$$

that is
$$\frac{xx_1}{a^2}+\frac{yy_1}{b^2} = 1$$

since $(x_1, y_1)$ lies on the curve. This gives the form found above for the tangent at $(a\cos\alpha, b\sin\alpha)$; it also accords with the mnemonic for the equations of tangents noticed on p. 138.

Finally, the equation of the normal at $(a\cos\alpha, b\sin\alpha)$ is

$$y-b\sin\alpha = (a/b)\tan\alpha(x-a\cos\alpha)$$

which reduces to
$$\frac{ax}{\cos\alpha}-\frac{by}{\sin\alpha} = a^2-b^2.$$

At $(x_1, y_1)$ the equation of the normal is

$$\frac{xa^2}{x_1}-\frac{yb^2}{y_1} = a^2-b^2.$$

Elementary problems on the ellipse often involve finding the locus of some point, and the following example illustrates a usual technique.

***Illustration 1.*** *The normal at a general point $P$ to an ellipse meets the major axis in $L$ and the minor axis in $M$. The point $Q$ divides $LM$ in the ratio $q:1$ (where $q+1 \neq 0$). Show that the locus of $Q$ is in general a second ellipse, and find the values of $q$ for which this ellipse is a circle.*

Referred to the principal axes as axes of coordinates, the ellipse has the equation $x^2/a^2+y^2/b^2 = 1$, and the general point $P$ is $(a\cos\theta, b\sin\theta)$. The equation of the normal at $P$ is

$$xa/\cos\theta - yb/\sin\theta = a^2-b^2 = a^2e^2,$$

and this meets the major axis, $y = 0$, in $L(ae^2\cos\theta, 0)$ and the minor axis, $x = 0$, in $M\{0, -(a^2e^2\sin\theta)/b\}$. $Q(X, Y)$ divides $LM$ in the ratio $q:1$, so that

$$X = \frac{ae^2\cos\theta}{q+1}, \quad Y = -\frac{qa^2e^2\sin\theta}{b(q+1)}.$$

Now $\cos^2\theta+\sin^2\theta = 1$, so that

$$\frac{(q+1)^2X^2}{a^2e^4}+\frac{b^2(q+1)^2Y^2}{q^2a^4e^4} = 1.$$

It was given that $q+1 \neq 0$, so this is the equation of an ellipse.

If the lengths of the axes are $2A$ and $2B$,

$$A^2 = a^2e^4/(q+1)^2, \quad B^2 = q^2a^4e^4/b^2(q+1)^2$$

and
$$A = \pm ae^2/(q+1), \quad B = \pm qa^2e^2/b(q+1),$$

where the signs are to be chosen so that $A$ and $B$ are positive (this will depend on the sign of $q$ and of $q+1$).

The ellipse is a circle provided the lengths of the major axis and the minor axis are equal. In this case

$$\frac{ae^2}{q+1} = \pm\frac{qa^2e^2}{b(q+1)}$$

and
$$q = \pm b/a.$$

It is interesting to notice that if $q = e^2-1$, the locus of $Q$ coincides with the original ellipse.

The use of the substitution $\tan\frac{1}{2}\theta = t$ [$T$ 6], is shown in the next example. The result should be compared with the equivalent results given previously, for the parabola (p. 88) and the rectangular hyperbola (p. 105).

***Illustration 2.*** *Show that through a general point $(h, k)$ there are four normals to an ellipse whose general point is $(a\cos\theta, b\sin\theta)$, and that, if the parameters of the feet of these normals are $\alpha, \beta, \gamma, \delta$, then $\alpha+\beta+\gamma+\delta = (2k+1)\pi$, where $k$ is any integer.*

The normal at the point with parameter $\theta$ has equation

$$\frac{xa}{\cos\theta}-\frac{yb}{\sin\theta} = a^2-b^2 = a^2e^2.$$

This contains the point $(h, k)$ if and only if

$$ha\sin\theta - kb\cos\theta = a^2e^2\cos\theta\sin\theta.$$

Substituting $\tan\frac{1}{2}\theta = t$, and clearing of fractions, this gives

$$2hat(1+t^2)-kb(1-t^2)(1+t^2) = 2a^2e^2t(1-t^2),$$

or $$t^4kb+t^3(2ha+2a^2e^2)+t(2ha-2a^2e^2)-kb = 0.$$

This is a quartic equation in $t$, so there are at most four roots, and so at most four normals can be drawn through $(h, k)$.

If the roots of the equation are $p, q, r, s$, among the identities [A 7] holding are two which are independent of $h$ and $k$; these are

$$pq+pr+ps+rs+sq+qr = 0$$

and $$pqrs = -kb/kb = -1.$$

Now the parameters of the feet of the normals are $\alpha, \beta, \gamma, \delta$; let $\tan\frac{1}{2}\alpha = p$, and so on. In the formula [T 8]

$$\tan\frac{\alpha+\beta+\gamma+\delta}{2} = \frac{\Sigma p-\Sigma pqr}{1-\Sigma pq+pqrs}$$

the denominator vanishes, and so [T 2]

$$\frac{\alpha+\beta+\gamma+\delta}{2} = k\pi+\tfrac{1}{2}\pi,$$

where $k$ is any integer, and

$$\alpha+\beta+\gamma+\delta = (2k+1)\pi.$$

## EXAMPLES VI A

1. Find the equation of an ellipse, if the focus $S$ is taken as origin, and the directrix is the line $x = d$. By changing the origin to the point $(-ae, 0)$ and putting $d = a/e-ae$, recover the standard form of the equation.

2. Deduce the equation of the ellipse in standard form if it is defined as the locus of $P(x, y)$ when $SP + S'P = 2a$ and $S$, $S'$ are the points $(\pm ae, 0)$.

3. Find the eccentricity of an ellipse if the lengths of its axes are 10 and 6. What is the distance between the foci?

4. Find the equation of an ellipse with eccentricity $\frac{1}{4}$ whose foci are the points $(\pm\sqrt{2}, 0)$.

5. Find the equation of the ellipse whose foci are at the points $(\pm 12, 0)$ and whose minor axis is 10 units long. Find also the eccentricity and the equations of the directrices. One extremity of a latus rectum of the ellipse is $P$ and $O$ is the origin. The perpendicular from $P$ on to the corresponding directrix meets it in $N$ and the line through $P$ at right angles to $OP$ meets this directrix in $M$. Show that the distance $NM = 13$ units.

6. Show that, if an ellipse has $S$ as focus and $BB'$ as minor axis, then $SB = SB' = a$, where $2a$ is the length of the major axis.

7. Find the eccentricities, foci, and directrices of the ellipses whose equations are:

(i) $4x^2 + 9y^2 = 144$;

(ii) $x^2 + 3y^2 = 9$;

(iii) $9(x-1)^2 + 16(y+4)^2 = 144$.

8. What is the eccentricity of an ellipse if its latus rectum is equal to (i) its semi major axis, (ii) its semi minor axis?

9. Find the condition satisfied by the eccentricity $e$ of an ellipse, if the normal at the end of the latus rectum passes through an end of the minor axis.

10. The tangent to an ellipse at any point meets the tangents at the ends of the major axis in $P$ and $Q$. Show that the circle on $PQ$ as diameter passes through the foci.

11. Find the equations of the tangents of gradient $1/2$ to the ellipse $x^2 + 6y^2 = 15$.

12. The general point of an ellipse is

$$\{a(1-t^2)/(1+t^2),\quad 2bt/(1+t^2)\}.$$

Show that the equation of the chord joining the points with parameters $p$ and $q$ is

$$(1-pq)\frac{x}{a} + (p+q)\frac{y}{b} = (1+pq).$$

Deduce the equation of the tangent to the ellipse.

13. The foci of an ellipse are $S(4, 0)$ and $S'(-4, 0)$, and any point $P$ on the ellipse is such that $SP + S'P = 10$. Find the equation of the ellipse. If $Q$ is the point in the first quadrant which lies on this ellipse and has its

$x$-coordinate equal to 3, find the equation of the tangent to the ellipse at $Q$. Find also the coordinates of the point in which this tangent meets the directrix associated with the focus $S$.

14. A point $Q$ is taken on the circle $x^2+y^2 = a^2$ and a line is drawn through $Q$, parallel to the $y$-axis, to meet the ellipse $x^2/a^2+y^2/b^2 = 1$ at $P$ and *then* the $x$-axis at $N$. If the distance of $N$ from the origin is $a\cos\theta$ show that the coordinates of $P$ are $(a\cos\theta, b\sin\theta)$. If $S$ is a focus of the ellipse, show that the length of the perpendicular from $S$ on to the tangent at $Q$ to the circle is equal to $SP$.

15. The chord joining the points with eccentric angles $\alpha$ and $\beta$ on the ellipse $b^2x^2+a^2y^2 = a^2b^2$ passes through the focus $F(ae, 0)$. Show that $e\cos\frac{1}{2}(\alpha+\beta) = \cos\frac{1}{2}(\alpha-\beta)$. $PFQ$ and $PF'Q'$ are two focal chords of an ellipse, the foci being $F$ and $F'$, and the eccentric angles of $Q$ and $Q'$ are $\beta$ and $\beta'$. Show that the ratio $\tan\frac{1}{2}\beta:\tan\frac{1}{2}\beta'$ is constant for all positions of $P$.

16. If the tangent to the ellipse $x^2/a^2+y^2/b^2 = 1$ at the point

$$(a\cos\alpha, b\sin\alpha)$$

meets the coordinate axes at $Q$ and $R$, and $M$ is the mid-point of $QR$, find the coordinates of $M$, and hence show that as $\alpha$ varies, the locus of $M$ is the curve $a^2/x^2+b^2/y^2 = 4$.

17. The normal at $P$, a point on the ellipse $x^2/a^2+y^2/b^2 = 1$, meets the major axis at $G$ and the ordinate at $P$ meets this axis at $M$. Prove that $GM = (b^2/a)\cos\theta$, where $\theta$ is the eccentric angle of $P$. Prove also that the locus of the mid-point of $GP$ is a second ellipse concentric with the first, and find the eccentricity of the second ellipse if $a = 4$, $b = \sqrt{15}$.

18. The circle, having as diameter the line joining the foci of the ellipse whose major axis has length $2a$, meets the minor axis in $L$ and $M$. Prove that the sum of the squares of the perpendiculars from $L$ and $M$ on any tangent to the ellipse is $2a^2$.

19. The point $S$ is a focus of the ellipse $x^2/a^2+y^2/b^2 = 1$, and $P$ is a point on the ellipse such that $PS$ is perpendicular to the $x$-axis. The tangent and normal to the ellipse at $P$ meets the axis of $y$ in $Q$ and $R$ respectively. If $H$ is the other focus of the ellipse, prove that $QR = HP$.

20. $P$ is the point $(a\cos\theta, b\sin\theta)$ on the ellipse $x^2/a^2+y^2/b^2 = 1$ and $Q$ is the corresponding point on the auxiliary circle. $O$ is the centre of the ellipse and the normal to the ellipse at $P$ meets $OQ$ in $R$. Find the coordinates of $R$ in terms of $a$, $b$ and $\theta$. Hence show that, as $\theta$ varies, the locus of $R$ is the circle $x^2+y^2 = (a+b)^2$.

21. The normal at $P(a\cos\theta, b\sin\theta)$ to the ellipse $x^2/a^2+y^2/b^2 = 1$ meets the major axis at $G$ and the tangent meets the minor axis at $T$. Find the coordinates of $G$ and $T$ and show that the locus of the circumcentre of the triangle $OGT$ is $16a^2x^2y^2-4(a^2-b^2)^2y^2+b^2(a^2-b^2)^2 = 0$.

22. The perpendicular from any point $P$ on the ellipse $x^2/a^2+y^2/b^2 = 1$ to the major axis meets the ellipse again at the point $Q$. $O$ is the centre of

the ellipse and the normal at $P$ meets $OQ$ at the point $R$. Find the coordinates of $R$ and show that, as $P$ varies, the locus of $R$ is the ellipse

$$\frac{x^2}{a^2}+\frac{y^2}{b^2}=\frac{(a^2-b^2)^2}{(a^2+b^2)^2}.$$

23. Show that the ellipse $x^2/a^2+y^2/b^2=1$ $(a>b)$ touches the rectangular hyperbola $2xy=ab$ at the point $P(a/\sqrt{2}, b/\sqrt{2})$. A straight line is drawn from the centre of the ellipse perpendicular to the common tangent at $P$ and is produced to meet the hyperbola in $Q$. If the tangent to the hyperbola at $Q$ passes through a focus of the ellipse, show that $a^2=3b^2$.

24. The normal at a point $P$ on the ellipse $x^2/a^2+y^2/b^2=1$ meets the major axis at $G$. If $Q$ is the point of intersection of the line through $G$ parallel to the $y$-axis and the line joining $P$ to the centre of the ellipse, show that the equation of the locus of $Q$ is

$$\frac{x^2}{a^2}+\frac{y^2}{b^2}=\frac{(a^2-b^2)^2}{a^4}.$$

25. A circle $x^2+y^2+2gx+2fy+c=0$ meets the ellipse $x^2/a^2+y^2/b^2=1$ in the points $(a\cos\theta, b\sin\theta)$ where $\theta=\alpha, \beta, \gamma, \delta$. Prove that

$$\alpha+\beta+\gamma+\delta=2k\pi,$$

where $k$ is an integer.

## 4. Geometrical properties of the ellipse.

**4. Geometrical properties of the ellipse.** Like the parabola, the ellipse possesses a wealth of interesting properties, which follow simply from the definition, sometimes with the assistance of easy algebra (or, with a little more trouble, without it). Many of these properties were known to the ancient Greeks, and appear in a work on conic sections (that is, sections of a cone, which are circles, ellipses, parabolas or hyperbolas) compiled by Apollonius (262–200 B.C.)

For the first sequence of properties we refer to Fig. 62. The ellipse has principal axes $AA'$ and $BB'$, of lengths $2a$ and $2b$, meeting in $O$, and foci $S$ and $S'$ with corresponding directrices $d$ and $d'$. These axes are taken as axes of coordinates when required. $P$ is a general point $(x_1, y_1)$ of the curve, and the tangent at $P$ meets the directrices in $D$ and $D'$ respectively, $AA'$ in $T$ and $BB'$ in $U$. The normal at $P$ meets $AA'$ in $G$, and the ordinate of $P$ meets $AA'$ in $N$. The line through $P$ parallel to $AA'$ meets $BB'$ in $L$ and the directrices in $M$ and $M'$.

We have already established the fundamental property of the focal distances,

$$SP+S'P=2a.$$

The first results now follow from a little elementary algebra, and are simple illustrations of the formulae obtained in the last section.

(1) $ON.OT = a^2$.

For if $P$ is $(x_1, y_1)$, $ON = x_1$, and the tangent $xx_1/a^2 + yy_1/b^2 = 1$ meets $y = 0$ in the point $T$ where $xx_1 = a^2$.

(This means that $N$ and $T$ are inverse points with respect to the auxiliary circle of the ellipse.) As a corollary, it follows that the points of contact of tangents to a system of ellipses with the same major axis, from any point on it all lie on a line; in particular the

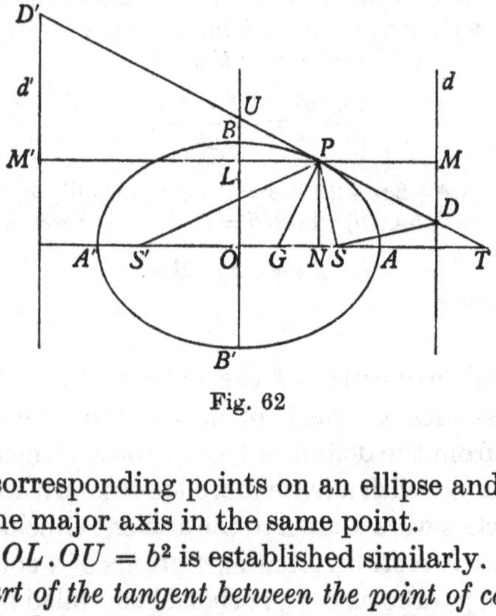


Fig. 62

tangents at corresponding points on an ellipse and its auxiliary circle meet the major axis in the same point.

The result $OL.OU = b^2$ is established similarly.

(2) *The part of the tangent between the point of contact and the directrix subtends a right angle at the corresponding focus*; that is, the angle $PSD$ is a right angle.

The tangent at $P(x_1, y_1)$ meets $d$ in $D(a/e, k)$ where

$$x_1/ae + ky_1/b^2 = 1,$$

that is $\quad aeky_1 = b^2(ae - x_1) = a^2(1-e^2)(ae - x_1).$

The gradient of $SP$ is $\quad y_1/(x_1 - ae)$

and of $SD$ is $\quad k/(a/e - ae),$

so that their product is

$$\frac{eky_1}{(x_1 - ae)\,a(1-e^2)} = -1.$$

It will perhaps be remembered that this result is also true for the parabola (p. 76); it is indeed true for any conic.

Two other results follow whose proof is exactly the same as for the parabola.

(3) *The points of contact of tangents from a point on the directrix are ends of a focal chord.*

(4) *Tangents at ends of a focal chord intersect on the directrix.*

(Such tangents are *not*, however, perpendicular; this part of the property is true only for parabolas. Indeed, for an ellipse the tangents which meet on the directrix can never be perpendicular (see p. 190, Example 7).)

The next results follow from a key result about the normal.

(5) $SG = e.SP$.

The normal at $P(x_1, y_1)$ is $xa^2/x_1 - yb^2/y_1 = a^2e^2$, and so meets the major axis, $y = 0$, in $G(e^2x_1, 0)$. Then

$$SG = OS - OG = ae - e^2x_1 = e^2(a/e - x_1)$$

$$= e^2.PM = e.SP.$$

(6) *The normal bisects the angle between the focal distances*; that is, the angles $SPG$, $S'PG$ are equal.

For $SG = e.SP$ and similarly $S'G = e.S'P$ so that

$$\frac{SG}{S'G} = \frac{SP}{S'P}.$$

Thus $PG$ bisects the angle $SPS'$.

(7) *The tangent at $P$ is equally inclined to the focal distances $SP$, $S'P$.*

The angles between the tangent at $P$ and $SP$, $S'P$ are the complements of the equal angles $SPG$, $S'PG$.

This is the fundamental 'reflecting' property of the ellipse; a ray of light from a focus $S$ is reflected at $P$ to pass through $S'$. It is this application to optics which gives the derivation of the word *focus*; the word comes from the Latin *focus* meaning a hearth, and was first introduced by Kepler (1571–1630), who is better known for recognizing the fundamental laws of planetary motion.

(8) *The feet $V$ and $V'$ of the perpendiculars from $S$ and $S'$ to the tangent at $P$ lie on the auxiliary circle.*

Produce $S'P$ and $SV$ to meet in $Q$ (Fig. 63). Then the angles

$S'PV'$, $SPV$ and $QPV$ are equal, so that the triangles $SPV$ and $QPV$ are congruent. Thus

$$S'Q = S'P + PQ = S'P + SP$$
$$= 2a \text{ (sum of focal distances).}$$

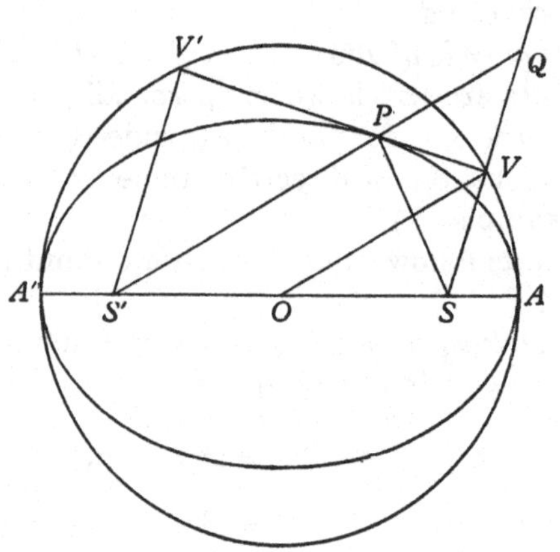


Fig. 63

Now $V$ is the mid-point of $SQ$ and $O$ is the mid-point of $S'S$, so that $S'Q = 2.OV$, and $OV = a$; $V$ thus lies on the auxiliary circle, and similarly so does $V'$.

(9) *If* $SV = p$ *and* $S'V' = p'$, $pp' = b^2$.

If $V'O$ meets the auxiliary circle again in $W$, $V'W$ is a diameter and so $WSV$ is a straight line. Also $SW = S'V'$ since the triangles $OS'V'$ and $OSW$ are congruent. Thus

$$pp' = SV.S'V'$$
$$= SV.SW = SA.SA' \text{ (intersecting chords of a circle)}$$
$$= (a-ae)(a+ae)$$
$$= a^2(1-e^2) = b^2.$$

(10) *The* $(p, r)$ *equation of the ellipse referred to a focus is* $b^2/p^2 = 2a/r - 1$.

The triangles $SPV$ and $S'PV'$ are similar, and so

$$p/SP = p'/S'P.$$

Then $$b^2/p^2 = p'/p = S'P/SP = (2a-r)/r,$$

where $r$ is the distance $SP$.

(11) *Perpendicular tangents to an ellipse meet in a point R whose locus is a circle called the director circle, with equation*

$$x^2+y^2 = a^2+b^2.$$

Let the two perpendicular tangents from $R$ meet the auxiliary circle in $V$, $V'$ and $W$, $W'$ respectively (Fig. 64). Then $SV$, $S'V'$ are perpendicular to one tangent, and $SW$, $S'W'$ to the other. $SVRW$ and $S'V'RW'$ are thus both rectangles, and so

$$RV.RV' = SW.S'W' = b^2.$$

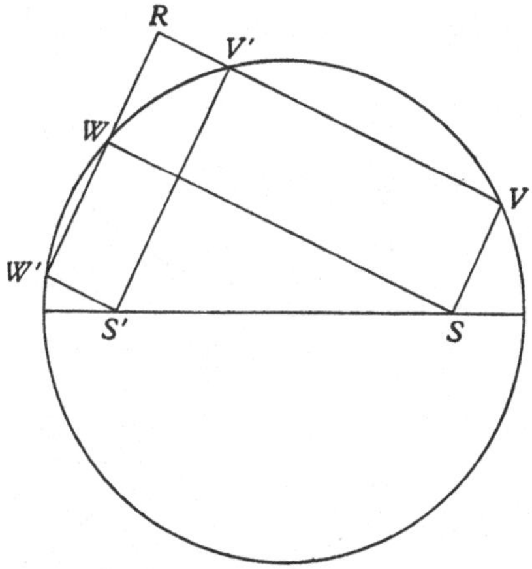


Fig. 64

Thus the length of the tangent from $R$ to the auxiliary circle is $b$ (Fig. 65), and so $OR^2 = a^2+b^2$, and the locus of $R$ is a circle whose radius is $\sqrt{(a^2+b^2)}$.

Comparison with the properties on the parabola (p. 77) shows that in this respect the director circle replaces the directrix which is the locus of intersections of perpendicular tangents of the parabola.

(12) *If TP and TQ are tangents to the ellipse, the angles TSP and TSQ are equal.*

Let the perpendiculars from $T$ to $SP$, $SQ$ meet them in $E$, $F$ respectively, and let the perpendiculars from $P$, $T$ to the directrix meet it in $M$, $H$. Let $PT$ meet the directrix in $R$ (Fig. 66).

Then

$$\frac{SE}{SP} = \frac{RT}{RP} \text{ (}TE\text{ and }SR\text{ being parallel, since } \angle RSP \text{ is a right angle)}$$

$$= \frac{TH}{PM} \text{ (since } TH \text{ and } PM \text{ are parallel).}$$

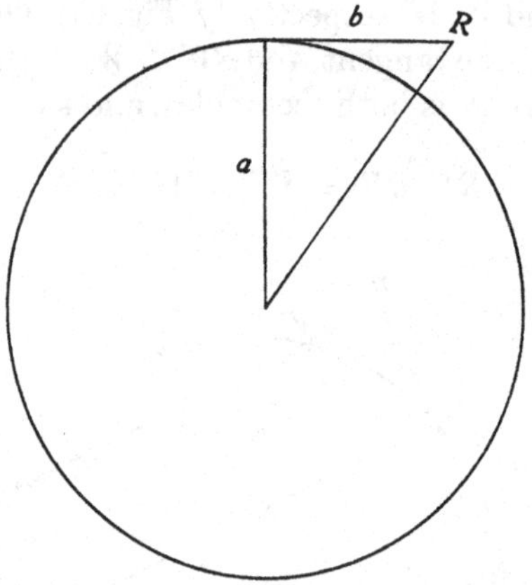


Fig. 65

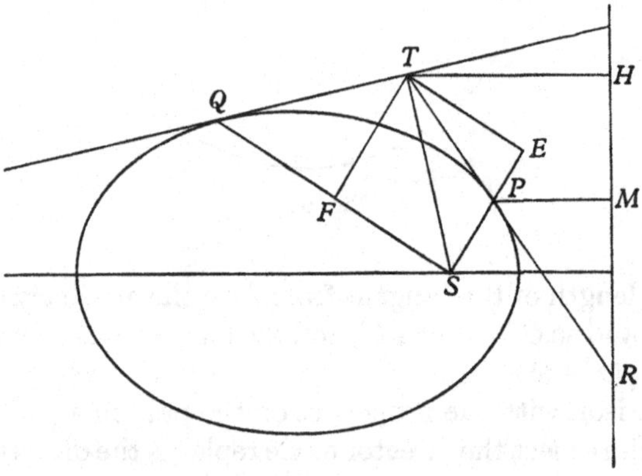


Fig. 66

But $SP = e.PM$ and so $SE = e.TH$. Similarly, $SF = e.TH$, and so the triangles $TSE$ and $TSF$ are congruent. It follows that $\angle TSP = \angle TSQ$.

(13) *If $TP$ and $TQ$ are tangents to the ellipse, the angles $STP$ and $S'TQ$ are equal.*

Referring to Fig. 67,

$$a_1 = a_2 \text{ (property 7)}$$

$$= a_3 \text{ (vertically opposite)};$$

$$b_1 = b_2 \text{ (property 12)};$$

$$a_1 + a_3 = b_1 + b_2 + c_1 \text{ (exterior angle)},$$

and $$a_1 = b_1 + d_1 \text{ (exterior angle)}.$$

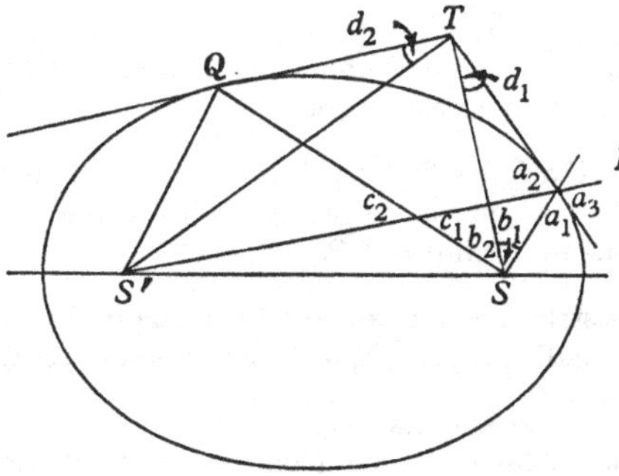


Fig. 67

Thus $c_1 = 2d_1$, and similarly $c_2 = 2d_2$. But $c_1 = c_2$, and so $d_1 = d_2$.

**EXAMPLES VI B**

In these questions the standard notation of Fig. 62 is used.

1. Let $MS$ and $M'S'$ meet in $H$, and $PH$ meet the major axis in $G_1$. Prove that (i) $H$ is on the minor axis; (ii) $SG_1 : MP = S'G_1 : M'P$; (iii) $PH$ is the normal at $P$, so that $G_1 = G$; (iv) $PG : PH = b^2 : a^2$ (cf. Illustration 1, p. 179).

2. Prove that $S$, $S'$, $P$, $H$ and $U$ are concyclic. Deduce that

$$OT \,.\, OG = OS^2.$$

3. If $AP$ and $A'P$ meet the directrix $d$ in $Z$ and $Z'$, show that the angle $ZSZ'$ is a right angle.

4. Show that $SP : S'P = PD : PD'$, and hence that the triangles $SPD$ and $S'PD'$ are similar. Deduce that the tangent at $P$ is equally inclined to the focal distances.

5. The perpendiculars from the foci $S$ and $S'$ to a tangent to an ellipse at $P$ are of length $p$ and $p'$, and the tangent is a perpendicular distance $q$ from the centre of the ellipse. Prove that $p/SP = p'/S'P = q/a$, and hence that $q = ab/\surd(SP.S'P)$.

6. Prove that $PG = (b/a)\surd(SP.S'P)$.

7. Show that if $K$ is any point of the directrix $d$, the length of the tangent from $K$ to the director circle is equal to $KS$. Deduce that tangents at ends of a focal chord of an ellipse can never be perpendicular (see p. 185).

8. If tangents at $P$ and $Q$ meet in $T$, and $PQ$ meets the directrix $d$ in $K$, show that $KS$ bisects the angle $PSQ$ externally. Deduce that angle $TSK$ is a right angle.

9. Tangents at ends of a focal chord meet on the directrix in $T$. Show that $TS$ is perpendicular to the chord.

10. Show that, if a focus $S$ of an ellipse, the length of the major axis, and one point $P$ of the curve are given, its centre lies on a fixed circle, whose centre is the mid-point of $SP$.

11. Give a construction for the second focus of an ellipse when one focus, the corresponding directrix, and one tangent are given.

12. Given an ellipse and its major axis, describe how to construct (i) the tangent at a point $P$, (ii) the foci, (iii) the directrices.

13. A series of ellipses is drawn with the same foci $S$ and $S'$. Show that there is just one ellipse of the system through any general point of the plane.

14. Given an ellipse and its foci and directrices, show how to construct the tangents to it from a given point $T$.

**5. Elementary properties of the hyperbola.** The hyperbola is the analogous locus to the ellipse, with the condition that $e > 1$ instead of $e < 1$. Applying the same method as for the ellipse (§1), we find two points $A$ and $A'$ of the locus on the line through $S$ perpendicular to $DD'$ and again choose the mid-point $O$ of $AA'$ as the origin, and $OS$ as $x$-axis. As before, $S$ is the point $(ae, 0)$ and $DD'$ the line $x = a/e$, though as $e > 1$ the length of $OS$ is now greater than that of $OA$, and that of $OZ$ less (Fig. 68).

If $P(x, y)$ is a general point on the locus, since $SP^2 = e^2.PM^2$

$$(x-ae)^2+y^2 = e^2(x-a/e)^2,$$

that is
$$x^2(e^2-1)-y^2 = a^2(e^2-1).$$

Now $e > 1$ so that $e^2 - 1 > 0$ and we may put $b^2 = a^2(e^2 - 1)$, with $b > 0$, reducing the equation of the hyperbola to the form

$$\frac{\mathbf{x}^2}{\mathbf{a}^2} - \frac{\mathbf{y}^2}{\mathbf{b}^2} = \mathbf{1}.$$

As an aid to sketching the curve, the following properties are noted.

(i) The curve is symmetrical about both axes.

(ii) $a^2y^2 = b^2(x^2 - a^2)$, so that there are no points on the curve if $-a < x < a$, and the points $(\pm a, 0)$ lie on it. There is no limitation on the possible values of $y$.

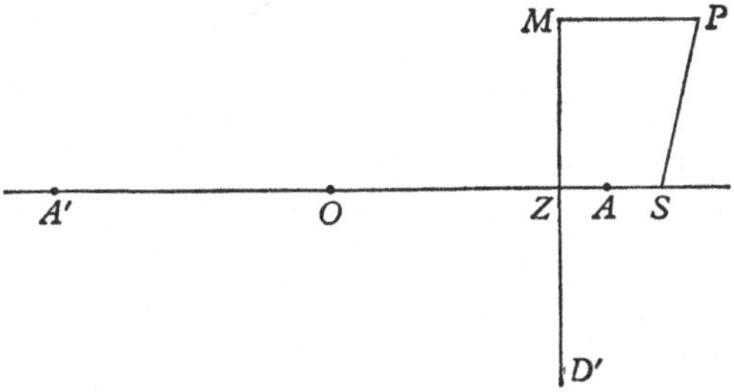


Fig. 68

(iii) A line $y = mx$ meets the curve where $x^2(b^2 - a^2m^2) = a^2b^2$. Thus for $m > b/a$ or $m < -b/a$ there are no points of the curve on $y = mx$; this gives two triangular 'forbidden regions'. The lines $y = \pm b/a.x$ are in fact asymptotes, as we shall show rigorously in §6.

The curve is shown in Fig. 69; the most obvious property is that it consists of two branches. $AA'$ is called the *transverse* axis, and the other axis of symmetry is called the *conjugate* axis; they meet in the *centre* $O$. The axes are also called *principal* axes.

Many of the properties of the hyperbola are exactly similar to those of the ellipse, and in §8 we shall group together many of the related formulae. To avoid repetition the more elementary facts about the hyperbola are here given without proof, whenever it can be supplied by analogy with work already done for the ellipse. The details may, and as a rule should, be supplied by the reader.

The latus rectum, the ordinate through $S$, is of length $2b^2/a$. There is a second focus $S'$ and a second directrix, symmetrically placed to $S$ and $DD'$ (Fig. 69). The difference of the focal distances is equal to $2a$; in fact if $P$ is on the branch near $S$, $S'P - SP = 2a$, and on the other branch $SP - S'P = 2a$. (These two formulae can be combined and written $SP \sim S'P = 2a$.)

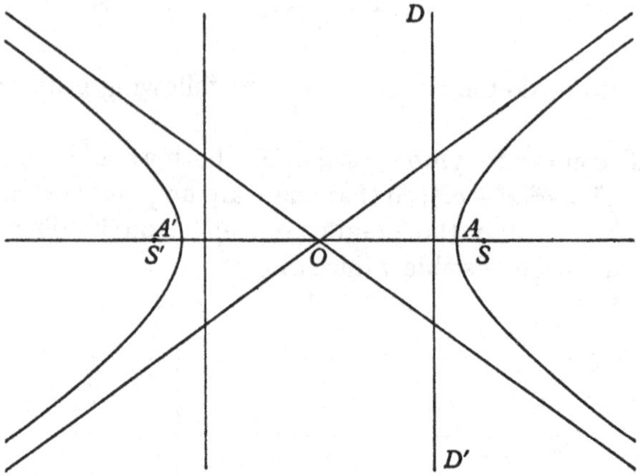


Fig. 69

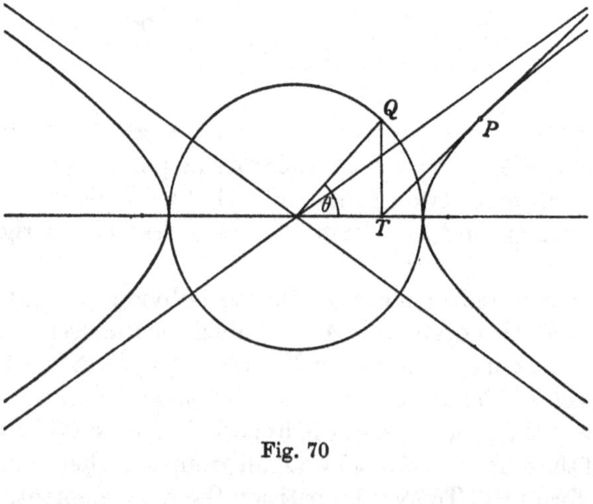


Fig. 70

For all values of $\theta$, the point $P(a \sec\theta, b\tan\theta)$ lies on the hyperbola. It can be verified from the equation of the tangent obtained below (p. 193) that if the tangent at $P$ meets the $x$-axis in $T$ and the ordinate through $T$ meets the circle $x^2+y^2 = a^2$ (the *auxiliary* circle) in $Q$, then $Q$ is $(a\cos\theta, a\sin\theta)$, so that $\angle QOT$ is the angle $\theta$ (Fig. 70).

Putting $\tan \frac{1}{2}\theta = t$, we have also an algebraic parametrization of the curve, $\{a(1+t^2)/(1-t^2),\ 2bt/(1-t^2)\}$. Thus if $(x, y)$ is any point of the curve,

$$\frac{x/a}{(1+t^2)} = \frac{y/b}{2t} = \frac{1}{(1-t^2)} = \frac{x/a-1}{2t^2} = \frac{x/a+1}{2}$$

or
$$t^2:t:1 = x/a-1:y/b:x/a+1.$$

Algebraic properties of the curve can be investigated from this point of view, by the methods of Chapter IV.

Another alternative is found by putting $\lambda = (1+t)/(1-t)$. Then

$$\lambda + \frac{1}{\lambda} = \frac{2(1+t^2)}{(1-t^2)} \quad \text{and} \quad \lambda - \frac{1}{\lambda} = \frac{4t}{1-t^2},$$

so that the general point on the curve is $\{\frac{1}{2}a(\lambda+\lambda^{-1}),\ \frac{1}{2}b(\lambda-\lambda^{-1})\}$. The reader should trace the variation of each of these parameters as $P$ moves on the curve.

Some readers will have met hyperbolic functions, $\cosh u$ and $\sinh u$, and will recognize that since $\cosh^2 u - \sinh^2 u = 1$, the point $(a \cosh u,\ b \sinh u)$ always lies on the hyperbola. Care must be taken when using this form, for since $\cosh u > 1$ for all values of $u$, the point $(a \cosh u,\ b \sinh u)$ lies only on one branch of the hyperbola, and the other branch is entirely lost.

The equations of the chord, tangent and normal can be deduced as in §3. The equation of the chord joining the points $(a \sec \alpha, b \tan \alpha)$ and $(a \sec \beta, b \tan \beta)$ is

$$x/a \cos \tfrac{1}{2}(\alpha-\beta) - y/b \sin \tfrac{1}{2}(\alpha+\beta) = \cos \tfrac{1}{2}(\alpha+\beta).$$

The equation of the tangent at $(x_1, y_1)$ is

$$xx_1/a^2 - yy_1/b^2 = 1,$$

or, at $(a \sec \alpha, b \tan \alpha)$, is

$$\frac{x \sec \alpha}{a} - \frac{y \tan \alpha}{b} = 1,$$

and the equations of the corresponding normals are

$$xa^2/x_1 + yb^2/y_1 = a^2 + b^2,$$

or
$$xa/\sec \alpha + yb/\tan \alpha = a^2 + b^2.$$

Proofs of these are left as an exercise for the reader.

The geometrical properties of the ellipse (§4) have analogues for the hyperbola, though in places equal angles are replaced by complementary angles and some results are different if two points are taken on different branches of the curve. The proof of property (11), p. 187, about the director circle, is not valid for a hyperbola; the director circle is referred to later (p. 199–200).

**6. Asymptotes of the hyperbola.** Consider the point $P(a \sec\theta, b \tan\theta)$, and suppose that $0 < \theta < \frac{1}{2}\pi$. Then both $x$ and $y$ are positive, and $P$ lies in the first quadrant. As $\theta \to \frac{1}{2}\pi$, $a \sec\theta \to \infty$, so that $P \to \infty$. Similarly, as $\theta \to \frac{3}{2}\pi$, $P \to \infty$.

The asymptote, if it exists, is the limiting form of the tangent as $\theta \to \frac{1}{2}\pi$. As $\theta \to \frac{1}{2}\pi$, $\sin\theta \to 1$ and $\cos\theta \to 0$, so that the tangent

$$\frac{x \sec\theta}{a} - \frac{y \tan\theta}{b} = 1,$$

or

$$\frac{x}{a} - \frac{y \sin\theta}{b} = \cos\theta$$

becomes $$x/a - y/b = 0 \quad \text{as} \quad \theta \to \tfrac{1}{2}\pi.$$

Similarly, as $\theta \to \frac{3}{2}\pi$, $\sin\theta \to -1$ and $\cos\theta \to 0$, and we have as the asymptote

$$x/a + y/b = 0.$$

These lines $$\mathbf{y = \pm bx/a}$$

are the asymptotes envisaged in the less rigorous discussion of §5.

Another hyperbola,

$$x^2/a^2 - y^2/b^2 = -1$$

which has the same asymptotes, is called the *conjugate* hyperbola.

***Illustration 3.*** *Through the point $P(x_0, y_0)$ on the hyperbola $x^2/a^2 - y^2/b^2 = 1$ are drawn two lines $l_1$ and $l_2$, the first parallel to the asymptote $bx = +ay$ and the second parallel to the asymptote $bx = -ay$. The line $l_1$ meets the asymptote $bx = -ay$ in the point $S$, and $l_2$ meets the asymptote $bx = +ay$ in the point $R$. Prove that the product of the distances $PR$ and $PS$ remains constant as $P$ moves on the hyperbola.*

This example shows the advantage of proceeding circumspectly when solving problems; the most obvious method is often by no means the simplest. In this case the unwary would probably find the equations of $l_1$ and $l_2$, then the coordinates of $S$ and of $R$, and then the distances $PR$ and $PS$; this procedure would lead to pages of algebra. It is more direct to take the coordinates of $R$ as $(ar, br)$ and $S$ as $(as, -bs)$. (In this way we ensure that $R$ and $S$ lie on the appropriate asymptotes.) From the geometry of Fig. 71 we see that $PR = OS$ and $PS = OR$, so that

$$PR.PS = OS.OR = r\sqrt{(a^2+b^2)}.s\sqrt{(a^2+b^2)}$$
$$= rs(a^2+b^2).$$

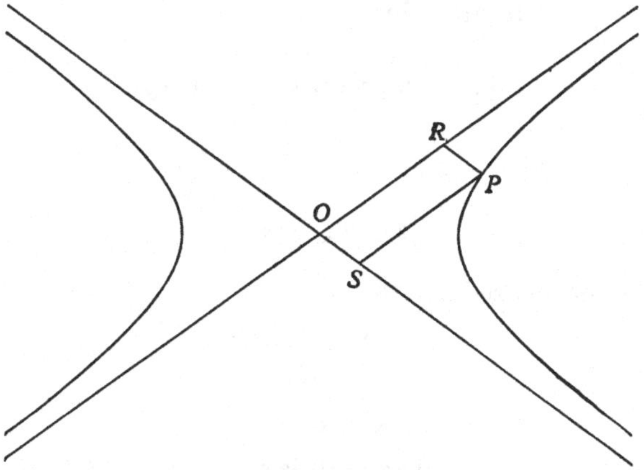


Fig. 71

Now the gradient of $PR$ is

$$(y_0 - br)/(x_0 - ar)$$

and this is $-b/a$, so

$$ay_0 - abr + bx_0 - abr = 0$$

and

$$2abr = ay_0 + bx_0.$$

Similarly (changing $b$ into $-b$, to deal with the other asymptote),

$$-2abs = ay_0 - bx_0,$$

and so

$$4a^2b^2rs = b^2x_0^2 - a^2y_0^2$$
$$= a^2b^2.$$

Thus $rs$ is a constant, and so $PR.PS$ is also a constant, for all positions of $P$ on the hyperbola.

**7. The rectangular hyperbola.** If the angle between the asymptotes is $2\phi$, $\tan\phi = b/a$. This can take all values (for $b$ is not, in the case of a hyperbola, restricted); in particular, when $b = a$, $\phi = \frac{1}{4}\pi$, and the asymptotes are perpendicular. We thus have the *rectangular hyperbola*

$$\mathbf{x^2 - y^2 = a^2.}$$

In order to reduce this equation to the more usual form considered in Chapter IV A, we rotate the axes, using the formula of Chapter II, p. 54. If $x^*$, $y^*$ are the new coordinates of a point, referred to axes rotated through an angle $\alpha = -\frac{1}{4}\pi$,

$$x^* = x\cos\alpha + y\sin\alpha = \frac{1}{\sqrt{2}}(x-y),$$

$$y^* = -x\sin\alpha + y\cos\alpha = \frac{1}{\sqrt{2}}(x+y).$$

Thus $$x^2 - y^2 = 2x^*y^*,$$

and the equation reduces to

$$\mathbf{x^*y^* = \tfrac{1}{2}a^2.}$$

## EXAMPLES VI C

1. Find the equation of the hyperbola whose foci are $(\pm 4, 0)$ and whose eccentricity is 2. What are the equations of the tangent and normal at $(4, 6)$?

2. A point $P$ moves so that the difference between its distances from the points $(1, 1)$ and $(-1, -1)$ is 2. Find the equation of the locus of $P$.

3. The asymptotes of a hyperbola are inclined at an angle of 60°. What is its eccentricity?

4. Find the coordinates of the centre and foci, and the eccentricity, of the hyperbola $16x^2 - 9y^2 = 225$.

5. Prove that the points where the asymptotes of a hyperbola meet the directrices lie on the auxiliary circle.

6. Find the equations of the tangent and normal at the point $A(9, -3)$ on the hyperbola $x^2 - 3y^2 = 54$. If the normal meets the curve again at $B$, find the coordinates of the point of intersection of the tangents at $A$ and $B$.

7. The line $y = 2x - 4$ meets the hyperbola $2x^2 - 3y^2 = 6$ at points $P$ and $Q$. Find the coordinates of the point of intersection of the tangents at $P$ and $Q$. Find also the angle between the normals at $P$ and $Q$.

8. A point moves on a hyperbola so that the difference of its distances from the fixed points $(4, 0)$ and $(-4, 0)$ is constant and equal to 6. Find

the equation of the hyperbola. Prove that the line $y = mx + c$ will be a tangent to the hyperbola if $c^2 = 9m^2 - 7$, and hence, or otherwise, obtain the equation of the locus of the meet of perpendicular tangents.

9. Prove that the equation of the normal to the rectangular hyperbola $x^2 - y^2 = a^2$ at the point $P(x_1, y_1)$ is $x/x_1 + y/y_1 = 2$. The normal meets the curve again at the point $Q$ and the mid-point of $PQ$ is $M$. Prove that the lines joining $P$ and $M$ to the origin are perpendicular.

10. $A$ and $A'$ are the vertices of the rectangular hyperbola $x^2 - y^2 = a^2$ and $P$ is any other point on the curve. The chords $PA$, $PA'$ meet one of the asymptotes in the points $Q$ and $R$. Prove that the length of the segment $QR$ is constant for all positions of the point $P$.

11. The tangent at a point $P$ on the rectangular hyperbola $x^2 - y^2 = a^2$ forms with the axes a triangle of area $A$; the normal at $P$ forms with the asymptotes a triangle of area $B$. Prove that $A^2B = a^6$.

12. The tangent at the point $P(a \sec \phi, b \tan \phi)$ to the hyperbola $x^2/a^2 - y^2/b^2 = 1$ meets the asymptotes at $Q$ and $Q'$. $O$ is the origin of coordinates. Prove that (i) $P$ is the mid-point of $QQ'$; (ii) the area of the triangle $OQQ'$ is $ab$.

13. The tangent at the point $(a \sec \theta, b \tan \theta)$ of the hyperbola

$$x^2/a^2 - y^2/b^2 = 1$$

meets the asymptotes at $P$ and $Q$. Find the coordinates of $P$ and $Q$. Prove that the locus of the point which divides $PQ$ in the ratio $k:1$ is

$$x^2/a^2 - y^2/b^2 = 4k/(1+k)^2.$$

14. Prove that the point $P$ whose coordinates are $(a \sec \theta, a \tan \theta)$ lies on the rectangular hyperbola $x^2 - y^2 = a^2$. If $Q$ is the point whose parameter is $\theta + \frac{1}{2}\pi$ and $R(x_1, y_1)$ is the mid-point of $PQ$, prove that

$$y_1/x_1 = \sin \theta + \cos \theta,$$

and find the locus of $R$.

15. Find the equation of the chord joining the points $(a \sec \theta, a \tan \theta)$ and $(a \sec \phi, a \tan \phi)$ on the hyperbola $x^2 - y^2 = a^2$. $P$ and $Q$ are points on the hyperbola whose parameters are $\alpha + \beta$ and $\alpha - \beta$ respectively, and $A$ and $A'$ are the points $(a, 0)$ and $(-a, 0)$. If $\beta$ varies, prove that the locus of the intersection of $AP$ and $A'Q$ is the circle

$$x^2 + y^2 - 2ay \tan \alpha = a^2.$$

16. Prove that the equation of the tangent to the hyperbola

$$x^2/a^2 - y^2/b^2 = 1 \quad \text{at the point} \quad \{\tfrac{1}{2}a(t+t^{-1}),\ \tfrac{1}{2}b(t-t^{-1})\}$$

is

$$\frac{x}{a}(t^2+1) - \frac{y}{b}(t^2-1) = 2t.$$

A variable tangent to the above hyperbola cuts the asymptotes in the points $L$ and $M$. Prove that the locus of the centre of the circle $OLM$, where $O$ is the origin of coordinates, is given by the equation

$$4(a^2x^2-b^2y^2) = (a^2+b^2)^2.$$

**8. Tangents to the general central conic.** It is often convenient to treat the ellipse and the hyperbola, the two conics which have a centre and so are called *central conics*, together. To do so, we write the equation in the form

$$\frac{x^2}{\alpha}+\frac{y^2}{\beta} = 1,$$

where $\alpha = a^2$, and $\beta = a^2(1-e^2)$, so that $\beta = +b^2$ when the curve is an ellipse, and $\beta = -b^2$ when the curve is a hyperbola. The results for either conic can at any stage be deduced by substituting these values for $\alpha$ and $\beta$.

We begin by recollecting the fundamental property, that the *tangent* at $(x_1, y_1)$ is

$$\mathbf{\frac{xx_1}{\alpha}+\frac{yy_1}{\beta} = 1.}$$

To find the condition for a line to touch the conic, we express the fact that the general line

$$lx+my+n = 0$$

should have the form of the tangent. Comparing coefficients,

$$\frac{x_1/\alpha}{l} = \frac{y_1/\beta}{m} = \frac{-1}{n},$$

so that
$$x_1 = \frac{-\alpha l}{n}, \quad y_1 = \frac{-\beta m}{n}.$$

Now $(x_1, y_1)$ lies on the tangent, so that

$$l.\frac{-\alpha l}{n}+m.\frac{-\beta m}{n}+n = 0$$

and the condition reduces to

$$\mathbf{\alpha l^2+\beta m^2 = n^2.}$$

If the equation of the tangent is expressed in the gradient form, $y = mx+c$, the condition becomes

$$\alpha m^2+\beta = c^2,$$

so that the line $$y = mx \pm \sqrt{(\alpha m^2 + \beta)}$$

is always a tangent to the conic.

We deduce two results.

(i) *There are two tangents to the conic through a general point outside the conic.*

For if $(h, k)$ is a general point, a tangent with gradient $m$ passes through it if

$$k = mh \pm \sqrt{(\alpha m^2 + \beta)},$$

or $$(k - mh)^2 = (\alpha m^2 + \beta).$$

This is a quadratic equation in $m$,

$$m^2(h^2 - \alpha) - 2mhk + (k^2 - \beta) = 0,$$

so there will be two roots which are real provided

$$(hk)^2 - (h^2 - \alpha)(k^2 - \beta) \geqslant 0,$$

which, if $\beta > 0$, reduces to $h^2/\alpha + k^2/\beta \geqslant 1$. This is the condition for $(h, k)$ to be outside the curve. Similar reasoning applies, with a change of sign, to the hyperbola.

(ii) *Points of intersection of perpendicular tangents lie on a circle*

Suppose $lx + my + n = 0$ is a tangent to the conic, then

$$\alpha l^2 + \beta m^2 = n^2,$$

so the ratio $l/m$ is a root of the equation

$$(\alpha l^2 + \beta m^2) = (lx + my)^2,$$

or $$l^2(\alpha - x^2) - 2lmxy + m^2(\beta - y^2) = 0.$$

If the two tangents through $(x, y)$ are perpendicular, the product of the roots of this equation will be $-1$, so that

$$\frac{\beta - y^2}{\alpha - x^2} = -1,$$

or $$x^2 + y^2 = \alpha + \beta.$$

This is the *director circle* whose equation, for an ellipse, was found earlier.

The same result may be obtained alternatively, by writing the equation of a tangent with gradient $m$ as

$$y - mx = \sqrt{(\alpha m^2 + \beta)}.$$

The perpendicular tangent has gradient $-1/m$, so that its equation is

$$y+\frac{1}{m}x = \sqrt{\left(\frac{\alpha}{m^2}+\beta\right)},$$

or

$$my+x = \sqrt{(\alpha+\beta m^2)}.$$

Squaring and adding these two equations, we find that the point of intersection of the tangents satisfies

$$(1+m^2)(x^2+y^2) = (1+m^2)(\alpha+\beta),$$

and since $(1+m^2) \neq 0$, the same equation of the director circle is obtained.

For the ellipse, the equation becomes

$$\mathbf{x^2+y^2 = a^2+b^2}.$$

For the hyperbola, it is

$$\mathbf{x^2+y^2 = a^2-b^2},$$

and here a further distinction arises. When $a^2 > b^2$ (that is, when $1 < e < \sqrt{2}$) the director circle exists, concentric with the auxiliary circle and inside it. When the hyperbola is rectangular, and $a^2 = b^2$, the director circle reduces to a point $(0, 0)$; the only perpendicular tangents from this point are the perpendicular asymptotes. Lastly, when $a^2 < b^2$ there are no points on the locus, and no tangents to such a hyperbola can be perpendicular.

**9. Chord of contact.** Suppose the points of contact of tangents drawn to the conic from $(h, k)$ are $(x_1, y_1)$ and $(x_2, y_2)$. The tangent at $(x_1, y_1)$ then contains $(h, k)$, so that

$$\frac{hx_1}{\alpha}+\frac{ky_1}{\beta} = 1,$$

and similarly, since the tangent at $(x_2, y_2)$ contains $(h, k)$,

$$\frac{hx_2}{\alpha}+\frac{ky_2}{\beta} = 1.$$

Looked at another way, these two equations express the fact that each of the points $(x_1, y_1)$ and $(x_2, y_2)$ lies on the line

$$\frac{hx}{\alpha}+\frac{ky}{\beta} = 1,$$

which is therefore the equation of the *chord of contact*, or *polar line*, of $(h, k)$.

This method can evidently be used for the finding of the equation of any chord of contact, when we know the equation of the tangent; a similar method was used for the circle (p. 138). The reader should master the technique by applying the same argument to find the chord of contact in the case of a parabola, or a rectangular hyperbola.

As in each previous case, we notice that the equation is formed in accordance with the general principle formulated on p. 138.

From the equation we can deduce the coordinates of the point of intersection of two tangents. Consider, for instance, the ellipse $x^2/a^2+y^2/b^2 = 1$. The polar line of $(h, k)$ is

$$\frac{hx}{a^2}+\frac{ky}{b^2} = 1.$$

If this is the chord joining the points $(a\cos\theta, b\sin\theta)$ and $(a\cos\phi, b\sin\phi)$, the equation is

$$\frac{x}{a}\cos\tfrac{1}{2}(\theta+\phi)+\frac{y}{b}\sin\tfrac{1}{2}(\theta+\phi) = \cos\tfrac{1}{2}(\theta-\phi),$$

so that
$$\frac{h/a}{\cos\frac{1}{2}(\theta+\phi)} = \frac{k/b}{\sin\frac{1}{2}(\theta+\phi)} = \frac{1}{\cos\frac{1}{2}(\theta-\phi)}$$

and so
$$h = \frac{a\cos\frac{1}{2}(\theta+\phi)}{\cos\frac{1}{2}(\theta-\phi)},$$

$$k = \frac{b\sin\frac{1}{2}(\theta+\phi)}{\cos\frac{1}{2}(\theta-\phi)}.$$

These are thus the coordinates of the point of intersection of the two tangents; they could also be found, with more trouble, by solving the two equations of the tangents simultaneously.

Rather than memorizing the result, it is better to notice the *method* used here, which can be applied to the ellipse and the hyperbola, and indeed to any other conic, in any of their parametric forms.

## EXAMPLES VI D

1. Write down the equation which gives the $x$-coordinates of the points where the line $y = mx+c$ meets the conic $x^2/\alpha+y^2/\beta = 1$. Hence show that the line touches the conic if and only if $c^2 = \alpha m^2+\beta$.

2. Prove that the line $x+y = 5$ touches the ellipse $9x^2+16y^2 = 144$, and find the coordinates of the point of contact.

3. Write down the equation of the polar line of $P$ with respect to the conic $S$, in each of the following cases:

(i) $S$ is $3x^2+4y^2-12=0$, $P$ is $(2, -1)$;

(ii) $S$ is $2x^2-3y^2-24=0$, $P$ is $(-6, 4)$;

(iii) $S$ is $x^2/a^2+y^2/b^2-1=0$, $P$ is $(a, b)$.

In which cases (if any) does $P$ lie on $S$?

4. Write down the equations of the director circles of the conics given in the last example.

5. Use one of the methods of § 8 to find the equation of the director circle of the ellipse $9(x-5)^2+16(y+3)^2=144$.

6. Show that the line $lx+my+n=0$ is a tangent to the ellipse $x^2/a^2+y^2/b^2=1$, if $a^2l^2+b^2m^2=n^2$, and find the coordinates of the point of contact when this condition is satisfied. Find the points on the ellipse $4x^2+9y^2=1$ at which the tangents are parallel to the line $8x=9y$.

7. Prove that the line $y=mx+c$ touches the ellipse $x^2/a^2+y^2/b^2=1$ if $c^2=a^2m^2+b^2$, and the distance between the tangents so obtained is $2\sqrt{\{(a^2m^2+b^2)/(1+m^2)\}}$. If this is equal to the distance between the pair of tangents perpendicular to the first pair, show that this distance is $\sqrt{\{2(a^2+b^2)\}}$. An ellipse in which the semi-axes $a$, $b$ are in the ratio $2:1$ touches the four sides of a square. Prove that $a$ is $\sqrt{(2/5)}$ times the length of a side of the square.

8. The coordinates of two points $P$ and $Q$ on the ellipse $x^2/a^2+y^2/b^2=1$ are $\{a\cos(\alpha-\beta), b\sin(\alpha-\beta)\}$ and $\{a\cos(\alpha+\beta), b\sin(\alpha+\beta)\}$ respectively. $S$ and $S'$ are the foci $(ae, 0)$ and $(-ae, 0)$ respectively. If $SP$ is parallel to $S'Q$, prove that $\sin\beta=e\sin\alpha$. Prove that the tangents at $P$ and $Q$ meet at the point $T$ $(a\cos\alpha\sec\beta, b\sin\alpha\sec\beta)$, and that, if $SP$ is parallel to $S'Q$, $T$ lies on the circle $x^2+y^2=a^2$.

9. Find the equation of the normal to the ellipse $x^2/a^2+y^2/b^2=1$ at the point $P(a\cos\theta, b\sin\theta)$. This normal meets the ellipse again at $Q$ and the tangents at $P$ and $Q$ meet at $R(h, k)$. By comparing the two forms of the equation of $PQ$, in terms of $\theta$ and in terms of $h$, $k$, or otherwise, prove that the locus of $R$ when $P$ varies on the ellipse, is the curve

$$x^2y^2(a^2-b^2)^2=a^6y^2+b^6x^2.$$

10. The eccentric angles of two points $P$, $Q$ on the ellipse $x^2/a^2+y^2/b^2=1$ are $\theta$ and $\theta+\frac{1}{2}\pi$, and $\alpha$ is one of the angles between the tangents at $P$ and $Q$. If $e$ is the eccentricity of the ellipse, prove that $e^2\sin 2\theta\tan\alpha=2\sqrt{(1-e^2)}$. If the tangents at $P$ and $Q$ intersect in $R$, prove that the locus of $R$ as $\theta$ varies is an ellipse.

11. Prove that, if $p^2=a^2\cos^2\alpha+b^2\sin^2\alpha$, the line $x\cos\alpha+y\sin\alpha=p$ is a tangent to the ellipse $x^2/a^2+y^2/b^2=1$. $H$ is the point $(0, ae)$ and $K$ is the point $(0, -ae)$ where $e$ is the eccentricity of the ellipse. Prove that the sum of the squares of the perpendiculars from these points on to any tangent is equal to $2a^2$.

12. Prove that, if $PSQ$ and $PS'R$ are focal chords of an ellipse whose foci are $S$ and $S'$, the tangents to the ellipse at $Q$ and $R$ meet on the normal at $P$.

13. A point $P$ moves so that the chord of contact of the tangents from $P$ to the ellipse $b^2x^2+a^2y^2 = a^2b^2$ touches the ellipse $4(b^2x^2+a^2y^2) = a^2b^2$. Find the locus of $P$.

14. A point $P$ of an ellipse $x^2/a^2+y^2/b^2 = 1$ is joined to the points $(\pm c, 0)$, and the joins meet the ellipse again in $Q$ and $R$. Show that the point of intersection of the tangents to the ellipse at $Q$ and $R$ lies on the ellipse

$$\frac{x^2}{a^2}+\frac{(a^2-c^2)^2y^2}{(a^2+c^2)^2b^2} = 1.$$

15. Find the coordinates of the point of intersection of the tangents to the ellipse $x^2/a^2+y^2/b^2 = 1$ at the points

$$(a\cos\theta,\, b\sin\theta) \quad\text{and}\quad (a\cos\phi,\, b\sin\phi).$$

The tangents at a variable pair of points $P$, $Q$ on the ellipse intersect on one of the directrices in the point $T$. If $PT$ meets the major axis in $R$, show that the locus of the point of intersection of $QT$ with the other tangent from $R$ to the ellipse is a straight line parallel to the directrix and through the corresponding focus.

16. Prove that the tangents at the points with parameters $\theta, \phi$, on the ellipse $x^2/a^2+y^2/b^2 = 1$ meet at the point given by

$$x = \frac{a(1-t_1t_2)}{1+t_1t_2},\quad y = \frac{b(t_1+t_2)}{1+t_1t_2},$$

where $t_1 = \tan\frac{1}{2}\theta$, $t_2 = \tan\frac{1}{2}\phi$. A tangent from a point $P$ to the ellipse meets the tangent at $A$, the end $(a, 0)$ of the major axis, in $Q$ and the other tangent from $P$ meets the tangent at $B$, the end $(0, b)$ of the minor axis, in $R$. Find the equation of the locus of $P$ if $AQ/BR$ is equal to $b/a$.

17. Write down the equation of the polar of the point $P(x', y')$ with respect to the hyperbola $x^2/a^2-y^2/b^2 = 1$. Prove that, if $P$ moves on the fixed straight line $px+qy = r$, its polar passes through a fixed point $Q$, and find the coordinates of $Q$. Prove also that, if the fixed straight line is parallel to either asymptote of the hyperbola, $Q$ lies on that asymptote.

18. Prove that the equation of any tangent to the hyperbola

$$x^2/a^2-y^2/b^2 = 1$$

may be written in the form $y = mx\pm\sqrt{(a^2m^2-b^2)}$. Find the equation of the locus of the foot of the perpendicular from the origin to a variable tangent to the hyperbola.

19. The general point of the rectangular hyperbola $x^2-y^2 = a^2$ is $(a\sec\theta, a\tan\theta)$. Find the equation of locus of the point of intersection of tangents to the hyperbola at points with parameters $\theta$, $\theta+\alpha$, as $\theta$ varies. What is the locus when $\alpha = \pi$?

20. The polar lines of a point $P$, with respect to a hyperbola and with respect to its auxiliary circle, are perpendicular. Show that $P$ lies on one of the asymptotes of the hyperbola.

**10.* Conjugate diameters of a central conic.** We now consider a locus problem of fundamental importance; we shall treat it in such a way that several other problems are solved at the same time. All these are concerned with the mid-point of a chord.

We first obtain a connection between the mid-point $M(h, k)$ and the gradient $\tan\theta$ of the chord. The general point $P$ of this chord through $M$ is

$$(h+r\cos\theta,\ k+r\sin\theta)$$

(see p. 32), where $r$ is the distance $PM$, the sign of $r$ indicating on which side of $M$ the point $P$ lies. $P$ will lie on the conic

$$x^2/\alpha+y^2/\beta = 1$$

if and only if

$$\frac{(h+r\cos\theta)^2}{\alpha}+\frac{(k+r\sin\theta)^2}{\beta} = 1.$$

The two roots of this equation in $r$ give the two positions of $P$ on the conic. Now $M$ is the mid-point of the chord, so these extremities of the chord are equidistant from $M$, and they are therefore given by values of $r$ which are equal and opposite. The condition for this [A 4] is that the sum of the roots of the quadratic in $r$ is zero; that is, the coefficient of $r$ vanishes. Thus

$$\frac{2h\cos\theta}{\alpha}+\frac{2k\sin\theta}{\beta} = 0,$$

or

$$\tan\theta\,.\,k/h = -\beta/\alpha.$$

We make use of this result to prove several properties.

(i) *To find the equation of the chord of the conic* $x^2/\alpha+y^2/\beta = 1$ *whose mid-point is* $(h, k)$.

In this case the mid-point $(h, k)$ is given, and we have to find the gradient $m$ and so the equation. We have found that

$$m = \tan\theta = -\beta h/\alpha k,$$

and so the equation of the chord is

$$y-k = -\frac{\beta h}{\alpha k}(x-h),$$

or
$$\frac{h(x-h)}{\alpha}+\frac{k(y-k)}{\beta}=0.$$

(ii) *To find the locus of the mid-points of chords parallel to a fixed diameter, $y = mx$.*

In this case the gradient $\tan\theta = m$ is given, and we find the locus of $(h, k)$. The fundamental relation is

$$m \,.\, k/h = -\beta/\alpha$$

and so the locus of $(h, k)$ is

$$y = -(\beta/\alpha m)\, x.$$

This we see is another diameter of the conic, as it goes through the centre. The locus of mid-points of chords parallel to a diameter $y = mx$ is therefore a diameter $y = m'x$ where

$$m' = -\beta/\alpha m,$$

or
$$mm' = -\beta/\alpha.$$

From the symmetry of this result we deduce that each of the two diameters $y = mx$, $y = m'x$ arises, by a similar construction, from the other; they are called *conjugate diameters*. The property they show is that if a line parallel to either meets the curve in $P$ and $P'$, the mid-point of $PP'$ will lie on the other.

(iii) *The tangents drawn from a point on a diameter to the conic meet the conic in points on a chord parallel to the conjugate diameter.*

Suppose $(h, k)$ is a point on $y = mx$, so that $m = k/h$. The chord of contact of tangents from $(h, k)$ to the conic is

$$hx/\alpha + ky/\beta = 1 \qquad (\S 9)$$

and the gradient of this chord is

$$-\beta h/\alpha k = -\beta/\alpha m;$$

this is the gradient of the diameter conjugate to $y = mx$.

(iv) For the ellipse, we can find a particularly simple relation between extremities of conjugate diameters, in terms of the eccentric angles of the points. When $\alpha = a^2$, $\beta = b^2$, the relation for conjugate diameters is

$$mm' = -b^2/a^2.$$

If the ends of these diameters have eccentric angles $\phi, \phi'$, then

$$m = \frac{b \sin \phi}{a \cos \phi} = \frac{b}{a} \tan \phi,$$

and similarly $$m' = (b/a) \tan \phi'.$$

The condition for these diameters to be conjugate is

$$(b/a) \tan \phi . (b/a) \tan \phi' = -(b^2/a^2),$$

or $$\tan \phi . \tan \phi' + 1 = 0.$$

Thus $$\phi' = \phi \pm \tfrac{1}{2}\pi,$$

and we have the result that *the eccentric angles of conjugate diameters of an ellipse differ by a right angle.*

(v) A fundamental difference between the ellipse and the hyperbola is that for the hyperbola only one of two conjugate diameters actually meets the curve. For suppose, if possible, that ends of conjugate diameters of a hyperbola are $(a \sec \phi,\ b \tan \phi)$ and $(a \sec \phi',\ b \tan \phi')$. The gradients of these diameters are $(b/a) \sin \phi$ and $(b/a) \sin \phi'$;

but $$mm' = b^2/a^2$$

and so $$\sin \phi . \sin \phi' = 1.$$

Now both $\sin \phi$ and $\sin \phi'$ are less than or equal to 1, and since for a general value of $\phi$, $\sin \phi < 1$, we should have $\sin \phi' > 1$, which is impossible. The reader may consider what happens when $\sin \phi = 1$.

(vi) Finally, we prove a fundamental property of the hyperbola. *If a chord PQ of the hyperbola meets the asymptotes in $A_1$ and $A_2$, the mid-point of PQ is the same as the mid-point of $A_1A_2$.*

The fundamental property, used once again, gives that, for the hyperbola, the gradient $\tan \theta$ and the mid-point $(h, k)$ of a chord satisfy

$$\tan \theta . \frac{k}{h} = \frac{b^2}{a^2},$$

or $$a^2 k \sin \theta = b^2 h \cos \theta.$$

The general point on the chord is

$$(h + r \cos \theta,\ k + r \sin \theta)$$

and this is $A_1$, on the asymptote $y = bx/a$, if $r = r_1$, say, where

$$k + r_1 \sin\theta = b(h + r_1 \cos\theta)/a;$$

thus
$$r_1 = -\frac{ak - bh}{a\sin\theta - b\cos\theta}.$$

Similarly, the point of intersection $A_2$ of the chord with the other asymptote $y = -bx/a$ occurs when $r = r_2$, and the expression for $r_2$ will be the same as that of $r_1$, except that the sign of $b$ is changed. Therefore

$$r_2 = -\frac{ak + bh}{a\sin\theta + b\cos\theta}.$$

Now
$$\begin{aligned} r_1 + r_2 &= -\frac{ak - bh}{a\sin\theta - b\cos\theta} - \frac{ak + bh}{a\sin\theta + b\cos\theta} \\ &= -\frac{2a^2k\sin\theta - 2b^2h\cos\theta}{a^2\sin^2\theta - b^2\cos^2\theta} \\ &= 0, \end{aligned}$$

so that $A_1$ and $A_2$ are equidistant from the mid-point $(h, k)$ of the chord. This proves the result.

As a corollary, we notice the fact that was proved in the special case of a rectangular hyperbola earlier; the portion of a tangent intercepted between the asymptotes of a hyperbola is bisected at the point of contact.

## EXAMPLES VI E

1. Prove that the mid-point of the chord joining points with eccentric angles $\theta$, $\phi$ on the ellipse $x^2/a^2 + y^2/b^2 = 1$ is $(h, k)$ where

$$h = a\cos\tfrac{1}{2}(\theta+\phi)\cos\tfrac{1}{2}(\theta-\phi), \quad k = b\sin\tfrac{1}{2}(\theta+\phi)\cos\tfrac{1}{2}(\theta-\phi).$$

Hence find the locus of the mid-points of chords with gradient $m$.

2. Show that, if $P_1(x_1, y_1)$ and $P_2(x_2, y_2)$ lie on a central conic

$$x^2/\alpha + y^2/\beta = 1 \quad \text{then} \quad \beta(x_1^2 - x_2^2) = -\alpha(y_1^2 - y_2^2).$$

By factorizing each side of this equation, find the locus of the mid-point of $P_1P_2$ if the gradient of $P_1P_2$ has a constant value $m$.

3. A system of chords of an ellipse all pass through a fixed point. Prove that their mid-points all lie on another ellipse.

4. Prove that the sum of the squares on two conjugate semi-diameters of an ellipse is a constant, $a^2 + b^2$, where $2a$, $2b$ are the lengths of the principal axes.

5. Show that the tangents at the ends of a diameter of a central conic are parallel to the conjugate diameter.

6. Prove that a parallelogram which touches an ellipse at ends of conjugate diameters has constant area.

7. If $P$, $Q$ are ends of conjugate diameters of an ellipse which has foci $S$, $S'$ and centre $O$, prove that $SP.S'P = OQ^2$.

8. Prove that the line joining the centre of an ellipse to the point of intersection of tangents at $P$ and $Q$ bisects the chord $PQ$. If a parallelogram $ABCD$ is inscribed in an ellipse, $AB$ being parallel to $CD$ and $BC$ parallel to $DA$, deduce that the tangents to the ellipse at $A$ and $B$ intersect on the line through the centre of the parallelogram parallel to $BC$.

9. Find the equation of the chord of the ellipse $3x^2+4y^2 = 28$ whose mid-point is the point $(1, 1)$. Find also the length of this chord.

10. Prove that the locus of the point of intersection of the tangents at the ends of conjugate diameters of the ellipse $x^2/a^2+y^2/b^2 = 1$ is the ellipse $x^2/a^2+y^2/b^2 = 2$.

11. Lines are drawn through the origin perpendicular to the tangents from a point $P$ to the ellipse $x^2/a^2+y^2/b^2 = 1$. If the lines are conjugate diameters of the ellipse, prove that $P$ lies on the curve $a^2x^2+b^2y^2 = a^4+b^4$.

12. The centre of the ellipse $x^2/a^2+y^2/b^2 = 1$ is $C$ and $CP$, $CQ$ are a pair of conjugate semi-diameters. If the eccentric angle at $P$ is $\theta$ and the chord $PQ$ passes through the point $(a\sqrt{2}, 0)$, prove that $\theta = \pm\frac{1}{12}\pi,\ \pm\frac{7}{12}\pi$.

13. $CP$ and $CD$ are conjugate semi-diameters of an ellipse whose foci are $S$ and $S'$. The tangents to the ellipse at $P$ and $D$ meet in $T$. Prove that $SP, S'P, SD, S'D$ touch a circle whose centre is $T$, and find the radius of the circle.

14. Find the equation of the chord of the hyperbola $x^2/a^2-y^2/b^2 = 1$ whose mid-point is the general point $(\alpha, \beta)$, and interpret your result in the special case when the point $(\alpha, \beta)$ lies on the hyperbola. A variable chord of the hyperbola is a tangent to the circle $x^2+y^2 = c^2$. Prove that the locus of the mid-point of the chord is the curve

$$(x^2/a^2-y^2/b^2)^2 = c^2(x^2/a^4+y^2/b^4).$$

15. Find the equation of the chord of the ellipse $b^2x^2+a^2y^2 = a^2b^2$ whose mid-point is $(\alpha, \beta)$. Deduce (i) the locus of the mid-points of chords of the ellipse parallel to the given line $lx+my = 0$; (ii) the locus of the mid-points of chords of the ellipse which touch the circle $x^2+y^2 = b^2$.

## MISCELLANEOUS EXAMPLES VI

1. $P$ and $Q$ are variable points of the ellipse $x^2/a^2+y^2/b^2 = 1$ such that the lines joining them to its centre are perpendicular, and the tangents at $P$ and $Q$ meet in $T$. Find the locus of $T$.

2. The normal at a point $P$ on the parabola $y^2 = 4ax$ meets the ellipse $2x^2+y^2 = c^2$ in the points $M, N$. Prove that $P$ is the mid-point of $MN$. Hence, or otherwise, prove that the curves cut at right angles.

3. Prove that the points $(\pm 3, 0)$ are the foci of the ellipse

$$x^2/25+y^2/16 = 1.$$

Find the length of the semi-major axis and the eccentricity of the hyperbola which has the same two foci and meets the ellipse at the point $(3, 3\frac{1}{5})$. Obtain the equations of the tangents to the ellipse and the hyperbola at the point $(3, 3\frac{1}{5})$ showing that they are at right angles.

4. The points $A$ and $B$ are $(1, 0)$ and $(-1, 0)$ respectively. Find equations for the loci of points $P$ and $Q$ which are such that

$$AP+BP = 4, \quad AQ-BQ = \pm 1.$$

Find the points of intersection of these loci, and show that they intersect orthogonally.

5. A circle is drawn on the semi-major axis of the ellipse $x^2/a^2+y^2/b^2 = 1$ as diameter, and the circle and the ellipse meet in three points. Show that the centroid of the triangle formed by these points is a distance

$$a(a^2+b^2)/3(a^2-b^2)$$

from the origin.

6. The point $(0, -b)$ is joined to the focus $(ae, 0)$ of the ellipse

$$x^2/a^2+y^2/b^2 = 1$$

to meet the ellipse in $P$. Prove that the gradient of the tangent at $P$ is $-2e/\sqrt{(1-e^2)}$.

7. By writing the equation $4x^2-8x = y^2$ in the form

$$(x-p)^2/a^2-y^2/b^2 = 1$$

show that it represents a hyperbola. Find the length of its transverse axis and the equations of its asymptotes.

8. The normal at a variable point on a hyperbola cuts the principal axes in $P, Q$. Prove that the locus of the mid-points of $PQ$ is a hyperbola. Can it coincide with the original hyperbola?

9. The product of the perpendiculars from the foci of the hyperbola $x^2/(a^2-2b^2)-y^2/b^2 = 1$ to the normal at a point $P$ on the hyperbola is equal to $b^2$. Find the coordinates of $P$, and hence show that $P$ also lies on the ellipse $x^2/a^2+y^2/b^2 = 1$.

10. The length of the minor axis $BB'$ of an ellipse is $2b$. $O$ is the centre of the ellipse and $e$ is its eccentricity. $C$ and $D$ are the two points on $BB'$ such that $OC = OD = eb$. Any pair of parallel lines through $C$ and $D$ meet the circle on $BB'$ as diameter at $P$ and $Q$ on the same side of $BB'$. Show that the tangents to this circle at $P$ and $Q$ intersect on the ellipse.

11. $A$ is one end of the major axis of an ellipse with semi-axes $a$ and $b$. Prove that chords of the ellipse that subtend a right angle at $A$ pass through a fixed point on the major axis, dividing it in the ratio $b^2/a^2$.

12. Prove that there are two points $P$ and $Q$ on the ellipse

$$x^2/a^2+y^2/b^2 = 1,$$

other than the point $(0, -b)$, the normals at which pass through the point $B(0, b)$, provided the eccentricity $e$ of the ellipse exceeds a certain value, and find this value. Prove that, when $P$ and $Q$ exist, $PB = QB = a/e$.

13. $P$, $Q$, $R$ are points on the ellipse $x^2/a^2+y^2/b^2 = 1$ with eccentric angles $\theta$, $\alpha$, $\beta$; $PQ$ goes through $(ae, 0)$ and $PR$ through $(-ae, 0)$. Prove that $\tan\frac{1}{2}\alpha/\tan\frac{1}{2}\beta = (1-e)^2/(1+e)^2$.

14. If the normal at any point $P$ on the hyperbola $x^2/a^2-y^2/b^2 = 1$ meets the $y$-axis at $Q$, prove that $(e^2-1)\,QS^2 = e^2.PS.PS'$, where $e$ is the eccentricity of the hyperbola, and $S$, $S'$ the foci.

15. Prove that the tangent at the point

$$\{a(1-t^2)/(1+t^2),\ 2bt/(1+t^2)\}$$

on the ellipse $x^2/a^2+y^2/b^2 = 1$ has the equation $x(1-t^2)/a+2ty/b = 1+t^2$. The tangents at the points $t = t_1$ and $t = t_2$ meet the line $x = a$ in the points $P$ and $Q$ such that $PQ = 2b$. Prove that the difference between $t_1$ and $t_2$ is 2. Prove also that the point of intersection of these tangents lies on the parabola $ay^2 = 2b^2(x+a)$.

16. Verify that the equation of the chord joining the points $(x_1, y_1)$ and $(x_2, y_2)$ on the ellipse $x^2/a^2+y^2/b^2 = 1$ is $x(x_1+x_2)/a^2+y(y_1+y_2)/b^2 = k$, and deduce the equation of the tangent at $(x_1, y_1)$. The circles with two parallel chords $x = x_1$, $x = x_2$ of the ellipse as diameters pass through the focus $(ae, 0)$; prove that $x_1+x_2 = 2ae/(2-e^2)$, where $e$ is the eccentricity. If one chord is $x = 0$, and the other is $PQ$, prove that the tangent at $P$ or $Q$ makes an intercept of length $3\sqrt{2}a/2$ on the $y$-axis.

17. Tangents are drawn to an ellipse of eccentricity $e$ from a point on a concentric circle. Prove that the chord of contact of the tangents touches a concentric coaxal ellipse of eccentricity $e\sqrt{(2-e^2)}$.

18. Find the equations of the real straight lines which touch the parabola $y^2 = 4ax$ and also the ellipse $\frac{1}{2}x^2+y^2 = \frac{1}{3}a^2$. Show that they intersect on the directrix of the parabola.

19. Prove that, in a hyperbola, that part of any tangent intercepted by the asymptotes is bisected at the point of contact. The tangent at a point $P$ of a rectangular hyperbola whose centre is $C$ meets the asymptotes in $H$ and $K$, and the normal at $P$ meets its real axis in $G$. Prove that $C$, $G$, $H$, $K$ lie on a circle whose centre is $P$.

20. $BCB'$ is the minor axis of an ellipse which has foci at $S$ and $H$. Prove that the circumcircle of the triangle $SBH$ touches the ellipse at $B$ and does not cut it again if the eccentricity is less than $1/\sqrt{2}$.

21. The tangent at $P$ to the ellipse $x^2/a^2+y^2/b^2 = 1$ meets the similar ellipse $x^2/a^2+y^2/b^2 = 2$ at the points $T$ and $T'$. Chords $TR$, $T'R'$ of the larger ellipse are drawn parallel to the chord $PQ$. Prove that $Q$ is the mid-point of $RR'$.

22. The normal at a point $P$ to a hyperbola meets the transverse axis in $G$; the line perpendicular to this axis through $P$ meets an asymptote in $Q$. Show that $GQ$ is perpendicular to the asymptote.

23. $P$ and $Q$ are the points $(x', y')$ and $(x', -y')$ on the hyperbola $x^2 - y^2 = a^2$. Show that the tangent at $P$ is perpendicular to the line joining $Q$ to the centre $C$ of the hyperbola. If the tangent at $P$ meets $CQ$ at $R$, prove that $CQ\,.\,CR = a^2$.

24. The tangent at any point $P$ on the hyperbola $x^2 - y^2 = a^2$ meets the asymptotes in $Q$ and $R$. $C$ is the centre of the hyperbola. Through $Q$ a line is drawn parallel to $CR$, through $R$ a line is drawn parallel to $CQ$. Prove that as $P$ varies on the hyperbola the locus of the intersection of these lines is a similar and similarly situated hyperbola whose dimensions are twice those of the given hyperbola.

25. A straight line cuts a hyperbola at $P$ and $Q$ and its asymptotes at $L$ and $M$. Prove that $LP = MQ$ whether the order of the letters on the line is $LPQM$ or $PLMQ$. Given the asymptotes of a hyperbola and one point on the curve, show how to construct a series of points on each branch.

---

26. The circle $x^2 + y^2 = c^2$ meets the ellipse $b^2x^2 + a^2y^2 = a^2b^2$ at four real points. Find the equations of the four common tangents, and prove that, if they are the sides of a square, then $2c^2 = a^2 + b^2$.

27. Prove that chords of an ellipse which are such that the tangents at their extremities are perpendicular touch an ellipse which has the same foci as the given ellipse.

28. Given that the line $x \cos\alpha + y \sin\alpha = p$ touches the ellipse

$$x^2/a^2 + y^2/b^2 = 1,$$

prove that $p^2 = a^2 \cos^2\alpha + b^2 \sin^2\alpha$. The ellipses

$$x^2/a^2 + y^2/b^2 = 1, \quad x^2/c^2 + y^2/d^2 = 1$$

have four real common tangents, and the four (finite) points of intersection of these tangents lie on a circle. Prove that $a^2 + b^2 = c^2 + d^2$ and that the radius of the circle is $\sqrt{(a^2 + b^2)}$.

29. Prove that the line $lx + my + n = 0$ touches the hyperbola

$$x^2/a^2 - y^2/b^2 = 1 \quad \text{if} \quad a^2l^2 - b^2m^2 = n^2.$$

The circle $x^2 + y^2 - 2\alpha y + \beta = 0$ cuts one asymptote of the hyperbola in the distinct points $P, P'$, and the other in $Q, Q'$, where $P$ and $Q$ are on the same side of the $y$-axis. If $PQ$ is a tangent to the hyperbola, prove that the circle passes through the foci of the hyperbola.

30. $P$ and $P'$ are any two points of an ellipse, and $A$ and $A'$ are the ends of its major axis. The chords $AP, A'P'$ meet in $M$, $AP', A'P$ meet in $N$, and $AA', PP'$ meet in $F$. Show that $FM, FN$ are perpendicular if and only if $F$ is a focus.

31. Given a central conic, $F$ is a point such that the tangents at the ends of any chord through $F$ meet on the line through $F$ perpendicular to the chord. Prove that $F$ is a focus.

32. The tangent to an ellipse at the point $P$ meets the directrices in $U$ and $V$. The normal at $P$ meets the minor axis in $M$, and $Q$ is the point on $PM$ produced such that $PM = MQ$. Prove that the lines $UQ$ and $VQ$ are tangents to the ellipse.

33. Prove that the point of contact of a tangent to a hyperbola bisects the part of the tangent which lies between the asymptotes. $P$ is the circumcentre of the triangle formed by a tangent and the two asymptotes of a given hyperbola. Prove that the locus of $P$ is a hyperbola which is similar to the hyperbola conjugate to the given one, but has its axis turned through 90°.

34. Prove that if a rectangle is inscribed in an ellipse, then the axes of the ellipse are parallel to the sides of the rectangle. If a square of side $2p$ is inscribed in an ellipse whose semi-axes are $a$, $b$ prove that

$$1/a^2 + 1/b^2 = 1/p^2.$$

35. What is the relation between the focal distances of a point on a central conic? Show that there is one ellipse and one hyperbola which have points $(\alpha, \beta)$ and $(-\alpha, -\beta)$ as foci and also pass through the point $(\alpha, -\beta)$. Prove that the points of intersection of the two conics lie on a circle.

36. The normal to the ellipse $b^2x^2 + a^2y^2 = a^2b^2$ at the point $(a\cos\theta, b\sin\theta)$ meets the ellipse again at the point $(a\cos\phi, b\sin\phi)$. Prove that $(a^2-b^2)\{\sin 2\theta - \sin(\theta+\phi)\} = (a^2+b^2)\sin(\theta-\phi)$. Find the range of values of the eccentricity $e$ for which two real normals can be drawn to the ellipse from an extremity of the minor axis in addition to the minor axis itself.

37. Two conjugate diameters $CP$, $CQ$ of the ellipse $x^2/a^2 + y^2/b^2 = 1$ meet the ellipse $x^2/a^2 + y^2/b^2 = 2x/a$ in $C$, the centre of the first ellipse, and also in the points $P'$ and $Q'$. Prove that $P'Q'$ passes through $C'$, the centre of the second ellipse. The diameter of the second ellipse, conjugate to $P'Q'$, meets the first ellipse in $C'$ and $R$. If the eccentric angles of $P$ and $R$ are $\theta$ and $\phi$, prove that $4\theta - \phi$ is zero or a multiple of $2\pi$.

38. Prove that the tangents at the points $\theta = \alpha$, $\beta$ on the ellipse $x = a\cos\theta$, $y = b\sin\theta$ meet at the point

$$\{a\cos\tfrac{1}{2}(\alpha+\beta)\sec\tfrac{1}{2}(\alpha-\beta),\ b\sin\tfrac{1}{2}(\alpha+\beta)\sec\tfrac{1}{2}(\alpha-\beta)\}.$$

Find the area of any parallelogram which is circumscribed to the ellipse and show that the area has a minimum value, $4ab$, when the sides of the parallelogram are parallel to a pair of conjugate diameters.

39. The points $A$ and $B$ are the vertices of a hyperbola. The straight lines $p$ and $q$ pass through $B$ and are parallel to the asymptotes. A straight line is drawn through $A$ to meet the hyperbola again in $R$ and to intersect the lines $p$ and $q$ at $P$ and $Q$ respectively. Prove that $R$ is the mid-point of $PQ$.

40. Prove that the line $lx+my+n = 0$ is a normal to the ellipse $x^2/a^2+y^2/b^2 = 1$ if, and only if, $n^2(a^2m^2+b^2l^2) = l^2m^2(a^2-b^2)^2$. If a variable chord of an ellipse is normal to the chord at one of its extremities, prove that the locus of its mid-point is given by

$$(b^2x^2+a^2y^2)^2(b^6x^2+a^6y^2) = a^4b^4(a^2-b^2)^2x^2y^2.$$

41. The polar lines of a point $P$ with respect to the two ellipses $x^2/a^2+y^2/b^2 = 1$ and $x^2/A^2+y^2/B^2 = 1$ meet in $Q$. If the locus of $P$ is the line $lx+my = n$, find the equation of the locus of $Q$.

42. Show that the feet of the four normals to the ellipse $x^2/a^2+y^2/b^2 = 1$ from the point $(\alpha, \beta)$ lie on a rectangular hyperbola which passes through the centre of the ellipse and the point $(\alpha, \beta)$ and whose asymptotes are parallel to the axes of the ellipse. If the normals at the points $A, B, D, E$ on the ellipse are concurrent, show that $DE$ and the diameter conjugate to $AB$ are equally inclined to the major axis of the ellipse.

43. Find the locus of the mid-points of chords of an ellipse which are the polar lines of points on the director circle.

44. Find the equation of the locus of the intersections of the normals at the ends of conjugate diameters of an ellipse.

45. $P$ and $Q$ are the points $(a\sec\alpha, b\tan\alpha)$ and $(a\sec\beta, b\tan\beta)$ on the hyperbola $x^2/a^2-y^2/b^2 = 1$. If the tangent at $P$ cuts the asymptote $ay-bx = 0$ at $T$, and $TQ$ is parallel to the other asymptote, prove that $\sec\beta+\tan\beta = 2(\sec\alpha+\tan\alpha)$, and that, if the chord $PQ$ meets the asymptotes at $R$ and $S$, then $P$ and $Q$ trisect $RS$.

46. $P$ is a point on the ellipse $x^2/a^2+y^2/b^2 = 1$. Its polar line with respect to the hyperbola $x^2/a^2-y^2/b^2 = 1$ meets the asymptotes of the hyperbola in $Q$ and $R$. If $QR$ is given equal to $c$ prove that the eccentric angle, $\theta$, of $P$ is given by $c^2\cos^2 2\theta+2(a^2-b^2)\cos 2\theta = 2(a^2+b^2)$, and deduce that if $a > b$ and $c > 2a$, this equation gives real values of $\theta$.

47. A chord $PQ$ of an ellipse, centre $C$, is drawn in a fixed direction and meets the minor axis in $Z$. $W$ is a point on the major axis equidistant from $P$ and $Q$. Show that $CW/CZ$ is constant.

48. Prove that the focal distances of a point $(a\cos\theta, b\sin\theta)$ on the ellipse $x^2/a^2+y^2/b^2 = 1$ are $a(1\pm e\cos\theta)$. If $a^2 > 2b^2$ show that there are four possible positions of a chord which subtends a right angle at each of the two foci; if $\alpha$ and $\beta$ $(\alpha < \beta)$ are the eccentric angles of the two extremities which are in the first quadrant, show that

$$\tan^2\alpha = \tan^2\beta/(1+2\tan^2\beta).$$

49. Find the equations of the conjugate diameters of the ellipse $x^2/a^2+y^2/b^2 = 1$ which are equal in length. Show that the locus of the intersections of normals at the extremities of a chord parallel to one of the equiconjugate diameters is the diameter perpendicular to the other equiconjugate diameter.

50. Show that the locus of the mid-points of the chords of the hyperbola $x^2/a^2 - y^2/b^2 = 1$ which are tangents to the ellipse $x^2/a^2 + y^2/b^2 = 1$ is $a^2b^2(b^2x^2 + a^2y^2) = (a^2y^2 - b^2x^2)^2$.

51. From an end $A$ of the major axis of an ellipse a line $APQ$ is drawn cutting the ellipse in $P$ and the minor axis in $Q$; prove that $AP \cdot AQ = 2CD^2$, where $CD$ is the semi-diameter parallel to $AP$.

52. A parallelogram circumscribes the ellipse $x^2/a^2 + y^2/b^2 = 1$, and two opposite vertices lie on the ellipse $a^2x^2 + b^2y^2 = a^2b^2$. Prove that the other pair of vertices lie on the hyperbola $x^2/a^4 - y^2/b^4 = 1/(a^2 - b^2)$.

53. Prove that four normals can be drawn to a central conic from any point in its plane. From the point $(a\cos\phi, b\sin\phi)$ of the ellipse

$$x^2/a^2 + y^2/b^2 = 1$$

three normals (other than the normal at the point) are drawn to the ellipse; verify that their feet lie on the circle

$$x^2 + y^2 - \frac{b^2}{a} x\cos\phi - \frac{a^2}{b} y\sin\phi - a^2 - b^2 = 0.$$

54. Points $P$, $Q$ are taken one on each of two similar and similarly situated ellipses with a common centre $O$ and the angle $POQ$ is bisected by the major axes of the ellipses. Prove that $PQ$ is always normal to a fixed ellipse.

55. Prove that if $PSP'$ and $QSQ'$ are two focal chords of an ellipse such that $PQ$ is a diameter, $P'Q'$ passes through a fixed point $O$ on the major axis such that $A'O/OA = (A'S/SA)^2$, where $A$ is the nearer vertex to $S$ and $A'$ is the other vertex.

56. A circle with fixed centre $(3h, 3k)$ and of variable radius cuts the rectangular hyperbola $x^2 - y^2 = 9a^2$ in the points $A, B, C, D$; prove that the locus of the centroid of the triangle $ABC$ is given by

$$(x - 2h)^2 - (y - 2k)^2 = a^2.$$

57. Any point $P$ is taken on an ellipse whose foci are $A, B$. $Q$ is a point on the bisector of the angle $APB$ and $L$ is the foot of the perpendicular from $P$ on to $AB$. From $Q$ the perpendicular $QM$ is drawn to $PA$. If $PL = QM$ prove that the length of $PM$ has the constant value $b^2/ae$ where $a$ and $b$ are the principal semi-axes and $e$ is the eccentricity.

58. Find the equation of the chord of the ellipse $x^2/a^2 + y^2/b^2 = 1$ which is bisected at the point $(\alpha, \beta)$. Chords of this ellipse are drawn to pass through a fixed point $Q$. Prove that the mid-points of the chords lie on an ellipse whose centre is the mid-point of the line joining $Q$ to the centre of the given ellipse.

59. Define the *eccentric angle* of a point on an ellipse, and prove that the eccentric angles of the extremities of two conjugate diameters differ

by an odd multiple of $\frac{1}{2}\pi$. Prove also that the sum of the eccentric angles of four concyclic points on the ellipse is a multiple of $2\pi$. $PCP'$ and $DCD'$ are conjugate diameters of an ellipse, and $\phi$ is the eccentric angle of $P$. Prove that $\frac{1}{2}\pi - 3\phi$ is the eccentric angle of the point where the circle $PP'D$ again cuts the ellipse.

60. Lines are drawn through the origin perpendicular to the tangents from a point $P$ to the ellipse $x^2/a^2 + y^2/b^2 = 1$. If the lines are conjugate diameters of the ellipse, prove that $P$ lies on the curve $a^2x^2 + b^2y^2 = a^4 + b^4$.

## METHODS OF DRAWING AN ELLIPSE

(1) Fix two drawing pins $A$ and $B$ to a sheet of paper on a drawing board, a distance $2c$ apart, and attach to them the ends of a piece of fine string of length $2a$ ($a > c$). With a pencil in the loop of the string keep the string taut, and the point $P$ of the pencil will trace an ellipse whose foci are $A$ and $B$, and whose eccentricity is $c/a$.

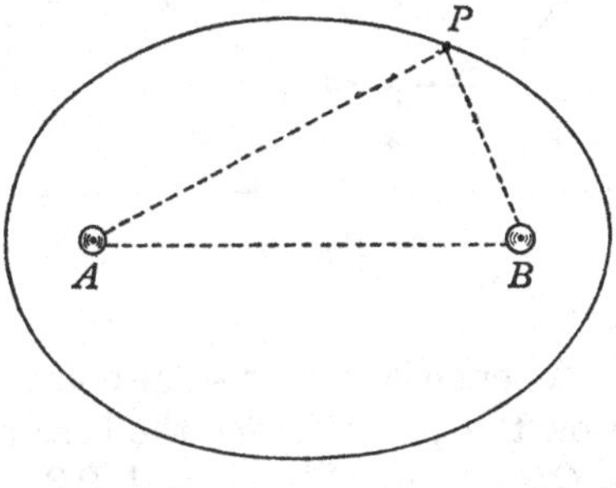


Fig. 72

Inaccuracies occur in this drawing, because the string gets wrapped round the pins and the effective length of it is then reduced. A better figure is obtained by using a loop of string of length $2(a+c)$ which passes round the pins but is not tied to them (Fig. 72).

(2) A further modification of this method produces a neat model. Fix the pins $A$ and $B$ to a board which can be rotated in a vertical plane. Pass the loop of thread through an eye, from which hangs a heavy bob, and to the eye attach a short horizontal bar, as a tangent. As the board is rotated the eye describes an

ellipse, whose path may be marked on the board (Fig. 73). The fact that the tangent is equally inclined to the focal distances is seen to follow from the principle of statics.

(3) A machine commonly used in drawing offices for constructing ellipses is the *Trammel of Archimedes*. Two perpendicular slots on a base plate meet in $O$, and two points $A$ and $B$ of a rod are fixed to blocks which are free to slide in these

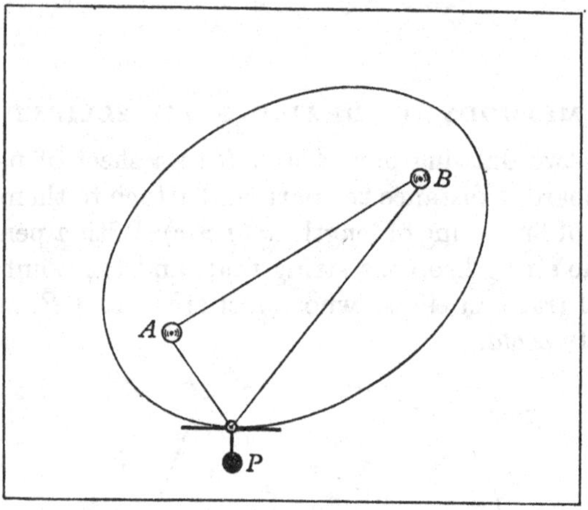


Fig. 73

grooves (Fig. 74). A pencil is fixed at some point $P$ of the rod, and it traces a path on the paper under the base plate, as the rod moves. If angle $OBA$ is $\theta$, $PA = a$ and $PB = b$, the point $P$ is $(a\cos\theta, b\sin\theta)$ and describes an ellipse with axes of length $2a$ and $2b$.

(4) Let $C$ be a fixed circle and $S$ a fixed point inside it (Fig. 75). Let $P$ be a variable point on $C$ and draw $PQ$ perpendicular to $SP$. The lines $PQ$ touch an ellipse with $S$ as focus and $C$ as auxiliary circle.

(5) This last construction can also be modified to give an envelope of creases on transparent paper. Draw the circle $C$ on the paper, and mark a point $S$ inside the circle. Fold the paper so that $S$ is on $C$, and mark the crease in the paper. As the position of $S$ varies on $C$, the creases are tangents of an ellipse with focus $S$.

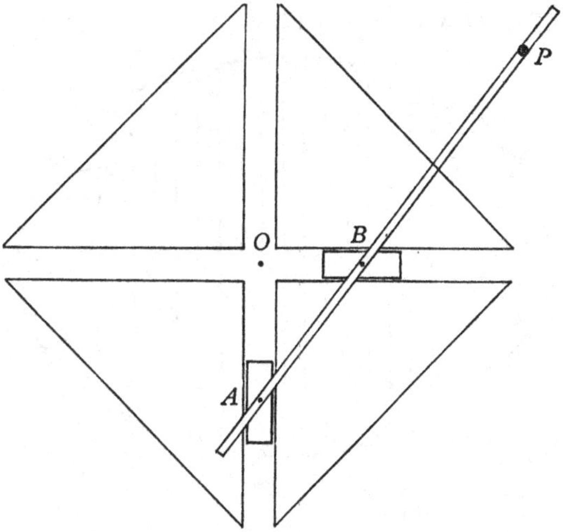


Fig. 74

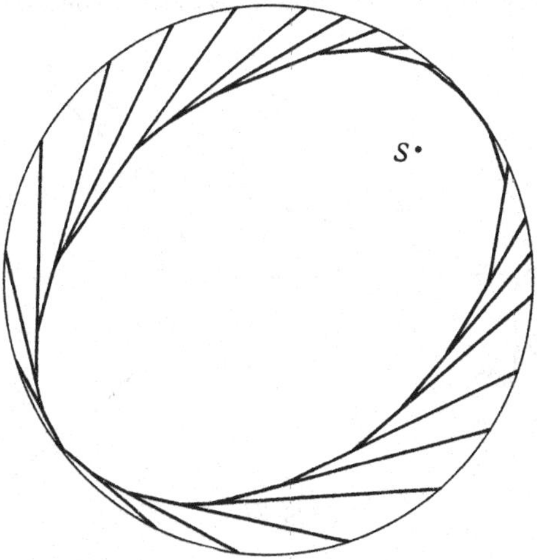


Fig. 75

(6) Draw a circle and a series of parallel chords of it. The envelope of circles drawn on these chords as diameters is a special ellipse (eccentricity $\frac{1}{2}\sqrt{2}$) (Fig. 76).

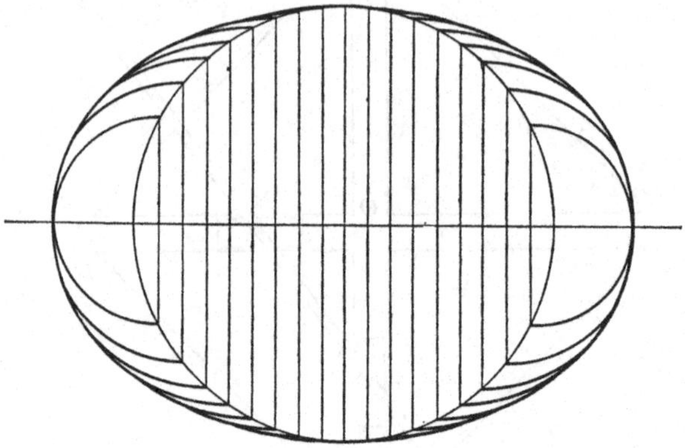

Fig. 76

## METHODS OF DRAWING A HYPERBOLA

(7) As in (1), pass a string round pins $A$ and $B$ (Fig. 77). Fix the pencil point $P$ to the string (passing a loop of the string round it),

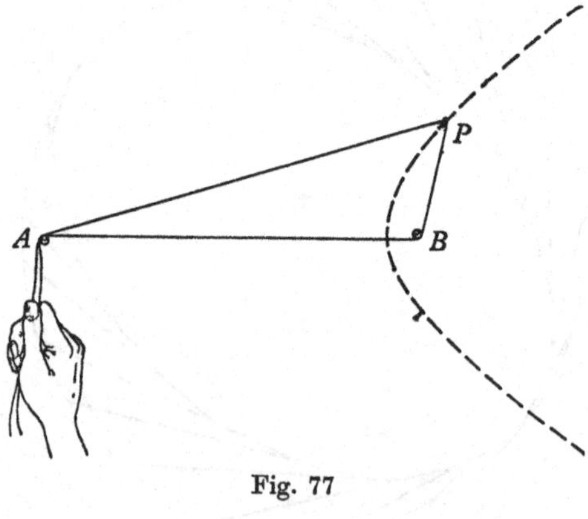


Fig. 77

and hold the free ends of the string. By pulling these ends, the lengths of $AP$ and $BP$ are each shortened by the same amount, so that $AP-BP$ is a constant, and the locus of $P$ is a hyperbola.

(8) The two methods, (4) and (5), of drawing an ellipse can be modified, taking $S$ outside $C$ instead of inside, to give a hyperbola in each case.

(9) Let $l$, $m$ be fixed lines meeting in $O$. Let $L$ be any point on $l$, where $OL = p$. For some fixed number $k$, construct the point $M$ on $m$ so that $OM = k/p$. As $L$ varies on $l$, $LM$ is a tangent to a hyperbola, and the mid-points of $LM$ lie on the hyperbola.

# VII

## LINE PAIRS

In the previous chapters we have discussed the properties of a number of curves, the parabola, circle, ellipse and hyperbola, which all have one thing in common. The equation of each of these curves is a quadratic in the variables $x$ and $y$, and this is true whether the axes are taken related in some special way to the curve, or completely generally. In this chapter we shall assemble results about the one remaining locus which is represented by a quadratic equation; in the next we shall prove that the general quadratic expression in $x$ and $y$ always represents one of these five loci.

In the work to come, much use will be made of the two results proved in earlier chapters (I, § 5, and II, § 14) on change of origin and rotation of axes, and these formulae should be revised now. By a combination of these two operations, the position of the axes may be changed to any other position in the plane. To put it another way, any two perpendicular lines in the plane may be taken as axes.

**1. The equation of a line pair through the origin.** Suppose

$$lx+my = 0$$

and

$$l'x+m'y = 0$$

are the equations of two lines $p$ and $p'$, each of which passes through the origin. Then the equation

$$(lx+my)(l'x+m'y) = 0$$

is satisfied by the coordinates of any point which lies on either $p$ or $p'$ or on both. That is, this equation represents the *pair* of lines $p, p'$. We refer to this combined locus briefly as a *line pair*.

When the two linear expressions are multiplied together, the combined equation of the line pair is

$$ll'x^2+(lm'+l'm)xy+mm'y^2 = 0.$$

This is a quadratic expression in $x$ and $y$; there are no terms in $x$

or $y$, and there is no constant term, independent of $x$ and $y$. Such an expression is called *homogeneous* of the second degree.

The general expression of this form is

$$ax^2 + 2hxy + by^2 = 0,$$

and we shall show that this always represents a pair of lines through the origin, provided that there are points other than $O$ whose coordinates satisfy the equation. If $a = 0$, the expression evidently represents a line pair,

$$y = 0 \quad \text{and} \quad 2hx + by = 0.$$

If $a \neq 0$, and $h^2 - ab \geqslant 0$, the equation can be factorized in the form

$$\frac{x}{y} = \frac{-h \pm \sqrt{(h^2 - ab)}}{a}.$$

Thus the equation represents two lines through the origin, which are distinct unless $h^2 = ab$.

If $a \neq 0$ and $h^2 - ab < 0$, we can write

$$a(ax^2 + 2hxy + by^2) = (ax + hy)^2 + (ab - h^2)\,y^2.$$

Thus $ax^2 + 2hxy + by^2$ is zero, if and only if each of the positive expressions

$$(ax + hy)^2 \quad \text{and} \quad (ab - h^2)\,y^2$$

is zero, and so $ax + hy = 0$ and $y = 0$, which gives $x = y = 0$. It follows that the only point on the locus in this case is the origin $(0, 0)$.

**2. Angle between the lines $ax^2 + 2hxy + by^2 = 0$.** Suppose the lines represented by the equation are

$$lx + my = 0 \quad \text{and} \quad l'x + m'y = 0.$$

Then

$$\frac{ll'}{a} = \frac{lm' + l'm}{2h} = \frac{mm'}{b}.$$

If the acute angle between these two lines is $\theta$,

$$\tan\theta = \frac{|l/m - l'/m'|}{1 + ll'/mm'} = \frac{|lm' - l'm|}{ll' + mm'}.$$

But

$$(lm' - l'm)^2 = (lm' + l'm)^2 - 4ll'mm'$$

and so
$$\tan\theta = \frac{\sqrt{(4h^2-4ab)}}{a+b},$$

or
$$\tan\theta = \frac{2\sqrt{(h^2-ab)}}{a+b}.$$

We notice three things about this expression:

(i) It is significant that $h^2-ab \geqslant 0$, for otherwise the expression would have no meaning.

(ii) If $h^2-ab=0$, $\theta=0$, and (as we already know) the lines coincide.

(iii) The lines are *perpendicular* if and only if

$$a+b=0,$$

that is, the sum of the coefficients of $x^2$ and $y^2$ is zero.

**3. Bisectors of the angle between two lines.** Let $p$ and $p'$ be the lines
$$lx+my=0 \quad \text{and} \quad l'x+m'y=0$$

forming the line pair $ax^2+2hxy+by^2=0$. As before

$$\frac{ll'}{a} = \frac{lm'+l'm}{2h} = \frac{mm'}{b}.$$

If $y=x\tan\alpha$ is the equation of a bisector, then

$$2\alpha = \theta+\theta' \quad \text{or} \quad \theta+\theta'+\pi,$$

where $\tan\theta$, $\tan\theta'$ are the gradients of the two given lines, so that

$$\tan\theta = -l/m, \quad \tan\theta' = -l'/m'.$$

Thus
$$\tan 2\alpha = \tan(\theta+\theta')$$

and so
$$\frac{2y/x}{1-y^2/x^2} = \frac{-l/m-l'/m'}{1-ll'/mm'}.$$

Substituting in terms of $a$, $h$, $b$, this becomes

$$\frac{x^2-y^2}{a-b} = \frac{xy}{h}.$$

As a check, we notice that the lines are perpendicular.

**4. Illustrations.** Problems about line pairs through the origin usually need the technique of introducing separate equations for the lines, as has been done in the last two paragraphs. In the following examples, the line pair $ax^2+2hxy+by^2=0$ is always taken to be

$$(lx+my)(l'x+m'y)=0,$$

so that

$$\frac{ll'}{a}=\frac{lm'+l'm}{2h}=\frac{mm'}{b}.$$

***Illustration 1.*** *Find the equation of the lines through the origin which are perpendicular to the lines $ax^2+2hxy+by^2=0$.*

The lines through the origin perpendicular to the given lines are

$$(mx-ly)(m'x-l'y)=0$$

or

$$mm'x-(lm'+l'm)xy+ll'y^2=0.$$

Substituting the values for these coefficients, we have the required line pair

$$bx^2-2hxy+ay^2=0.$$

***Illustration 2.*** *A triangle has two of its sides along the lines $ax^2+2hxy+by^2=0$, and its orthocentre is $H(p,q)$. Find the equation of the third side, and investigate the situation when $a+b=0$.*

The orthocentre lies on the altitude through each vertex. One vertex is the origin $O$, and the gradient of $OH$ is $q/p$. The gradient of the third side, opposite $O$, is thus $-p/q$, and the equation of the third side is

$$px+qy=k,$$

where the value of $k$ is to be found.

This line meets $lx+my=0$ in the vertex

$$\left(\frac{km}{pm-ql},\frac{-kl}{pm-ql}\right).$$

The gradient of the altitude joining this point to the orthocentre $(p,q)$ is

$$\left(q+\frac{kl}{pm-ql}\right)\Big/\left(p-\frac{km}{pm-ql}\right),$$

or

$$\frac{pqm-q^2l+kl}{p^2m-pql-km},$$

and this is perpendicular to the remaining edge $l'x+m'y=0$, so that

$$\frac{pqm-q^2l+kl}{p^2m-pql-km}\times\frac{-l'}{m'}=-1.$$

This gives

$$k(ll'+mm') = q^2ll'-pq(lm'+l'm)+p^2mm',$$

or

$$k(a+b) = aq^2-2hpq+bp^2.$$

Thus, if $a+b \neq 0$, the equation of the third side of the triangle is

$$(a+b)(px+qy) = aq^2-2hpq+bp^2.$$

If $a+b = 0$, the given sides of the triangle are perpendicular, and the orthocentre $(p, q)$ is at the origin, whatever line in the plane is the third side.

## EXAMPLES VII A

1. Find the angle between the lines

(i) $6x^2-13xy+5y^2 = 0$;

(ii) $x^2-2xy\sec\alpha+y^2 = 0$;

(iii) $3x^2+5xy-y^2 = 0$.

2. The angle between the lines

$$(3k+2)x^2+2(k+1)xy-ky^2 = 0$$

is $\theta$. Find the values of $k$ when $\theta = 0, \frac{1}{6}\pi, \frac{1}{4}\pi, \frac{1}{3}\pi, \frac{1}{2}\pi$.

3. Find the equation of the bisectors of the angles between the line pairs

(i) $3x^2+8xy+4y^2 = 0$;

(ii) $15x^2-xy-6x^2 = 0$;

(iii) $12x^2-7xy-y^2 = 0$.

4. Find the condition on $m$ if the bisectors of the angles between the line pair $m^2x^2+2(m+1)xy-y^2 = 0$ are the axes of coordinates.

5. Show that if the line $x+y = 1$ meets the lines of the line pair $ax^2+2hxy+by^2 = 0$ in the points $(\alpha, 1-\alpha)$ and $(\beta, 1-\beta)$, then $\alpha$ and $\beta$ are the roots of the equation

$$(a-2h+b)x^2+2x(h-b)+b = 0.$$

Hence, or otherwise, calculate the area of the triangle formed by the three lines.

6. Prove that the line pair $x^2+4xy+y^2 = 0$ and the line $x-y = \sqrt{6}$ form an equilateral triangle, and find the length of a side.

7. Find the perimeter of the triangle whose edges are given by

$$3x+4y = 2 \quad \text{and} \quad x^2-3xy+2y^2 = 0.$$

8. Two line pairs are given by $x^2+xy = 0$ and $6x^2-xy-y^2 = 0$. Show that the angle between the first pair is equal to the angle between the second pair.

**5. Line pair joining the origin to the points of intersection of a line and a conic.** We establish a useful technique by a method already used several times (pp. 41, 143), that of *inspection.*

Let $lx+my+n = 0$ be a general line, and $ax^2+by^2 = 1$ be a central conic. At the points where the two loci meet

$$\frac{lx+my}{-n} = 1 \quad \text{and} \quad ax^2+by^2 = 1,$$

so that the equation

$$ax^2+by^2 = \left(\frac{lx+my}{-n}\right)^2$$

is satisfied at these points. But this is a homogeneous equation of the second degree, so it represents a line pair through the origin, provided that the given line does meet the conic. It is thus the equation of the line pair joining the origin to the points of intersection of the line and the conic.

The same technique, that of making the equation homogeneous by substituting $(lx+my)/-n$ for 1, can be used for any conic; an example of its use follows.

***Illustration 3.*** *Find the gradient of a chord PQ of a parabola with focus S, which meets the directrix and the axis in the same point, and is such that SP is perpendicular to SQ.*

In order to use the method just described, we choose $S$ as origin and the axis of the parabola as the $x$-axis, so that the equation of the curve is

$$y^2 = 4a(x+a).$$

The axis meets the directrix in the point $(-2a, 0)$, and a chord $PQ$ through this point and with gradient $m$ has equation

$$y = mx+2am$$

so that

$$(y-mx)/2am = 1.$$

Making the equation of the conic homogeneous by means of this expression, we have

$$y^2 = 4ax\left(\frac{y-mx}{2am}\right)+4a^2\left(\frac{y-mx}{2am}\right)^2.$$

This is the equation of a line pair through the origin, and is satisfied by $P$ and $Q$, so it is the equation of the line pair $SP$, $SQ$. When simplified, the expression is

$$m^2x^2 + (m^2 - 1)\,y^2 = 0.$$

These lines are perpendicular if and only if

$$m^2 + m^2 - 1 = 0$$

so that $$m = \pm 1/\sqrt{2}.$$

**6. The condition for the general quadratic equation to represent a line pair.** We now consider the general quadratic equation

$$S \equiv ax^2 + 2hxy + by^2 + 2gx + 2fy + c = 0,$$

which we write more shortly $S = 0$. The main result of this paragraph is the following:

A necessary and sufficient condition for $S = 0$ to represent a line pair† is that

$$\Delta = \begin{vmatrix} a & h & g \\ h & b & f \\ g & f & c \end{vmatrix} = 0.$$

(This determinant is of great importance in the study of conics, and it will be useful to remember it. In particular, notice that the 'non-diagonal' terms, $f$, $g$ and $h$, are each *half* the corresponding coefficients of $y$, $x$ and $xy$.)

Further

(i) if $h^2 - ab > 0$, the lines are real and distinct, meeting in the point $\{(bg - hf)/(h^2 - ab), (af - gh)/(h^2 - ab)\}$;

(ii) if $h^2 - ab < 0$, there is only one real point on the locus;

(iii) if $h^2 - ab = 0$, the lines are parallel;

(iv) if $a{:}h{:}g = h{:}b{:}f = g{:}f{:}c$, the lines coincide.

It is necessary to consider separately the cases of intersecting and of parallel lines; let us first suppose that $S = 0$ represents a pair of lines which meet in the point $(\lambda, \mu)$.

*Change the origin to the point $(\lambda, \mu)$, by the transformation*

$$x = x^* + \lambda,$$

$$y = y^* + \mu.$$

† We use the term 'line pair' loosely, and include loci like $ax^2 + 2hxy + by^2 = 0$, with $h^2 < ab$, which was considered in §1, and which contains only one real point.

The equation becomes

$$a(x^*+\lambda)^2+2h(x^*+\lambda)(y^*+\mu)+b(y^*+\mu)^2$$
$$+2g(x^*+\lambda)+2f(y^*+\mu)+c = 0.$$

This represents a line pair through the new origin, so that terms in $x^*$ and $y^*$ and the constant term all vanish; that is

$$a\lambda+h\mu+g = 0,$$
$$h\lambda+b\mu+f = 0$$

and $$a\lambda^2+2h\lambda\mu+b\mu^2+2g\lambda+2f\mu+c = 0.$$

This last equation can be written

$$\lambda(a\lambda+h\mu+g)+\mu(h\lambda+b\mu+f)+(g\lambda+f\mu+c) = 0,$$

and since the first two brackets are zero, so is the third. Thus we have the three equations

$$a\lambda+h\mu+g = 0,$$
$$h\lambda+b\mu+f = 0,$$
$$g\lambda+f\mu+c = 0,$$

and they have a solution in $\lambda$, $\mu$, so [A 3]

$$\begin{vmatrix} a & h & g \\ h & b & f \\ g & f & c \end{vmatrix} = 0.$$

The equation of the line pair, referred to the new axes, is

$$ax^{*2}+2hx^*y^*+by^{*2} = 0,$$

so the effect of the change of origin is to remove the linear and constant terms, but *not to alter* the coefficients $a$, $b$ and $2h$. It follows from §1 that if $h^2-ab > 0$ the lines are real and distinct, and if $h^2-ab < 0$ there is only one point on the locus. The coordinates of the point of intersection

$$\{(bg-hf)/(h^2-ab), (af-gh)/(h^2-ab)\}$$

are found by solving the first two of the equations found above. These expressions for $\lambda$ and $\mu$ are hardly worth remembering, but the equations from which they are found are easily recalled from the form of the determinant $\Delta$. It also follows that the results of §2 hold for line pairs not through the origin, and in particular that the lines are perpendicular if $a+b=0$.

We now turn to the converse of the theorem, and assume that $\Delta = 0$ and that $h^2-ab \neq 0$. Since $\Delta = 0$, there are numbers $p, q, r$, not all zero, [A 3], such that

$$ap+hq+gr = 0,$$

$$hp+bq+fr = 0,$$

$$gp+fq+cr = 0.$$

Now $h^2-ab \neq 0$, so $r \neq 0$. Let $p/r = \lambda$, $q/r = \mu$, and change the origin to $(\lambda, \mu)$. Since the relations

$$a\lambda+h\mu+g = 0,$$

$$h\lambda+b\mu+f = 0,$$

$$g\lambda+f\mu+c = 0,$$

hold, the equation of $S$ becomes

$$ax^{*2}+2hx^*y^*+by^{*2} = 0,$$

and since $h^2-ab \neq 0$, this represents either a pair of real, distinct lines, or (when $h^2-ab<0$) a single point. This completes the proof in the cases (i) and (ii).

The proof given so far breaks down if $(\lambda, \mu)$ does not exist, so we consider separately the case when $S = 0$ represents a pair of parallel lines,

$$lx+my+n = 0 \quad \text{and} \quad lx+my+N = 0,$$

say. In this case, $S = 0$ is the same equation as

$$l^2x^2+2lmxy+m^2y^2+l(n+N)x+m(n+N)y+nN = 0,$$

so
$$\frac{l^2}{a} = \frac{lm}{h} = \frac{m^2}{b} = \frac{l(n+N)}{2g} = \frac{m(n+N)}{2f} = \frac{nN}{c}.$$

From these equations, we see that if $m \neq 0$,

$$\frac{a}{h} = \frac{h}{b} = \frac{g}{f} = \frac{l}{m},$$

so that two rows of $\Delta$ are proportional, and if $m = 0$,

$$h = b = f = 0,$$

and $\Delta$ has a row of zeros. Either way, $\Delta = 0$ and $h^2 - ab = 0$.

The investigation for reality conditions is a little tedious, but we give it for completeness. If the lines are real and distinct, $n$ and $N$ must be real and distinct, and so, putting $n + N = 2u$, $nN = v$, the roots $n$ and $N$ of the equation

$$t^2 - 2ut + v = 0$$

are real and distinct, and $\quad u^2 - v > 0.$

Now $$\frac{l^2}{a} = \frac{m^2}{b} = \frac{2lu}{2g} = \frac{2mu}{2f} = \frac{v}{c} = \rho, \quad \text{say} \quad (\rho \neq 0),$$

so that $$l^2(u^2 - v) = \rho^2(g^2 - ac)$$

and $$m^2(u^2 - v) = \rho^2(f^2 - bc).$$

Now the parallel lines exist, so not both of $l$ and $m$ can vanish, and since $u^2 - v > 0$, it follows that neither of

$$g^2 - ac \quad \text{and} \quad f^2 - bc$$

can be negative, and one at least must be strictly positive.

Conversely, suppose that $\Delta = 0$ and $ab - h^2 = 0$. Then, as before, there exist $p$, $q$, $r$, not all zero, such that

$$ap + hq + gr = 0,$$

$$hp + bq + fr = 0,$$

$$gp + fq + cr = 0.$$

If $r \neq 0$, from the first two equations $p$ and $q$ can be eliminated, giving $hg - af = 0$, and so

$$\frac{a}{h} = \frac{h}{b} = \frac{g}{f}.$$

If $r = 0$, these ratios are equal to $-q/p$, so that in either case we may put

$$\frac{a}{h} = \frac{h}{b} = \frac{g}{f} = \frac{l}{m}, \quad \text{say,}$$

and so $$\frac{a}{l^2} = \frac{h}{lm} = \frac{b}{m^2} = \frac{g}{lu} = \frac{f}{mu} = \frac{c}{v} = \frac{1}{\rho}, \quad \text{say.}$$

Thus the equation $S = 0$ is

$$l^2x^2 + 2lmxy + m^2y^2 + 2lux + 2muy + v = 0$$

or

$$(lx + my + n)(lx + my + N) = 0,$$

where

$$n + N = 2u, \quad nN = v.$$

The equation therefore represents a pair of parallel lines.

To deal with the reality of these lines, suppose the conditions found previously hold; that is $g^2 - ac$ and $f^2 - bc$ are not negative and at least one is positive. Now $\rho^2(g^2 - ac) = l^2(u^2 - v)$ and $\rho^2(f^2 - bc) = m^2(u^2 - v)$ and it follows that $u^2 - v > 0$. But since $n + N = 2u$ and $nN = v$, it follows that $n$ and $N$ are the roots of the quadratic $t^2 - 2ut + v = 0$, for which the condition for real and distinct roots is satisfied. Thus $n$ and $N$ are real and distinct, and $S = 0$ represents real and distinct parallel lines.

Lastly, if the lines coincide, so that $S \equiv (lx + my + n)^2 = 0$, it is a trivial matter to verify that $\Delta = 0$ and that

$$a:h:g = h:b:f = g:f:c;$$

conversely, if these conditions hold, the equation evidently represents a repeated line.

## EXAMPLES VII B

1. For what values of $\lambda$ do the following equations represent line pairs?

(i) $x^2 - y^2 + 2\lambda x + 4 = 0$;

(ii) $2x^2 + xy - 6y^2 + 2x + 5y + \lambda = 0$;

(iii) $x^2 + 2\lambda xy + y^2 + 6x + 2y + 9 = 0$;

(iv) $\lambda x^2 + 2xy + \lambda y^2 + 4x + 4y + 3 = 0$;

(v) $x^2 + xy + \lambda y^2 + 7x - 14y - 14 = 0$.

2. Show that each of the following expressions represents a line pair, and find the point of intersection of the lines when it exists:

(i) $x^2 + 8xy + y^2 + 16x + 4y + 4 = 0$;

(ii) $4x^2 + 12xy + 9y^2 - 10x - 15y + 6 = 0$;

(iii) $x^2 - 2xy + y^2 - 4x + 4y + 4 = 0$;

(iv) $39x^2 + 8xy - 7y^2 + 16x - 28y - 28 = 0$;

(v) $x^2 + 3xy + 2y^2 + 4x + 5y + 3 = 0$.

3. Show that the equation $x^2 - xy - 6y^2 - 7x + 31y - 18 = 0$ represents a pair of lines, and find the angle between them.

4. Show that the lines joining the origin to the points of intersection of the line $3x-2y=1$ and the conic $3x^2+5xy-3y^2+2x+3y=0$ are perpendicular.

5. For what value of $k$ does the equation

$$12x^2+7xy+ky^2+13x-y+3=0$$

represent a line pair, and what is then the angle between them?

6. For what values of $\lambda$ is the conic

$$\lambda(x^2+y^2+2gx+2fy+c)+2xy=0$$

a line pair? If two of these line pairs consist of parallel lines, what condition must be satisfied by $g, f$ and $c$?

7. Find the angle between the following pairs of lines:

(i) $2x^2-7xy+3y^2=0$;

(ii) $x^2-4xy+y^2+6x-3=0$;

(iii) $4x^2+5xy+y^2+3y-4=0$.

8. Find the angle between the lines joining the origin to the points of intersection of the line $x-3y+2=0$ and the curve

$$y^2-17xy+16y-12=0.$$

9. Find the equation of the line pair joining the origin to the intersections of the line $ax+by=a+2b$ and the locus $x^2-xy+y-1=0$. Interpret the result.

10. Find the equation of the bisectors of the angles between the lines:

(i) $3x^2+4xy+2y^2-2x+1=0$;

(ii) $x^2-3xy+2y^2=0$;

(iii) $2x^2-6xy+y^2+16x-10y+18=0$.

11. Find the condition on a chord $ax+by+c=0$ of the curve

$$x^2-3y^2-4x+2y=0$$

if it subtends a right angle at the origin.

12. Show that the equation $x^2+2xy\tan\theta-y^2-2ax\sec\theta+a^2=0$ represents a pair of lines, and find the locus of their point of intersection.

## MISCELLANEOUS EXAMPLES VII

1. A chord $PQ$ of a parabola with focus $S$ meets the axis in $C$ and $SP$ is perpendicular to $SQ$. Find what restriction (if any) this imposes on the position of $C$.

2. Show that the equation $x^2+4xy+4y^2=25$ represents a pair of parallel straight lines, and find their distance apart. Find the equation of the two circles which touch both these lines and pass through the origin.

3. Prove that the line $ax+by+c=0$ forms with the two lines

$$(ax+by)^2 = 3(ay-bx)^2$$

an equilateral triangle. Find the equations of the bisectors of the angles between the lines represented by the latter equation.

4. Prove that the equation of the pair of straight lines joining the origin to the points of intersection of the circle $x^2+y^2+2gx+2fy=0$ and the straight line $px+qy=r$ is

$$(2pg+r)\,x^2+2(pf+qg)\,xy+(2qf+r)\,y^2=0.$$

Hence, or otherwise, obtain the equation of the straight line joining the two points of intersection (other than the origin) of the circle

$$x^2+y^2+2x+2y=0$$

and the pair of straight lines $x^2-4xy+2y^2=0$.

5. Show that, for any value of $k$, the circumcircle of the triangle formed by the $x$-axis and the pair of perpendicular lines

$$x^2+kxy-y^2-kx+2y-1=0$$

passes through the points $(0, \pm 1)$.

6. $P$ is a variable point which moves on the lines of the pair

$$ax^2+2hxy+by^2=0,$$

and the perpendicular through $P$ to the line on which $P$ lies meets the other line in $Q$. Prove that the locus of the mid-point of $PQ$ is given by the equation

$$(b^3+2bh^2+ah^2)\,y^2+2h(a^2+b^2+ab+h^2)\,xy+(a^3+2ah^2+bh^2)\,x^2=0.$$

7. Prove that the equation of the pair of lines through the origin perpendicular to the pair $ax^2+2hxy+by^2=0$ is $bx^2-2hxy+ay^2=0$. Deduce the equation of the pair of lines through the point $(x', y')$ which are perpendicular to the pair $ax^2+2hxy+by^2=0$, and of the circle through the four points of intersection of these two pairs, showing that its centre is at the point

$$\left(\frac{bx'-hy'}{a+b}, \frac{-hx'+ay'}{a+b}\right).$$

8. Show that $3x^2+xy-3y^2-11x-8y+7=0$ is the equation of a pair of straight lines and find the coordinates of the point $A$ where they meet. These two lines form the sides of a rectangle $ABCD$ whose vertex $C$ is at the point $(-2, 1)$. Obtain the equation of the sides $BC$, $CD$ and the equation of the diagonals. Calculate the area of the rectangle.

9. Find for what values of $\lambda$ the equation

$$6x^2+8y^2-26xy-22y-6+\lambda(2x^2+4y^2-14xy-10y-2)=0$$

represents a pair of straight lines. Obtain the equation of each of the six lines of the line pairs.

10. The lines represented by the equation $ax^2+2hxy+by^2=0$ are turned in their own plane about the origin $O$ of coordinates through the acute angle $\tan^{-1}2$ in the sense from $Ox$ to $Oy$. Prove that, in their new position, their equation referred to the original axes is

$$(a-4h+4b)x^2+(4a-6h-4b)xy+(4a+4h+b)y^2=0.$$

Prove that the length of the chord joining the points of intersection of the above pair of lines in either position with the circle $x^2+y^2=r^2$ is

$$r\left[2-\frac{2(a+b)}{\sqrt{\{(a-b)^2+4h^2\}}}\right]^{\frac{1}{2}}.$$

11. Prove that the equation

$$my^2-m^3x^2-(m^2+1)y-m(m^2-1)x+m=0$$

represents two straight lines. Find the intersection of these straight lines, and show that for different values of $m$ the locus of the point of intersection is $(2x+1)y^2=(x+1)^2$.

12. Show that the lines $(m+1)y^2-(m^2+2m-1)xy+(m^2-m)x^2=0$ intersect each other at an angle of $\frac{1}{4}\pi$ radians for all values of $m$. Find the equations of the chords of the parabola $y^2=4ax$ which pass through the point $(-6a, 0)$ and which subtend an angle of $\frac{1}{4}\pi$ radians at the vertex.

13. Prove that, if $\mu \leqslant 169/56$, there are two finite real values of $\lambda$ for which the equation $9x^2+\lambda xy+\mu y^2-45x+13y+14=0$ represents a pair of lines. If $\mu=-1$, find the separate equations of the lines constituting one of these pairs.

14. Find the equation of the pair of lines joining the origin to the points in which the pair of lines

$$4x^2-15xy-4y^2+39x+65y-169=0$$

are met by the line $x+2y-5=0$. Show that the quadrilateral having the first pair and also the second pair as adjacent edges is cyclic, and find the equation of its circumcircle.

15. Find the equation of the pair of lines obtained by rotating the lines represented by $ax^2+2hxy+by^2=0$ about the origin through a positive angle of 60°. Write down the equation which corresponds to the case $a=0$, $b=1$, $2h=\sqrt{3}$, and sketch the two pairs of lines in this case.

16. Prove that the equation $32x^2+24xy-108y^2-52x+123y-24=0$ represents two straight lines. Find the coordinates of their common point and the tangent of the angle between them.

17. Prove that the equation $x^2-3xy+2y^2-3x+5y+2=0$ represents a pair of straight lines, finding their separate equations. Find the area of the parallelogram formed by these lines and the lines drawn parallel to them through the origin.

18. Find separately the equations of the bisectors of the angles between the straight lines drawn through the point $(-2, 1)$ parallel to the straight lines $3x^2+12xy-2y^2 = 0$.

19. If the lines $ax^2+2hxy+by^2 = 0$ meet the line $qx+py = pq$ in points which are equidistant from the origin, prove that

$$h(p^2-q^2) = pq(b-a).$$

20. Obtain the equation of the pair of lines through the point $(p, q)$ at right angles to the pair represented by $ax^2+2hxy+by^2 = 0$. Show that the two pairs of lines $2x^2-xy-y^2+7x+5y-4 = 0$, $AB$ and $AC$, and $x^2-xy-2y^2-5x-2y+4 = 0$, $DB$ and $DC$, form a cyclic quadrilateral and find the equation of the circle circumscribing $ABCD$.

---

21. Find the equations of the diagonals of the parallelogram of which the lines $ax^2+2hxy+by^2+2gx+2fy+c = 0$, where $c \neq 0$, are two adjacent sides, the other two sides meeting in the origin. Prove that, if this parallelogram is a rhombus, then $(a-b)fg+h(f^2-g^2) = 0$.

22. A variable line $lx+my = 1$ cuts the fixed circle $(x-a)^2+(y-b)^2 = r^2$ in points $A$ and $B$, such that $AB$ subtends a right angle at the origin, ($a$, $b$ and $r$ being such that this is possible). Show that the locus of the foot of the perpendicular from the origin on to $AB$ is a circle.

23. Two sides of a parallelogram lie along the lines

$$ax^2+2hxy+by^2 = 0$$

and the diagonal which does not pass through the origin lies along the line $lx+my+n = 0$. Find the coordinates of the vertex opposite the origin.

24. Show that the radius of the circumcircle of the triangle formed by the lines $x/\alpha+y/\beta = 1$ and $ax^2+2hxy+by^2 = 0$ is

$$\frac{\alpha\beta\,.\,(\alpha^2+\beta^2)^{\frac{1}{2}}\{(a-b)^2+4h^2\}^{\frac{1}{2}}}{2(a\alpha^2-2h\alpha\beta+b\beta^2)}.$$

25. From a point $P(p, q)$ perpendiculars $PM$, $PN$ are drawn to the lines given by the equation $ax^2+2hxy+by^2 = 0$. Show that if $O$ is the origin, the area of the triangle $OMN$ is

$$\frac{(aq^2-2hpq+bp^2)\,(h^2-ab)^{\frac{1}{2}}}{(a-b)^2+4h^2}.$$

26. Through a fixed point $(p, q)$ a variable line is drawn to cut the lines $ax^2+2hxy+by^2 = 0$ at $R$ and $S$. Show that the equation of the locus of the mid-point of $RS$ is $ax^2+2hxy+by^2 = p(ax+hy)+q(hx+by)$.

27. Find the area of the segment cut off by $x = y\tan\phi$ from the circle $x^2+y^2-2ax = 0$. Prove that the area of that part of the circle which lies between the lines $px^2+2qxy+ry^2 = 0$ is

$$a^2\lambda + \frac{2a^2(r-p)\sqrt{(q^2-pr)}}{(r-p)^2+4q^2} \quad \text{where} \quad \tan\lambda = \frac{2\sqrt{(q^2-pr)}}{r+p}.$$

28. Prove that the product of the perpendicular distances from $(x', y')$ to the lines $ax^2+2hxy+by^2 = 0$ is $(ax'^2+2hx'y'+by'^2)/\{(a-b)^2+4h^2\}^{\frac{1}{2}}$.

29. Two sides of a triangle are the fixed lines $ax^2+2hxy+by^2 = 0$ and the coordinates of its centroid $G$ are $(p, q)$. Find the equation of the third side. If the third side is variable but always passes through the point $(u, v)$ show that $G$ lies on the conic

$$3(ax^2+2hxy+by^2)-2(au+hv)x-2(hu+bv)y = 0.$$

30. If one of the lines $ax^2+2hxy+by^2 = 0$ and one of the lines

$$Ax^2+2Hxy+By^2 = 0$$

coincide, show that

$$(Ab-aB)^2+4aAbB\left(\frac{H}{B}-\frac{h}{b}\right)\left(\frac{H}{A}-\frac{h}{a}\right) = 0$$

and that the equation of the bisectors of the angles between the other two lines is

$$\left(\frac{aB}{Ab}-\frac{Ab}{aB}\right)(x^2-y^2) = 4xy\left\{\left(\frac{h}{b}-\frac{H}{B}\right)+\left(\frac{h}{a}-\frac{H}{A}\right)\right\}.$$

# VIII

## THE GENERAL CONIC

We now complete the study of the conic in Cartesian coordinates, by a full discussion of the general quadratic equation

$$S \equiv ax^2 + 2hxy + by^2 + 2gx + 2fy + c = 0.$$

This, as we already know, represents a line pair if and only if $\Delta = 0$; if $\Delta \neq 0$, we shall show that it represents either an ellipse, circle, hyperbola or parabola, the curves called conics which we have already studied separately. Much of the preceding theory will be unified in this chapter.

We refer to the locus whose equation is $S = 0$ as the conic $S$, and we shall assume that $S$ is not a line pair, so that $\Delta \neq 0$.

**1. Invariance of $ab - h^2$, $a + b$.** Consider the effect of a rotation through an angle $\alpha$,

$$x = X\cos\alpha - Y\sin\alpha,$$

$$y = X\sin\alpha + Y\cos\alpha,$$

on a quadratic expression

$$ax^2 + 2hxy + by^2.$$

It becomes

$$a(X\cos\alpha - Y\sin\alpha)^2 + 2h(X\cos\alpha - Y\sin\alpha)(X\sin\alpha + Y\cos\alpha)$$
$$+ b(X\sin\alpha + Y\cos\alpha)^2$$
$$\equiv AX^2 + 2HXY + BY^2, \quad \text{say}.$$

Now the distance of a point from the origin is unchanged by this rotation, so

$$x^2 + y^2 \equiv X^2 + Y^2,$$

and so, for any value of $\lambda$,

$$ax^2 + 2hxy + by^2 + \lambda(x^2 + y^2) \equiv AX^2 + 2HXY + BY^2 + \lambda(X^2 + Y^2).$$

If the left-hand side,

$$(a + \lambda)x^2 + 2hxy + (b + \lambda)y^2 = 0$$

represents a pair of coincident lines, then

$$(a+\lambda)(b+\lambda) = h^2,$$

and then the right-hand side,

$$(A+\lambda)X^2+2HXY+(B+\lambda)Y^2 = 0,$$

which is the equation in the new coordinate system, will also represent a pair of coincident lines, so that

$$(A+\lambda)(B+\lambda) = H^2.$$

These two quadratics in $\lambda$ are

$$\lambda^2+(a+b)\lambda+(ab-h^2) = 0$$

and

$$\lambda^2+(A+B)\lambda+(AB-H^2) = 0,$$

and since they give the same values of $\lambda$,

$$\mathbf{a+b \equiv A+B}$$

$$\mathbf{ab-h^2 \equiv AB-H^2}.$$

We have thus established that the two expressions $a+b$ and $ab-h^2$ are *invariant* under any rotation of the axes; in particular, if one or other vanishes in one coordinate system, it will vanish in any other coordinate system. An example of this has already appeared in the previous chapter, when we saw that the lines given by

$$ax^2+2hxy+by^2 = 0$$

are perpendicular if $a+b = 0$, and coincident if $ab-h^2 = 0$. These properties of the line pair are naturally independent of the particular coordinate axes chosen.

The identities may also be established by plain substitution, since

$$A = a\cos^2\alpha+2h\cos\alpha\sin\alpha+b\sin^2\alpha,$$

$$H = -a\cos\alpha\sin\alpha+h(\cos^2\alpha-\sin^2\alpha)+b\cos\alpha\sin\alpha,$$

$$B = a\sin^2\alpha-2h\sin\alpha\cos\alpha+b\cos^2\alpha,$$

but the algebra is too laborious to be worth exhibiting.

**2. Centre of the conic $S$.** Let $(\lambda, \mu)$ be any point in the plane, and consider a line through it whose gradient is $\tan\alpha$. Then the general point of the line (see p. 32) is

$$(\lambda + r\cos\alpha,\ \mu + r\sin\alpha),$$

where $r$ is the distance of the general point from $(\lambda, \mu)$. This point lies on the conic $S$, if and only if it satisfies the equation

$$ax^2 + 2hxy + by^2 + 2gx + 2fy + c = 0,$$

so that

$$a(\lambda + r\cos a)^2 + 2h(\lambda + r\cos\alpha)(\mu + r\sin\alpha) + b(\mu + r\sin\alpha)^2$$
$$+ 2g(\lambda + r\cos\alpha) + 2f(\mu + r\sin\alpha) + c = 0,$$

or

$$r^2\{a\cos^2\alpha + 2h\cos\alpha\sin\alpha + b\sin^2\alpha\}$$
$$+ 2r\{(a\lambda + h\mu + g)\cos\alpha + (h\lambda + b\mu + f)\sin\alpha\}$$
$$+ \{a\lambda^2 + 2h\lambda\mu + b\mu^2 + 2g\lambda + 2f\mu + c\} = 0.$$

This important equation is called the *r-quadratic*, and a number of results can be deduced from it. First, as the name suggests, it is a quadratic equation in $r$, so, unless it is satisfied for all $r$, there are at most two roots in $r$. Since $r$ is the distance of a point on the line from $(\lambda, \mu)$, it follows that the line meets the conic $S$ in at most two points (unless the line is part of the conic, when $S$ is a line pair, contrary to our assumption).

Now suppose $(\lambda, \mu)$ is the mid-point of the chord through it. Then the roots of the quadratic are equal and opposite, and so the coefficient of $r$ vanishes. We can then find the gradient $\tan\alpha$ of a chord whose mid-point is $(\lambda, \mu)$; this idea is followed up in Examples VIII A, 2.

We now go further, and suppose that $(\lambda, \mu)$ is the *centre* of $S$, so that it is the mid-point of *every* chord through it. This means that

$$(a\lambda + h\mu + g)\cos\alpha + (h\lambda + b\mu + f)\sin\alpha = 0$$

for *all* angles $\alpha$, so that

$$a\lambda + h\mu + g = 0,$$
$$h\lambda + b\mu + f = 0.$$

If $ab - h^2 \neq 0$, these equations have a solution

$$\lambda = \frac{hf - bg}{ab - h^2}, \quad \mu = \frac{gh - af}{ab - h^2},$$

and, conversely, if there is a solution, $ab-h^2 \neq 0$. In this case, $S$ is called a *central* conic.

Two things should be noticed about this result. First, the expression $ab-h^2$, which is one of the invariants found in §1, is important here; when it does not vanish, $S$ has a centre. Secondly, the equations which give the centre are easy to remember, because the coefficients

$$\begin{matrix} a & h & g \\ h & b & f \end{matrix}$$

are the first two rows of the determinant $\Delta$.

**3. Types of central conic.** Suppose $S$ has a centre $(\lambda, \mu)$, so that $ab-h^2 \neq 0$ and

$$a\lambda + h\mu + g = 0,$$
$$h\lambda + b\mu + f = 0.$$

Transfer the origin to $(\lambda, \mu)$, putting

$$x = x^* + \lambda,$$
$$y = y^* + \mu.$$

The equation of $S$ becomes

$$ax^{*2} + 2hx^*y^* + by^{*2} + 2x^*(a\lambda + h\mu + g) + 2y^*(h\lambda + b\mu + f)$$
$$+ (a\lambda^2 + 2h\lambda\mu + b\mu^2 + 2g\lambda + 2f\mu + c) = 0,$$

or

$$ax^{*2} + 2hx^*y^* + by^{*2} = k, \quad \text{say}.$$

Now

$$\begin{aligned} -k &= \lambda(a\lambda + h\mu + g) + \mu(h\lambda + b\mu + f) + (g\lambda + f\mu + c) \\ &= g\lambda + f\mu + c \\ &= \frac{1}{ab-h^2}\{g(hf-bg) + f(gh-af) + c(ab-h^2)\} \\ &= \Delta/(ab-h^2), \end{aligned}$$

since

$$\Delta = \begin{vmatrix} a & h & g \\ h & b & f \\ g & f & c \end{vmatrix} = abc + 2fgh - af^2 - bg^2 - ch^2.$$

Since we have assumed that $\Delta \neq 0$, it follows that $k \neq 0$.

Having moved the origin to the centre of the conic, we now rotate the axes through an angle $\alpha$, putting

$$x^* = X\cos\alpha - Y\sin\alpha,$$
$$y^* = X\sin\alpha + Y\cos\alpha.$$

The equation of $S$ becomes

$$X^2\{a\cos^2\alpha + 2h\cos\alpha\sin\alpha + b\sin^2\alpha\}$$
$$+2XY\{(b-a)\cos\alpha\sin\alpha + h(\cos^2\alpha - \sin^2\alpha)\}$$
$$+Y^2\{a\sin^2\alpha - 2h\cos\alpha\sin\alpha + b\cos^2\alpha\} = k.$$

Choose $\alpha$ so that the coefficient of $XY$ is zero; this is always possible, for then

$$2h\cos 2\alpha - (a-b)\sin 2\alpha = 0$$

or
$$\tan 2\alpha = 2h/(a-b),$$

and, for any values of $a$, $b$ and $h$, the angle $\alpha$ can be found.

The equation of the conic is now

$$AX^2 + BY^2 = k, \quad k = -\Delta/(ab-h^2) \neq 0.$$

Also $A$ and $B$ cannot vanish, or the equation would represent a line pair. The equation thus represents either

(i) an ellipse, if $A$, $B$ and $k$ are all positive or all negative, and in particular a circle when $A = B$; or

(ii) nothing, if $A$ and $B$ are of the same sign, and $k$ has the opposite sign; or

(iii) a hyperbola, if $A$ and $B$ have opposite signs.

From § 1, we know that $AB = ab - h^2$ (since $H = 0$), so we conclude that the central conic $S$ either

(i) is an ellipse, if $ab - h^2 > 0$ and $\Delta < 0$,

or (ii) has no points, if $ab - h^2 > 0$ and $\Delta > 0$,

or (iii) is a hyperbola, if $ab - h^2 < 0$.

**4. Principal axes of a central conic.** The process which has been outlined, that of changing the origin and then rotating the axes, has the effect of referring the conic to its principal axes as axes of coordinates. Once the equation of the conic is written in this standard form, the lengths of its principal axes can be deduced, and so its eccentricity.

There are several practical methods of finding the lengths, and equations, of the principal axes of a central conic. In each case the first step is to change the origin to the centre, and we suppose this already done, the equation of the conic then being

$$ax^2 + 2hxy + by^2 = 1$$

(where we have divided through by the non-zero $k$ to get 1 as the constant term).

*Method 1.* If only the lengths of the principal axes are wanted, the invariant theory of §1 can be used. If the equation of the conic referred to its principal axes is

$$AX^2+BY^2 = 1,$$

then
$$a+b = A+B$$

and
$$ab-h^2 = AB-H^2 = AB, \quad \text{since} \quad H = 0.$$

Thus $A$ and $B$ are the roots of the quadratic equation

$$r^2-(a+b)r+ab-h^2 = 0.$$

If the roots of this equation are both positive, it is usual to take $A$ as the smaller, and then if $A = 1/\alpha^2$, $B = 1/\beta^2$, with $x > \beta$, the equation of the resulting ellipse is

$$x^2/\alpha^2+y^2/\beta^2 = 1,$$

and the lengths of the major and minor axes are $2\alpha$ and $2\beta$ respectively.

If the roots of the quadratic in $r$ have opposite signs, let $A$ be the positive root. If then $A = 1/\alpha^2$ and $B = -1/\beta^2$, the equation of the hyperbola is
$$x^2/\alpha^2-y^2/\beta^2 = 1$$

and the length of the transverse axis is $2\alpha$.

Lastly, if the roots of the quadratic are both negative, the conic has no real points.

This method does not find the directions of the principal axes.

*Method 2.* Suppose the principal axes make angles $\alpha_1$ and $\alpha_2$ with the $x$-axis, so that a rotation

$$x = X\cos\alpha-Y\sin\alpha,$$

$$y = X\sin\alpha+Y\cos\alpha,$$

where $\alpha = \alpha_1$ or $\alpha_2$, reduces the equation of the conic to the form in which the term in $XY$ vanishes. Then $\alpha_1$ and $\alpha_2$ satisfy the equation
$$(b-a)\cos\alpha\sin\alpha+h(\cos^2\alpha-\sin^2\alpha) = 0.$$

Now
$$X = x\cos\alpha+y\sin\alpha,$$

$$Y = -x\sin\alpha+y\cos\alpha,$$

and the equations of the principal axes are $X = 0$ and $Y = 0$; together, then, the equation of the two axes is

$$(x\cos\alpha + y\sin\alpha)(-x\sin\alpha + y\cos\alpha) = 0$$

or
$$(y^2 - x^2)\cos\alpha\sin\alpha + xy(\cos^2\alpha - \sin^2\alpha) = 0.$$

Comparing these two equations for $\alpha_1$ and $\alpha_2$,

$$\frac{x^2 - y^2}{a - b} = \frac{xy}{h},$$

and this is the equation of the principal axes. We notice as a check that the lines are perpendicular. This method finds only the equation of the principal axes, not their lengths.

*Method 3.* From first principles, we can find the values of $\tan\alpha_1$ and $\tan\alpha_2$ from the equation

$$(b - a)\cos\alpha\sin\alpha + h(\cos^2\alpha - \sin^2\alpha) = 0$$

or
$$h\tan^2\alpha + (a - b)\tan\alpha - h = 0.$$

It is then easy to calculate $\sin\alpha_1$ and $\cos\alpha_1$, say, and substitute the values in the equation

$$X^2\{a\cos^2\alpha + 2h\cos\alpha\sin\alpha + b\sin^2\alpha\}$$
$$+ Y^2\{a\sin^2\alpha - 2h\cos\alpha\sin\alpha + b\cos^2\alpha\} = 1.$$

If $\alpha_2$ is used instead of $\alpha_1$, the coefficients of $X^2$ and $Y^2$ are interchanged.

Two other methods apply most suitably to ellipses.

*Method 4.* Put $x = r\cos\theta$, $y = r\sin\theta$, in the equation of $S$. The equation satisfied by $r$ and $\theta$ is then

$$r^2(a\cos^2\theta + 2h\cos\theta\sin\theta + b\sin^2\theta) = 1,$$

where $r$ is the distance of a point $P$ on the conic from the centre $O$, and $\theta$ is the angle which $OP$ makes with the $x$-axis (Fig. 78). $P$ will lie at an end of the major axis when $r$ is a maximum, and at an end of the minor axis when $r$ is a minimum. For this, the expression

$$E = a\cos^2\theta + 2h\cos\theta\sin\theta + b\sin^2\theta$$

is a minimum or a maximum, and the values of $\theta$ for which this happens are found by putting $dE/d\theta = 0$. The roots in $\theta$ of this equation give the directions of the major and minor axes, and then the values of $r$ can be calculated.

If the conic is a hyperbola, this method will give a minimum value of $r$, corresponding to a maximum of $E$, but no maximum of $r$, since a hyperbola extends to infinity, and a point on it can be arbitrarily far from the centre.

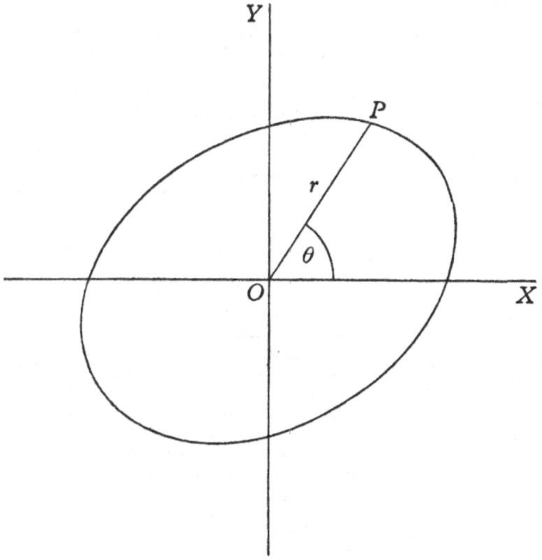


Fig. 78

*Method 5.* The four points where the conic

$$ax^2 + 2hxy + by^2 = 1$$

meets the circle
$$x^2 + y^2 = r^2$$

satisfy
$$ax^2 + 2hxy + by^2 = \frac{x^2 + y^2}{r^2},$$

or
$$\left(a - \frac{1}{r^2}\right)x^2 + 2hxy + \left(b - \frac{1}{r^2}\right)y^2 = 0.$$

This is a homogeneous equation in $x$ and $y$, so it represents a line pair through the origin, $POP'$, $QOQ'$ (Fig. 79). For varying $r$, different line pairs arise; in particular, when the lines coincide, they do so in either the major axis $OA$ or the minor axis $OB$. This happens when

$$\left(a - \frac{1}{r^2}\right)\left(b - \frac{1}{r^2}\right) = h^2,$$

or
$$\frac{1}{r^4}-\frac{1}{r^2}(a+b)+(ab-h^2) = 0,$$

and the roots of this equation in $r^2$ give the squares of the lengths of the semi-axes, $OA^2$ and $OB^2$. Then, if $r = r_1$ gives $OA$, the equation of $OA$ is
$$\left(a-\frac{1}{r_1^2}\right)x+hy = 0,$$

and similarly the equation of $OB$ can be found.

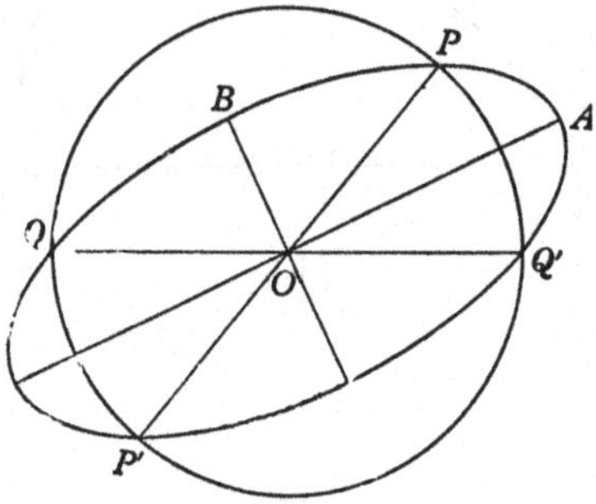


Fig. 79

The reader should consider why this method is unsuitable for use with hyperbolas.

We illustrate the theory, by finding the centre, principal axes, and eccentricity, of the conic
$$8x^2-4xy+5y^2-8x-16y-16 = 0.$$

First, $ab-h^2 = 8.5-(-2)^2 = 36$, so the equation represents an ellipse. The equations for the centre $(\lambda, \mu)$ are
$$8\lambda-2\mu-4 = 0,$$
$$-2\lambda+5\mu-8 = 0,$$
so
$$\lambda = 1, \quad \mu = 2.$$

Transfer the centre to (1, 2), putting
$$x = x^*+1,$$
$$y = y^*+2;$$

the equation becomes

$$8x^{*2}-4x^*y^*+5y^{*2} = 36.$$

If the equation of the ellipse referred to its principal axes is

$$AX^2+BY^2 = 36,$$

$A$ and $B$ are the roots of

$$r^2-(8+5)r+36 = 0,$$

so $$A = 4, \quad B = 9.$$

The reduced equation is thus

$$\tfrac{1}{9}X^2+\tfrac{1}{4}Y^2 = 1,$$

and so the lengths of the major and minor axes are 6 and 4, and the eccentricity $e$ is given by

$$9-4 = 9e^2,$$

$$e = \sqrt{5}/3.$$

(As a check, we notice that $e < 1$.)

The equation of the principal axes is

$$\frac{x^{*2}-y^{*2}}{8-5} = \frac{x^*y^*}{-2}$$

or $$2(x^{*2}-y^{*2})+3x^*y^* = 0;$$

separately the axes are

$$2x^*-y^* = 0 \quad \text{and} \quad x^*+2y^* = 0.$$

The equations of the axes referred to the original origin are then

$$2x-y = 0 \quad \text{and} \quad x+2y = 5.$$

(As a check, notice that these lines do meet in the centre $(1, 2)$, and they are perpendicular, as axes should be.)

The reader should work this example by at least one of the remaining methods already discussed.

**5. Asymptotes of a hyperbola.** Suppose the equation

$$AX^2+BY^2 = k$$

represents a hyperbola, so that $k/A = \alpha^2$, $-k/B = \beta^2$. The asymptotes of this hyperbola are

$$Y = \pm \beta/\alpha . X$$

or
$$\alpha^2 Y^2 = \beta^2 X^2;$$

that is
$$AX^2 + BY^2 = 0.$$

In the previous coordinate system this is

$$ax^{*2} + 2hx^*y^* + by^{*2} = 0;$$

referred to the original origin it is

$$a(x-\lambda)^2 + 2h(x-\lambda)(y-\mu) + b(y-\mu)^2 = 0.$$

Expanding this, and remembering that

$$a\lambda + h\mu + g = 0 \quad \text{and} \quad h\lambda + b\mu + f = 0,$$

we obtain the equation of the asymptotes

$$ax^2 + 2hxy + by^2 + 2gx + 2fy + C = 0, \quad \text{say},$$

where $C = a\lambda^2 + 2h\lambda\mu + b\mu^2$.

There are two things to notice here. First, the asymptotes of $S$ are lines through the centre parallel to the line pair

$$ax^2 + 2hxy + by^2 = 0.$$

In particular, they are perpendicular, so that the hyperbola is rectangular, if and only if $a + b = 0$. The axes, which we showed in §4 to be parallel to

$$h(x^2 - y^2) = (a-b)\,xy$$

are bisectors of the angles between the asymptotes.

Secondly, the equations of the conic and of its asymptotes differ only in the constant term. This means that if $S = 0$ represents a hyperbola, its asymptotes are given by $S + p = 0$, where $p$ is chosen so that $S + p = 0$ represents a line pair.

**6. The conic without a centre.** We now return to the case which has so far been omitted, and see what happens if $ab - h^2 = 0$. There is no centre of the conic in this case, so we cannot as a start simplify by transferring the origin as we have done before. We therefore begin by rotating the axes through a suitable angle $\alpha$

(given by $\tan\alpha = 2h/(a-b)$ as in §3), to remove the term in $XY$, so that the equation becomes

$$AX^2+BY^2+2GX+2FX+C = 0.$$

Now $H = 0$, and $AB = ab-h^2 = 0$, so that either $A = 0$ or $B = 0$. Not both of $A$ and $B$ are zero, or the equation represents a single line. Suppose, without loss of generality, that $A = 0$, $B \neq 0$. The equation can then be written

$$\left(Y+\frac{F}{B}\right)^2 = -2\frac{G}{B}X+\frac{F^2}{B^2}-\frac{C}{B},$$

and $G \neq 0$, since the equation does not represent a line pair. Now transfer the origin, putting

$$x^* = X-\frac{F^2-BC}{2BG},$$

$$y^* = Y+F/B$$

and

$$a^* = -G/2B.$$

The equation then becomes

$$y^{*2} = 4a^*x^*,$$

which represents a parabola.

This completes the proof that the general quadratic equation in $x$ and $y$ represents one of the conics we have studied in previous chapters.

It is always worth using all the information already available, when tackling problems on the general conic, as the following example shows.

***Illustration 1.*** *Show that the equation*

$$9x^2-24xy+16y^2+20x-10y+24 = 0$$

*represents a parabola, and find the equation of its axis, its latus rectum, and its vertex.*

In this equation $ab-h^2 = 9.16-(-12)^2 = 0$, so the equation represents a parabola. The second-degree terms form a perfect square, so the equation can be written

$$(3x-4y)^2+20x-10y+24 = 0.$$

Now we introduce $p$, so that

$$(3x-4y+p)^2 = 2x(3p-10)-2y(4p-5)+p^2-24.$$

Choose $p$ so that the lines

$$3x-4y+p = 0$$

and $$2x(3p-10)-2y(4p-5)+p^2-24 = 0$$

are perpendicular. Then

$$\frac{3}{4}\cdot\frac{3p-10}{4p-5} = -1,$$

and so $$p = 2.$$

The equation of the parabola is thus

$$\frac{(3x-4y+2)^2}{25} = -\frac{2}{5}\left(\frac{4x+3y+10}{5}\right),$$

arranging the coefficients so that when

$$X = \frac{4x+3y+10}{5},$$

$$Y = \frac{-3x+4y-2}{5},$$

the coefficients 4/5 and 3/5 are respectively the cosine and sine of an angle. The equation of the parabola is then

$$Y^2 = -2/5.X.$$

From this we see that the axis of the parabola is

$$Y = 0 \quad\text{or}\quad 3x-4y+2 = 0,$$

and the latus rectum is $\frac{2}{5}$. The vertex is the point where

$$X = Y = 0,$$

and this is the point $x = -\frac{46}{25}$, $y = -\frac{22}{25}$.

### EXAMPLES VIII A

1. $S = 0$ represents an ellipse, which referred to its principal axes is $AX^2+BY^2 = k$. From the condition $A = B$ for this to be a circle, deduce the usual condition $a = b$, $h = 0$, for $S = 0$ to represent a circle.

2. If $(\lambda, \mu)$ is the mid-point of a chord of the conic

$$8x^2-4xy+5y^2-8x-16y-16 = 0,$$

which has gradient $\tan\alpha$, show that $8\lambda-2\mu-4 = (2\lambda-5\mu+8)\tan\alpha$. Hence find the locus of the mid-points of chords of the conic which are parallel to $y+2x = 0$.

3. Show that the conic $29x^2+24xy+36y^2-34x+48y-84 = 0$ is an ellipse, and find the equations and lengths of its principal axes.

4. Find the lengths and directions of the principal axes of the conics

(i) $62x^2-20xy+83y^2 = 174$;

(ii) $19x^2-16xy+49y^2 = 85$;

(iii) $14x^2-4xy+11y^2 = 25$.

5. Find the equation of each asymptote and each principal axis of the hyperbola $32x^2+60xy+7y^2 = 52$. Determine which of the axes meets the curve in real points.

6. Find the centre and the lengths of the principal axes of the conic whose equation is $13x^2-12xy+4y^2-16x = 0$. Show that the axes are inclined at an angle $\tan^{-1}2$ to the coordinate axes, and make a rough sketch of the conic.

7. Find the centre of the conic $9x^2-4xy+6y^2-14x-8y+1 = 0$. Show that this conic is an ellipse, and find its eccentricity.

8. Find the eccentricity of the conic $2x^2+2\sqrt{18}xy+5y^2 = 1$.

9. The equation of an ellipse is $8x^2+12xy+17y^2+2gx+2fy-27 = 0$. Determine $g$ and $f$ so that the centre is the point $(2, 1)$. Further, show that in this case the major axis is parallel to the line $x+2y = 0$ and that the lengths of the axes are $4\sqrt{5}$ and $2\sqrt{5}$.

10. Find the centre of the conic $13x^2-18xy+37y^2+2x+14y-2 = 0$ and show that the curve is an ellipse of semi-axes $\sqrt{(\frac{2}{5})}$ and $\frac{1}{2}\sqrt{(\frac{2}{5})}$.

11. Find the equations and the lengths of the principal axes of the conic $73x^2-72xy+52y^2 = 100$. What is the value of its eccentricity?

12. Show that the conic $x^2+y^2-4xy-2x-20y-11 = 0$ is a hyperbola. Find the coordinates of the centre, and show that the distance between the vertices of the two branches of the hyperbola is 12.

13. Find the points where the hyperbola $11x^2-16xy-y^2 = 75$ cuts its axis, and sketch the curve.

14. Prove that the latus rectum of the parabola

$$9x^2+6xy+y^2+2x+3y+4 = 0$$

is $7\sqrt{10}/100$, and sketch the curve.

15. Prove that the equation $13(x^2+y^2) = (3x+2y-6)^2$ represents a parabola, and find the length of its latus rectum.

16. Show that the principal axes of the conic

$$11x^2+6xy+19y^2-2x-26y+3 = 0$$

are $3x-y+1=0$ and $x+3y-2=0$. Determine $\alpha$, $\beta$, $\gamma$ so that

$$\alpha(3x-y+1)^2+\beta(x+3y-2)^2+\gamma=0$$

is identical with the given conic, and hence (or otherwise) show that the equation of the curve referred to its principal axes is $5x^2+10y^2=3$.

**7. Polar properties of a conic.** There is an alternative way of examining the general conic $S=0$, and this gives a neat synthesis of the properties of polar lines and tangents.

Consider

$$S \equiv ax^2+2hxy+by^2+2gx+2fy+c=0,$$

and two points $P_1(x_1, y_1)$, $P_2(x_2, y_2)$ in the plane. The general point of $P_1P_2$ is

$$\left(\frac{x_1+\lambda x_2}{1+\lambda}, \frac{y_1+\lambda y_2}{1+\lambda}\right)$$

dividing $P_1P_2$ in the ratio $\lambda:1$. This point lies on $S$ if and only if

$$a(x_1+\lambda x_2)^2+2h(x_1+\lambda x_2)(y_1+\lambda y_2)+b(y_1+\lambda y_2)^2$$
$$+2g(x_1+\lambda x_2)(1+\lambda)+2f(y_1+\lambda y_2)(1+\lambda)+c(1+\lambda)^2=0.$$

Such an expression urgently demands a workable shorthand notation, and we write

$$S_{11} \equiv ax_1^2+2hx_1y_1+by_1^2+2gx_1+2fy_1+c,$$

$$S_{12} \equiv ax_1x_2+h(x_1y_2+x_2y_1)+by_1y_2+g(x_1+x_2)+f(y_1+y_2)+c,$$

$$S_{22} \equiv ax_2^2+2hx_2y_2+by_2^2+2gx_2+2fy_2+c.$$

The equation for the intersections of $P_1P_2$ with $S$ then becomes

$$\mathbf{S_{11}+2\lambda S_{12}+\lambda^2 S_{22}=0.}$$

This important equation, which plays a part in the present theory similar to the part played in §2 by the $r$-quadratic, is called *Joachimsthal's ratio equation.*

The equation is quadratic in $\lambda$, so it follows that $P_1P_2$ meets $S$ in at most two points, which will coincide if the equation has coincident roots.

Let us examine the intersections of $P_1P_2$ with the conic $S$ more closely. Let $P_1P_2$ meet $S$ in $A_1$ and $A_2$, where $A_1$ is given by $\lambda=\lambda_1$, and $A_2$ by $\lambda=\lambda_2$. Then

$$P_1A_1:A_1P_2=\lambda_1:1 \quad \text{and} \quad P_1A_2:A_2P_2=\lambda_2:1.$$

Suppose now that $\lambda_1 = -\lambda_2$, so that $A_1$ and $A_2$ divide $P_1P_2$ internally and externally in the same ratio (Fig. 80). The condition for this is that the roots of

$$S_{11} + 2\lambda S_{12} + \lambda^2 S_{22} = 0$$

should be equal and opposite; that is, that

$$S_{12} = 0.$$

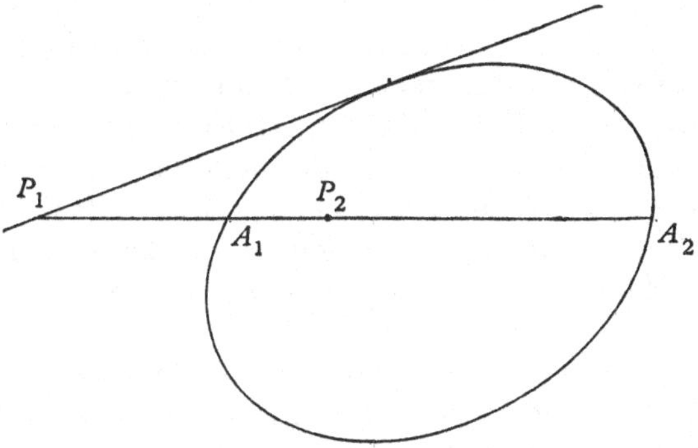


Fig. 80

The argument used so far breaks down if $P_1P_2$ does not meet the conic, but we can generalize the idea so as to include this case. We *define* two points to be *conjugate* with respect to $S$, and only if

$$S_{12} = 0;$$

from the foregoing discussion this means that if $P_1P_2$ meets $S$, the points of intersection divide $P_1P_2$ internally and externally in the same ratio.

As a special case, we see (Fig. 80) that if $P_1$ and $P_2$ are conjugate, and if $A_1$ coincides with $P_1$, then $\lambda_1 = 0$ so that $\lambda_2 = -\lambda_1 = 0$ and so $A_2$ also coincides with $P_1$. Thus if $P_1$ is any general point and a tangent is drawn from $P_1$ to the conic, the point of contact of this tangent with the conic is conjugate to $P_1$.

We now deduce a series of results.

(i) *The polar line of* $P_1$. Consider a fixed point $P_1$. A conjugate point $P_2$ is such that $S_{12} = 0$; that is

$$ax_1x_2 + h(x_1y_2 + x_2y_1) + by_1y_2 + g(x_1 + x_2) + f(y_1 + y_2) + c = 0.$$

The *locus* of conjugate points of $P_1$ is called the *polar* of $P_1$ with respect to $S$; it is given by

$$ax_1x+h(x_1y+xy_1)+by_1y+g(x_1+x)+f(y_1+y)+c = 0,$$

(simply dropping the suffix 2). We write this, more shortly, as

$$S_1 = 0.$$

An immediate result is that the polar of $P_1$ with respect to $S$ is a *line*. Moreover, as it is the locus of conjugate points of $P_1$, it contains the points of contact of tangents drawn from $P_1$ to $S$, and so it is the chord of contact. The terminology used constantly since Chapter III is at last justified, as is the mnemonic inferred but not proved on p. 138. It must, however, be realized that the polar line as now defined exists for *any* position of $P_1$ in the plane, whereas the chord of contact can only be found when $P_1$ is outside the conic. If the polar line of $P_1$ is $l_1$, the point $P_1$ is called the *pole* of $l_1$.

From the definition of polar lines an important corollary follows. If $P_2$ lies on the polar line of $P_1$, then $P_1$ lies on the polar line of $P_2$. For the condition for the first is $S_{12} = 0$ and since $S_{12} = S_{21}$, it follows that $S_{21} = 0$ and so $P_1$ lies on the polar line of $P_2$ (Fig. 81).

(ii) *The tangent at a point* $P_1$. In the particular case when $P_1$ lies on the conic, the polar line of $P_1$ is the tangent to the conic. For then

$$S_{11} = 0$$

(since $P_1$ lies on the conic), and

$$S_{12} = 0,$$

if $P_2$ is a point on the polar line of $P_1$. The ratio equation then gives

$$\lambda^2 S_{22} = 0,$$

of which the roots are $\lambda = 0, 0$. It follows that the polar line of $P_1$ meets $S$ in the point $P_1$ repeated; that is, the line is a tangent to $S$ at $P_1$.

(iii) *The pair of tangents from* $P_1$ *to* $S$. Let $P_1$ now be a general point of the plane. If $P_2$ is any point (Fig. 82) on a tangent from $P_1$ to $S$, the ratio equation

$$S_{11}+2\lambda S_{12}+\lambda^2 S_{22} = 0,$$

which gives the intersections of $P_1P_2$ with $S$, has coincident roots in $\lambda$. Thus

$$S_{12}^2 - S_{11}.S_{22} = 0$$

and the locus of $P_2$ is $S_1^2 - S_{11}.S = 0.$

This is the equation of the pair of tangents from $P_1$ to $S$.

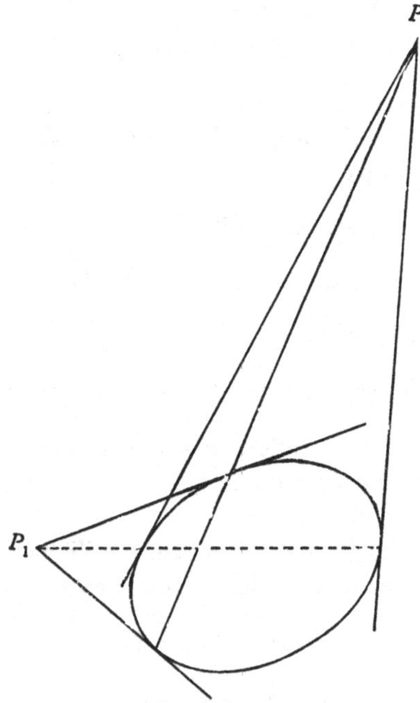


Fig. 81

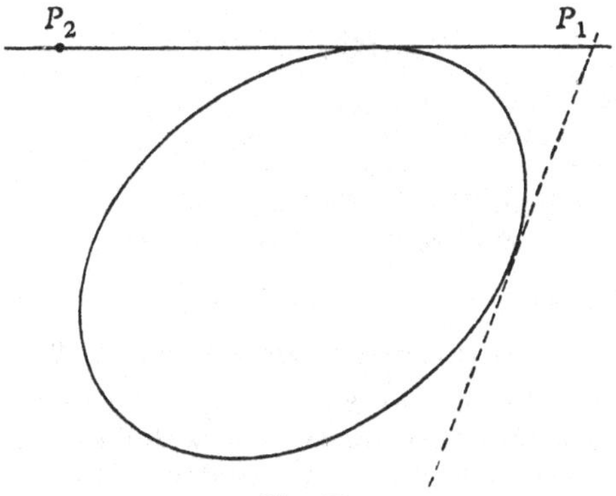


Fig. 82

(iv) *The director circle of S.* Suppose $P_1$ is a point of the director circle of $S$. The equation of the pair of tangents from $P_1$ to $S$ is, as found above,

$$\{axx_1+h(xy_1+yx_1)+byy_1+g(x+x_1)+f(y+y_1)+c\}^2$$
$$-\{(ax_1^2+2hx_1y_1+by_1^2+2gx_1+2fy_1+c)$$
$$\times(ax^2+2hxy+by^2+2gx+2fy+c)\}=0.$$

Since $P_1$ is on the director circle, the tangents from $P_1$ to $S$ are perpendicular and so the sum of the coefficients of $x^2$ and $y^2$ in this equation is zero. Thus

$$(ax_1+hy_1+g)^2+(hx_1+by_1+f)^2$$
$$-(a+b)(ax_1^2+2hx_1y_1+by_1^2+2gx_1+2fy_1+c)=0.$$

The locus of $P_1$ is therefore

$$(h^2-ab)(x^2+y^2)+2(hf-bg)x+2(gh-af)y+g^2-ac+f^2-bc=0,$$

which is indeed a circle, unless $ab-h^2=0$, in which case $S$ is a parabola and the director circle is actually a straight line, the directrix.

The reader would be wise to become familiar with the work of this section by repeating it with a conic in its simple form, say

$$x^2/a^2+y^2/b^2=1,$$

where the algebraic details are less heavy.

## EXAMPLES VIII B

1. Find the equation for the ratio in which the line joining the points $(x_1, y_1)$, $(x_2, y_2)$ is cut by the conic $ax^2+2hxy+by^2+2gx+2fy+c=0$, and deduce the equation of the polar of a point $(X, Y)$ and the equation of the pair of tangents from $(X, Y)$. Find the pole of the line $y+2=0$ with respect to the conic $x^2+2xy-y^2-4x-6y-1=0$.

2. Show that the conic whose equation is

$$a^2x^2+2hxy+b^2y^2-2acx-2bcy+c^2=0$$

touches both coordinate axes. Find the equation of the circle touching the line $bx+ay=0$ whose centre coincides with the centre of the conic. Prove that the two tangents that can be drawn to the conic from any point of this circle are at right angles.

3. Prove that the equation of the tangent at the point $(x', y')$ on the conic

$$ax^2 + 2hxy + by^2 + 2gx + 2fy + c = 0$$

is
$$axx' + h(xy' + yx') + byy' + g(x + x') + f(y + y') + c = 0.$$

A conic touches the $y$-axis at the origin and passes through the point $(1, 0)$, where the tangent is $y + 2x = 2$. Prove that the conic is a member of the family of conics $2x^2 + xy + \lambda y^2 - 2x = 0$. If the conic passes through the point $(1, -\frac{1}{2})$, prove that it is an ellipse.

4. If the line through $A(x_1, y_1)$ and $B(x_2, y_2)$ cuts the conic $S = 0$, where $S \equiv ax^2 + 2hxy + by^2 + 2gx + 2fy + c$ in the points $P$ and $Q$, find an equation whose roots are the ratios $AP/PB$ and $AQ/QB$. Prove that, in general, there is one and only one point $C$ other than $B$ such that the conic divides $AC$ in the same ratio as it divides $AB$. Determine the coordinates of $C$ in the special case when $A$ is $(2, 1)$, $B$ is $(-1, 2)$ and the conic is

$$3x^2 + 5y^2 + 18x = 0.$$

5. Write down the equations of the polars of the point $P(h, k)$ with respect to the circles $x^2 + y^2 + 2\lambda x + c = 0$, $x^2 + y^2 + 2\lambda y - c = 0$. Show that, as $\lambda$ varies, the point of intersection of the polars describes a conic. If $P$ is the point $(1, 1)$ show that this conic is a rectangular hyperbola, and find its equation referred to its principal axes.

6. Prove that the locus of the points of intersection of the polars of points on the line $y = c$ with respect to the parabolas $y^2 = 4ax$, $x^2 = 4ay$ is the hyperbola $2ax^2 - cxy + 4a^2(y + c) = 0$. Find the separate equations of the asymptotes of this hyperbola.

**8. Pencils of conics.** We now turn to a system of conics

$$S + \lambda S' = 0,$$

determined by two conics

$$S \equiv ax^2 + 2hxy + by^2 + 2gx + 2fy + c = 0,$$

$$S' \equiv a'x^2 + 2h'xy + b'y^2 + 2g'x + 2f'y + c' = 0,$$

where $\lambda$ is here an arbitrary parameter.

By the method of inspection (used before in similar circumstances, for $l + \lambda l'$, p. 41, and for circles with a common chord, p. 143), we affirm that:

(i) $S + \lambda S' = 0$ represents a conic, for it is a quadratic equation in $x$ and $y$;

(ii) the conic passes through any point which lies on both $S$ and $S'$.

Such a system is called a *pencil* of conics. The arbitrary parameter $\lambda$ can be chosen so as to satisfy one additional linear condition. In particular, there is just one conic belonging to a given pencil and passing through a given point in the plane, provided the given point does not lie on all the conics of the pencil.

Points which lie on *all* the conics of a pencil are called *base-points* of the pencil. Two conics meet in at most four points, a result which is obvious intuitively and easy to prove in any particular case (taking one of the conics in parametric form, when $x$ and $y$ are quadratic functions of the parameter).

Among the results about a general pencil of conics, we prove that in such a pencil there are at most three line pairs. For the general conic

$$S_\lambda \equiv S + \lambda S' = 0$$

has equation

$$(a+\lambda a')x^2 + 2(h+\lambda h')xy + (b+\lambda b')y^2 + 2(g+\lambda g')x + 2(f+\lambda f')y + (c+\lambda c') = 0,$$

and the condition for this to be a line pair is

$$\begin{vmatrix} a+\lambda a' & h+\lambda h' & g+\lambda g' \\ h+\lambda h' & b+\lambda b' & f+\lambda f' \\ g+\lambda g' & f+\lambda f' & c+\lambda c' \end{vmatrix} = 0.$$

This, when evaluated, will be a cubic equation in $\lambda$, so there are at most three roots in $\lambda$ and at most three line pairs in the pencil.

When we are considering the conics through four points, it is easy to see that there are exactly three line pairs in the pencil, as in Fig. 83. Use of this fact simplifies problems on pencils of conics, as the following example shows.

***Illustration 2.*** *A family of conics is such that each curve passes through the points $A(1,0)$, $B(-1,0)$, $C(0,1)$ and $D(0,2)$. Find the equation of the family. A hyperbola of the family has its asymptotes parallel to the lines $x-2y=0$, $4x+y=0$. Find the separate equations of these asymptotes.*

One line pair in the pencil is $AB$, $CD$ whose equation is

$$xy = 0.$$

Another line pair is $AC$, $BD$ whose separate equations are

$$x+y=1 \quad \text{and} \quad \frac{x}{-1}+\frac{y}{2}=1,$$

so that the combined equation is

$$(x+y-1)(2x-y+2) = 0$$

or
$$2x^2+xy-y^2+3y-2 = 0.$$

Thus the general equation of the family is

$$(2x^2+xy-y^2+3y-2)+\lambda(xy) = 0$$

or
$$2x^2+(1+\lambda)xy-y^2+3y-2 = 0.$$

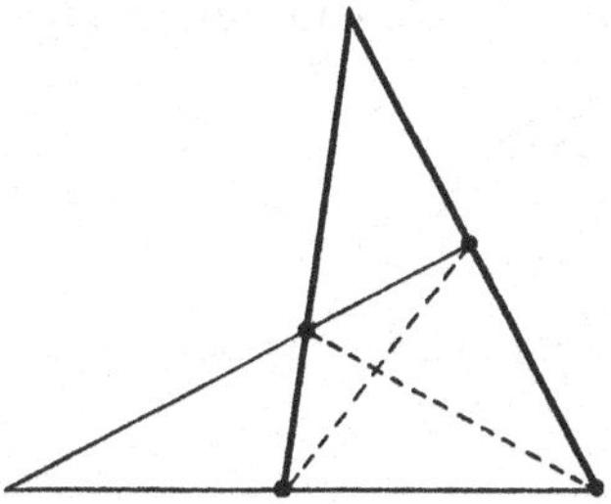

Fig. 83

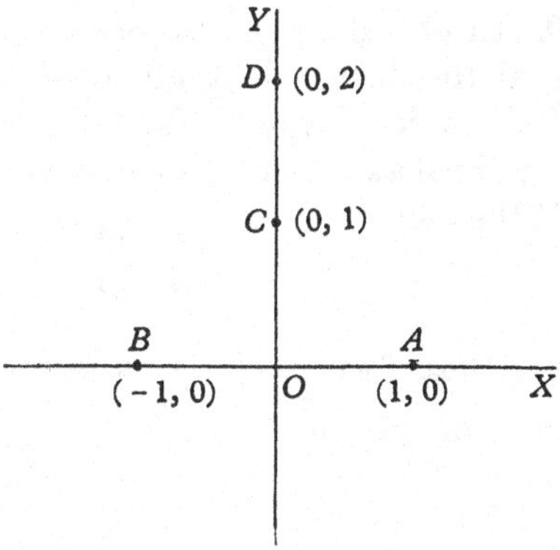


Fig. 84

Suppose for some value of $\lambda$ this equation represents a hyperbola whose asymptotes are

$$x-2y+p = 0 \quad \text{and} \quad 4x+y+q = 0.$$

The combined equation of the asymptotes is

$$(x-2y+p)(4x+y+q) = 0,$$

and so the equation of the hyperbola (which differs from the equation of the asymptotes only in the constant term, p. 246) is

$$4x^2-7xy-2y^2+x(4p+q)+y(p-2q)+pq+r = 0.$$

This is to be the same as the equation above, so

$$\frac{2}{4} = \frac{1+\lambda}{-7} = \frac{-1}{-2} = \frac{0}{4p+q} = \frac{3}{p-2q} = \frac{-2}{pq+r}.$$

From these equations we want values for $p$ and $q$, so we extract

$$4p+q = 0,$$

$$p-2q = 6,$$

giving
$$p = \tfrac{2}{3}, \quad q = -\tfrac{8}{3}.$$

Thus the separate equations of the asymptotes are

$$3x-6y+2 = 0 \quad \text{and} \quad 12x+3y-8 = 0.$$

**9. Confocal conics.** Another one-parameter system of conics is that to which belong all conics with the same foci, called *confocal* conics. Suppose the foci are $(p, 0)$ and $(-p, 0)$, so that the principal axes of all the conics coincide with the coordinate axes. If the ellipse

$$\frac{x^2}{a^2}+\frac{y^2}{b^2} = 1 \quad (a^2 > b^2)$$

has these foci,
$$p^2 = a^2e^2 = a^2-b^2,$$

and similarly if

$$\frac{x^2}{A}+\frac{y^2}{B} = 1$$

is any other central conic with the same foci,

$$A-B = p^2.$$

If
$$A = a^2+\lambda, \quad B = b^2+\lambda,$$

this equation is satisfied, and so all the conics

$$\frac{x^2}{a^2+\lambda}+\frac{y^2}{b^2+\lambda} = 1$$

have the same foci. Since $a^2 > b^2$, it is clear that for $\lambda > -b^2$, the conic is an ellipse and when $-a^2 < \lambda < -b^2$ the conic is a hyperbola.

Through a general point $(h, k)$ of the plane there are just two of the confocal conics. For then

$$\frac{h^2}{a^2+\lambda}+\frac{k^2}{b^2+\lambda} = 1$$

or
$$(a^2+\lambda)(b^2+\lambda)-h^2(b^2+\lambda)-k^2(a^2+\lambda) = 0,$$

which is a quadratic in $\lambda$. Put $b^2+\lambda = \mu$; since $a^2-b^2 = p^2$ the equation becomes

$$(p^2+\mu)\,\mu - h^2\mu - k^2(p^2+\mu) = 0$$

or

$$\mu^2 + \mu(p^2-h^2-k^2) - k^2p^2 = 0.$$

The product of the roots of this equation is negative, so that the values of $\mu = b^2+\lambda$ satisfying it have opposite signs; thus through the point there is one ellipse and one hyperbola (Fig. 85). This basic fact about confocal conics can also be deduced simply from the property $SP+S'P = 2a$ for ellipses, and the analogous result for hyperbolas.

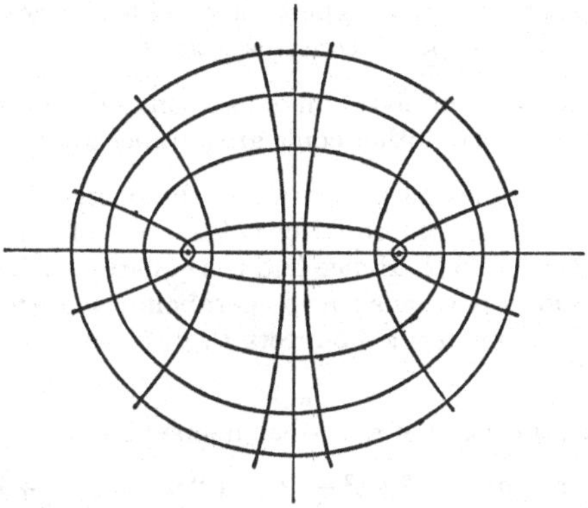

Fig. 85

Suppose the two conics through $(h, k)$ are given by $\lambda = \rho$ and $\lambda = \sigma$. Then

$$\frac{h^2}{a^2+\rho}+\frac{k^2}{b^2+\rho} = 1 = \frac{h^2}{a^2+\sigma}+\frac{k^2}{b^2+\sigma},$$

and so

$$\frac{h^2(\sigma-\rho)}{(a^2+\rho)\,(a^2+\sigma)} = \frac{k^2(\rho-\sigma)}{(b^2+\rho)\,(b^2+\sigma)}.$$

Now the tangents to the conics at $(h, k)$ are

$$\frac{xh}{a^2+\rho}+\frac{yk}{b^2+\rho} = 1$$

and

$$\frac{xh}{a^2+\sigma}+\frac{yk}{b^2+\sigma} = 1,$$

and the product of the gradients of these lines is

$$\frac{h^2}{(a^2+\rho)\,(a^2+\sigma)}\Bigg/\frac{k^2}{(b^2+\rho)\,(b^2+\sigma)} = -1.$$

Thus the two confocals through any point of the plane cut orthogonally.

This is yet another example of two orthogonal systems of curves (cf. p. 157).

## EXAMPLES VIII C

1. Find the equation of the general conic through the four points $(3, 2)$, $(-3, 2)$, $(2, -3)$, $(-2, -3)$. Show that the family of conics so defined contains one parabola, and find its equation.

2. Show that a general pencil of conics $S + \lambda S' = 0$ contains two parabolas and one rectangular hyperbola, and find the condition satisfied by the coefficients if the pencil contains a circle.

3. The four points of intersection of two conics lie upon a circle; prove that the principal axes of either conic are parallel to those of the other.

4. A system of conics all pass through the points $(3, \pm 4)$ and touch $x = -5$ at $(-5, 0)$. Find the equations of the pair of parallel lines in the system, and of the circle $C$. Hence find the equation of the general conic of the system, and in particular of the parabola $P$ and the ellipse $E$ with centre $(2, 0)$. If the line $y = x + 5$ meets $P$, $E$, $C$ again in $L$, $M$, $N$ show that $LM = MN$.

5. Prove that the conic of the pencil defined by

$$xy = c^2 \quad \text{and} \quad x^2 + y^2 - 2ax/k + 2aky = c^2(k^2 + 1/k^2)$$

which passes through the origin consists of one diameter of each conic. Find the equations of the two parabolas passing through the four points of intersection of the conics.

6. Consider the conic $\lambda(2x^2 + 2y^2 - 6xy - 6y - 2) - 2y^2 + 8xy + 4y = 0$. For what real values of $\lambda$, or range of real values, is this conic (i) an ellipse, (ii) a parabola, (iii) a hyperbola, (iv) a degenerate conic consisting of two straight lines? Find the equations of the six lines forming the three degenerate conics.

7. By considering the equation

$$x^2/a^2 + y^2/b^2 - 1 = (lx + my + n)(l'x + m'y + n')$$

or otherwise, show that if four points $A, B, C, D$ of an ellipse are concyclic the lines $AB$, $CD$ are equally inclined to the axes. The point $P(a, b)$ lies on the ellipse $x^2/2a^2 + y^2/2b^2 = 1$, the normal at $P$ meets the ellipse again at $Q$, and the circle on $PQ$ as diameter meets the ellipse again in $R$. Show that the equations of $PR$ and $QR$ are respectively $ax + by = a^2 + b^2$ and $x/a - y/b = 2(a^4 - b^4)/(a^4 + b^4)$.

8. A system of conics is given by the equation $x^2/(a^2 + \lambda) + y^2/(b^2 + \lambda)$, where $\lambda$ is a variable parameter. Prove that (i) the conics have the same

foci, (ii) two conics of the system pass through a general point of the plane, (iii) the poles of a given line $p$ with respect to the conics of the system lie on another line $p'$ perpendicular to $p$. Deduce that there is one conic of the system to which $p$ and $p'$ are tangent and normal.

## MISCELLANEOUS EXAMPLES VIII

1. Find new coordinate axes $Ox'$, $Oy'$ so that the equation of the ellipse $73x^2+72xy+52y^2 = 1$ takes the form $Ax'^2+By'^2 = 1$. Find the lengths of the axes of the ellipse.

2. Find the equations of the principal axes of the rectangular hyperbola $8x^2+12xy-8y^2+6x+2y = 9$, and find the length of the transverse axis.

3. Find the equation of the directrix and the coordinates of the focus of the parabola $9x^2+24xy+16y^2-26x+7y-9 = 0$.

4. Show that the equation $15x^2+34xy+15y^2-64x-64y+63 = 0$ represents a hyperbola. Find the equations of its asymptotes and axes. By a suitable choice of the coordinate axes reduce the equation of the hyperbola to its simplest form.

5. Find the equation of the directrix and the coordinates of the focus and of the vertex of the parabola $16x^2-24xy+9y^2-4x-12y+4 = 0$.

6. Find the coordinates of the centre, the lengths and equations of the axes of the conic $5x^2+8xy+5y^2 = 12x+6y$.

7. If the equation of a central conic $S$, referred to rectangular axes $Ox, Oy$ through the centre, is $7x^2+12xy+2y^2 = 1$, determine the angle $\theta$ so that the new axes $Ox'$, $Oy'$ are the principal axes of $S$, where $\theta$ is the angle $xOx'$. By transforming the equations of $S$, find whether $S$ is an ellipse or a hyperbola and find its eccentricity.

8. The equation of a conic, referred to rectangular axes, is

$$x^2+24xy-6y^2-22x+36y-28 = 0.$$

Find its equation referred to parallel axes through its centre. In terms of the new coordinates, find the equation of the principal axes of the conic. Find also the lengths of the semi-axes.

9. Find the equations of the principal axes of the conic

$$2x^2+12xy-7y^2-16x+2y-3 = 0.$$

10. Find the lengths of the principal axes of the ellipse

$$5x^2+4xy+8y^2 = 36a^2.$$

A rectangle whose sides touch the ellipse is drawn with its sides parallel to the coordinate axes. Prove that its area is $8a^2\sqrt{10}$.

11. Show that the conic

$$(16-4k)x^2+(-2+8k)\,2xy+(-56+59k)y^2+(128-38k)x$$
$$+(114-54k)y+(17+11k)=0$$

is a parabola if and only if $k$ has one of two values. Find the axes of these two parabolas, and show that the axes intersect in the point $(-\frac{9}{2}, 1)$.

12. The asymptotes of a hyperbola which passes through the origin are parallel to the lines $x = 0$, $x+y = 0$, and the line $3x-y-5 = 0$ touches it at (2, 1). Find the equations of the hyperbola and of its asymptotes.

13. Show that the conic $(2x+y-3)^2-(2x-4y+2)^2 = 3$ is a hyperbola, and find the equations of its asymptotes and axes. Obtain the equation of the hyperbola having the same asymptotes and passing through the origin, and show that this hyperbola meets the parabola $y^2 = 2(y-2x)$ in three coincident points at the origin.

14. The coordinates of a point on a curve are given by

$$x = (2t+1)/(t-1), \quad y = (1-2t)/(t+1).$$

Show that the curve is a rectangular hyperbola, and find the coordinates of its vertices and of its foci.

15. Prove that the curve for which the equations, in terms of a parameter $\lambda$, are $x = \lambda(\lambda-1)/(\lambda+1)$, $y = (2\lambda+3)/(\lambda+1)$ is a hyperbola. Show that the equations of the asymptotes are $y = 2$ and $2y-x = 7$, and find the equations of its axes.

16. Show that the equations $x = t^2+t-2$, $y = 1+t-t^2$ represent a parabola, and find the coordinates of its focus and the equation of its directrix.

17. Show that the curve given by the parametric equations

$$x = \cos\lambda+2\sin\lambda, \quad y = \cos\lambda-2\sin\lambda$$

is an ellipse, and find the lengths and the directions of its axes.

18. Find the lengths and positions of the axes of the ellipse described by a point $P$ whose coordinates at time $t$, referred to rectangular axes, are $x = 3\sin\omega t$, $y = 2\cos(\omega t+\frac{1}{6}\pi)$.

19. Find the Cartesian equation of the rectangular hyperbola whose vertices are at the points (5, 4), (−3, −2).

20. Show that the lengths of the semi-axes of the conic

$$ax^2+2hxy+by^2 = 1$$

are given by the equation $(ab-h^2)r^4-(a+b)r^2+1 = 0$. Find the equation of the ellipse having the following properties: (i) the $y$-axis and the line $x = y$ are conjugate diameters; (ii) the product of the lengths of the semi-axes is unity; (iii) it passes through the point $(1/\sqrt{2}, 0)$.

21. If an ellipse is represented by the equation

$$ax^2+2hxy+by^2+2gx+2fy+c = 0$$

prove that: (i) the product $ab$ is necessarily positive; (ii) the origin is outside or inside the ellipse according as the product $ac$ is positive or negative. A diameter of the conic $2x^2+xy+3y^2 = 6x+7y+44$ is parallel to the line $y = 2x$. Find where this diameter meets the $y$-axis.

22. Show that the equation $(x/a)^{\frac{1}{2}}+(y/b)^{\frac{1}{2}} = 1$ represents a parabola touching the coordinate axes, and that the line $y = mx+abm/(am-b)$ is a tangent to the curve for all values of $m$. Hence, or otherwise, prove that the equation of the directrix is $ax+by = 0$, and obtain the coordinates of the focus.

23. Find the equations of the axes of the ellipse $x^2+xy+7y^2/4 = 1$, and find the equation of its director circle. Show that the locus of mid-points of chords of this ellipse which pass through the point $(1,0)$ is $x^2+xy+7y^2/4-x-y/2 = 0$.

24. Prove that the points of intersection of the polars of points on the line given by $y = 0$, with respect to the circles defined by

$$x^2+(y-2)^2 = 1, \quad (x-3)^2+(y-3)^2 = 1$$

lie on a hyperbola.

25. The conics

$$ax^2+2hxy+by^2+2gx+2fy+c = 0, \quad \alpha x^2+\beta y^2 = 1 \quad (\alpha^2 \neq \beta^2),$$

intersect in the points $A, B, C, D$. Find the relations which hold between the coefficients $\alpha, \beta, a, h, \ldots, c$ (i) when $A, B, C, D$ are concyclic, (ii) when the join of two of the points passes through the origin, (iii) when the rectangular hyperbola through $A, B, C, D$ touches the $x$-axis.

---

26. The coordinates of a point $P$ are given parametrically in terms of a real variable $t$ by the relations

$$x = a+b\cosh t+c\sinh t, \quad y = p+q\cosh t+r\sinh t;$$

show that, in general, the locus of $P$ is one branch of a hyperbola. Prove that $q^2-r^2 = c^2-b^2$ is the condition that the hyperbola should be rectangular.

27. Find the general equation of a conic which passes through the origin $O$ and has the axes $OX$, $OY$ as tangent and normal at $O$. If $O$ is a fixed point of a conic $S$, and $P, Q$ are variable points such that the chord $PQ$ subtends a right angle at $O$, prove that, in general, $PQ$ meets the normal at $O$ in a fixed point. Show that, in the particular case in which $S$ is a rectangular hyperbola, the chords are all parallel to the normal at $O$.

28. Show that the equation of any conic may be taken as

$$ax^2+2hxy+by^2+2fy = 0$$

the origin $O$ being any point on the conic. $PQ$ is a variable chord of a conic subtending a right angle at a fixed point $O$ on the conic. Prove that $PQ$ passes through a fixed point $F$, which lies on the normal at $O$ and also on the diameter $HK$, where $OH$, $OK$ are chords parallel to the axes of the conic. Hence, or otherwise, prove that if $R$ is a variable point on the conic $ax^2+by^2=1$ then the corresponding point $F$ (through which chords subtending a right angle at $R$ pass) moves on the conic

$$ax^2+by^2=k^2, \quad \text{where} \quad k=(a-b)/(a+b).$$

29. Find the equation of the conic passing through the points $(a,0)$, $(b,0)$, $(0,c)$, $(0,d)$, $(b,d)$, where $b>a$, $d>c$, and $a$, $b$, $c$, $d$ are all positive. Prove that the conic is an ellipse and that, if $b=2a$ and $d=2c$, its centre is at the point $(6a/5,\ 6c/5)$.

30. Prove that the equation

$$25x^2-120xy+144y^2-2x-29y-1=0$$

represents a parabola. Find the equations of the axis and the tangent at the vertex, the coordinates of the vertex, and the length of the latus rectum.

31. A circle touches the ellipse $x^2/a^2+y^2/b^2=1$ at the point $(x_1,y_1)$ and cuts it at $Q$ and $R$. If the circle passes through the origin prove that the equation of $QR$ is $b^2(a^2-b^2)\,x_1x-a^2(a^2-b^2)\,y_1y=b^4x_1^2+a^4y_1^2$.

32. Show that $\lambda$ can be found so that

$$x^2/a^2+y^2/b^2-1+\lambda(lx+my-1)\,(lx-my-c)=0$$

represents, for any value of $c$, a circle through the intersection of the ellipse $x^2/a^2+y^2/b^2=1$ and the line $lx+my=1$. Show that, if $P$ and $Q$ are the points of contact of the tangents to an ellipse drawn from a point $L$ on an equi-conjugate diameter produced, then the circle through $P$, $Q$ and the centre of the ellipse also passes through $L$.

33. Prove that the polar lines of a fixed point $(h,k)$ with respect to the conics $x^2/(a^2+\lambda)+y^2/(b^2+\lambda)=1$ of a confocal system all touch a parabola with focus $\{h(a^2-b^2)/(h^2+k^2),\ k(b^2-a^2)/(h^2+k^2)\}$ and directrix $kx-hy=0$.

34. Prove that the conics $ax^2+2hxy+by^2=1+k(a'x^2+2h'xy+b'y^2)$ and $ax^2+2hxy+by^2=1$ have the same axes if $(a-b)/h=(a'-b')/h'$. Find the value of $k$ in order that the conic

$$4x^2+3xy+y^2=1+k(3x^2+5xy-2y^2)$$

may be confocal with the conic $4x^2+3xy+y^2=1$.

35. Find the locus of points whose polar lines with respect to the conics $x^2/\alpha+y^2/\beta=1$ $(\alpha+\beta\neq 0)$ and $ax^2+2hxy+by^2=1$ are perpendicular. Hence, or otherwise, show that the conics having centre at the origin which meet the first conic orthogonally at four points are either confocal with it, or else pass through four fixed points on its principal axes.

# IX

## POLAR COORDINATES

An alternative system of coordinates in a Cartesian plane was mentioned in Chapter IV, and a short account is now given. Polar coordinates are useful in their applications in other branches of mathematics, particularly mechanics.

**1. Definition of polar coordinates.** In this system a fundamental point $O$, called the *pole*, and a line through it, $OX$, called the *initial line*, are given (Fig. 86). The position of a point $P$ is specified by giving the length $r$ of $OP$ and the angle $\theta$ which $OP$ makes with the initial line. It is the custom to take $r$ as always positive, and $0 \leqslant \theta < 2\pi$.

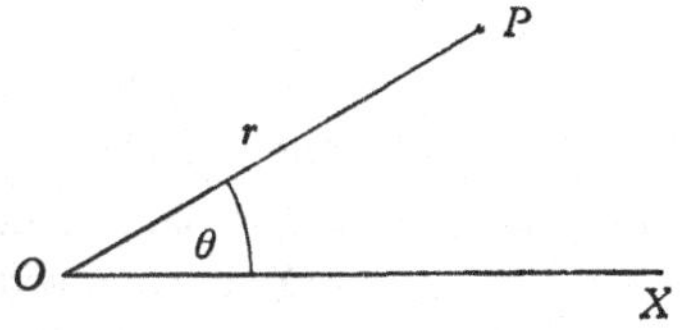

Fig. 86

The basic equations which connect the polar coordinates $(r, \theta)$ with Cartesian coordinates $(x, y)$ of $P$ (referred to $O$ as origin and $OX$ as $x$-axis) are

$$x = r\cos\theta,$$

$$y = r\sin\theta,$$

and in reverse

$$r = +\sqrt{(x^2+y^2)}$$

$$\cos\theta = x/\sqrt{(x^2+y^2)}, \quad \sin\theta = y/\sqrt{(x^2+y^2)}.$$

**2. The equations of a line and of a circle.** Suppose the perpendicular distance $OM$ from the pole $O$ to the line is $p$ (Fig. 87) and $\angle MOX = \alpha$. In Cartesian coordinates the equation of the line is

$$x\cos\alpha + y\sin\alpha = p,$$

and so in polar coordinates the equation is

$$r(\cos\theta\cos\alpha + \sin\theta\sin\alpha) = p,$$

or

$$\mathbf{r\cos(\theta-\alpha) = p}.$$

This can also be deduced direct from the geometry of the figure.

For a line through $O$, when $p = 0$, the equation is $\theta$ = constant.

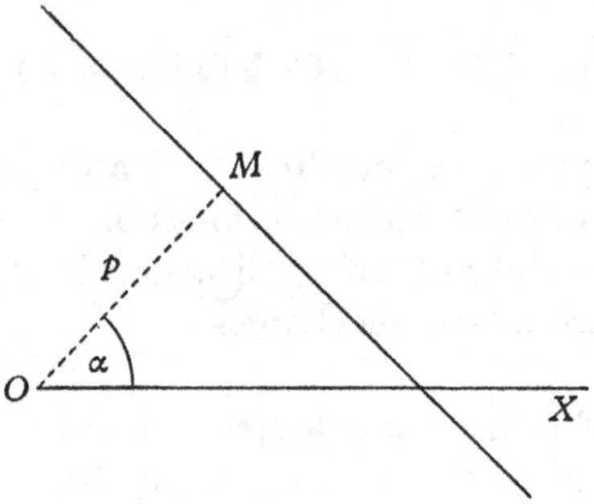


Fig. 87

The equation of a circle whose centre is the pole $O$ is evidently $r$ = constant. When the circle, of radius $a$, passes through $O$ (Fig. 88), take $OX$ along a diameter. Then the equation is

$$\mathbf{r = 2a \cos \theta.}$$

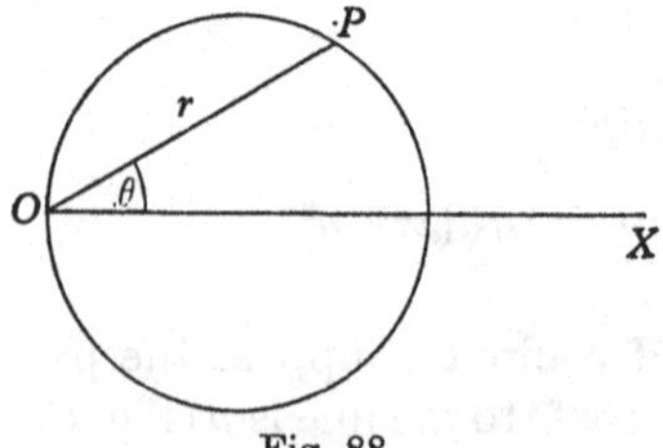


Fig. 88

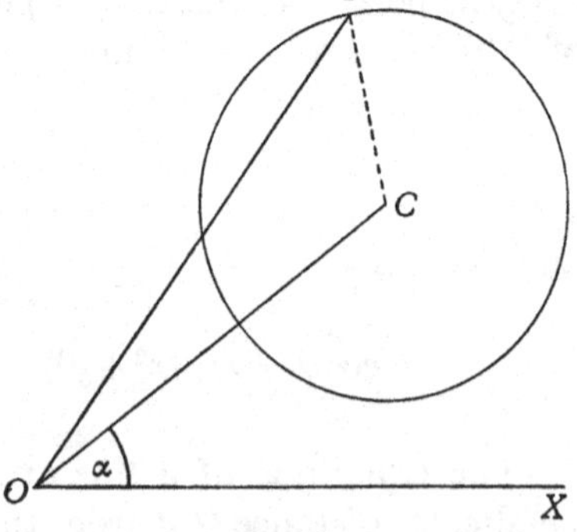


Fig. 89

In general (Fig. 89), if the centre of the circle is $C(d, \alpha)$,

$$CP^2 = OC^2 + OP^2 - 2.OC.OP \cos(\theta - \alpha),$$

so
$$a^2 = d^2 + r^2 - 2dr \cos(\theta - \alpha),$$

or
$$\mathbf{r^2 - 2rd \cos(\theta - \alpha) + d^2 - a^2 = 0.}$$

**3. The equation of a conic with its focus at the pole.** Take the initial line through the focus $S$ perpendicular to the directrix $d$. Let $PM$, $SD$ be perpendicular to $d$, and let $SD = k$. Then $SP = r$ and $\angle PSD = \theta$. For any point $P$ on the conic

$$SP = e.PM$$

so $$r = e(k - r\cos\theta)$$

or $$r(1 + e\cos\theta) = ek.$$

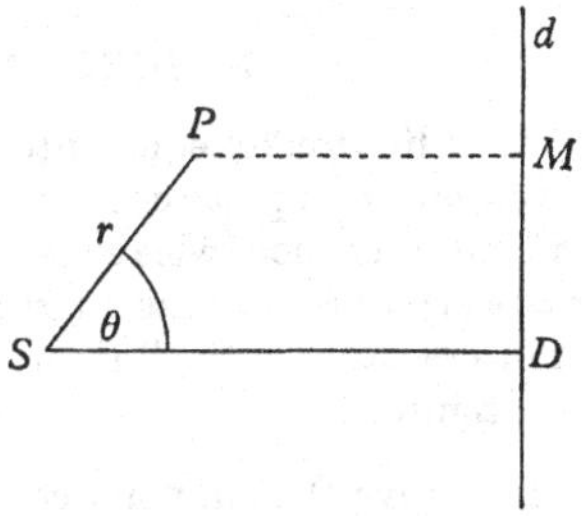

Fig. 90

It is usual to give the result in terms of the length $l$ of the semi-latus rectum. When $\theta = \frac{1}{2}\pi$, $\cos\theta = 0$ and $r = l$, so $l = ek$. The equation is thus

$$\mathbf{1/r = 1 + e\cos\theta.}$$

This form is often of use in calculating results about orbits of particles (or planets) in mechanics.

Properties of tangents are much less neat when expressed in polar coordinates, and few theorems that we have proved so far benefit from being transferred to this new coordinate system.

**4. Other curves in polar coordinates.** A number of curves, including some mentioned in Chapter IV, are most easily investigated in terms of their polar equations. In order to draw the graphs of such curves, it is useful to have a formula of the calculus which gives the angle $\phi$ between the tangent and the radius vector (Fig. 91). The formula is

$$\tan\phi = r\frac{d\theta}{dr}.$$

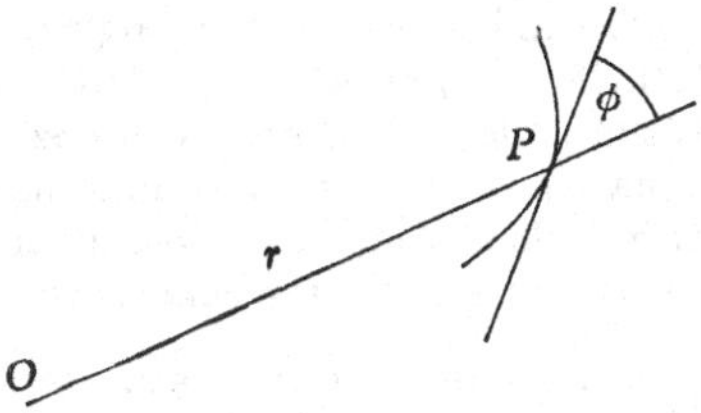

Fig. 91

By using this result, the directions of tangents at a few obvious points can be calculated, and the graph sketched.

Among the curves which are interesting to draw are the following, of which the first three have already been mentioned in Chapter IV.

| | |
|---|---|
| The Cardioid | $r = 2a(1 + \cos\theta)$. |
| The Cissoid | $r = 2(a+b)\sec\theta - 2a\cos\theta$. |

| | |
|---|---|
| The Lemniscate of Bernouilli | $r^2 = a^2 \cos 2\theta$. |
| The Limaçon of Pascal | $r = 2a \cos\theta + k$. |
| The equiangular spiral | $r = ae^{\theta \cot\alpha}$. |

## MISCELLANEOUS EXAMPLES IX

1. A line makes equal intercepts of $2a$ on the axes of $x$ and $y$. Prove that its equation in polar coordinates is $r\cos(\theta - \frac{1}{4}\pi) = a\sqrt{2}$, $\theta$ being measured in a counter-clockwise sense from the axis of $x$. Prove that the equation $r = 4a\cos\theta$ represents a circle. Obtain the values of $\theta$ for the two points of intersection of the line and the circle, and illustrate the result by means of a figure.

2. Prove that in polar coordinates the equation $r\cos(\theta - \alpha) = p$ is the equation of a line and $r = a\cos\theta$ that of a circle. Obtain in polar coordinates the equation of the tangent to the circle $r = 2\cos\theta$ at the point $(1, \frac{1}{3}\pi)$.

3. Find the equation of the tangent to the curve $x = a(2\cos t - \cos 2t)$, $y = a(2\sin t - \sin 2t)$ at the point with parameter $t$, and prove that one of the tangents parallel to the $y$-axis touches the curve at two points. Prove that the equation of the curve may be expressed in the form

$$r = 2a(1 - \cos\theta),$$

where $(r, \theta)$ are polar coordinates referred to the point with Cartesian coordinates $(a, 0)$ as origin and the $x$-axis as initial line. Give a rough sketch of the curve.

4. Show that the polar equation of a conic, with its focus at the origin, and its axis inclined at an angle $\alpha$ to the initial line, can be written in the form $l/r = 1 + e\cos(\theta - \alpha)$. Find the equation of the directrix of this conic. If the circle $r + 2a\cos\theta = 0$ intersects this conic in four real points find the equation in $r$ which determines the distances of these four points from the origin. Show that if the algebraic sum of these four distances is equal to $2a$, the eccentricity of the conic is equal to $2\cos\alpha$.

5. An ellipse of semi-latus rectum $l$, eccentricity $e$ and foci $S$, $S'$ is intersected at $P$ and $Q$ by a parabola of semi-latus rectum $el$, whose axis lies along $SS'$ and whose focus is at $S$. The vertex of the parabola lies on the same side of $S$ as does $S'$. Prove that the area of the triangle $PSQ$ is $l^2(1-e)\{e(2-e+e^2)\}^{\frac{1}{2}}/(1+e)^{\frac{3}{2}}$.

# X

## WHAT IS A CONIC?

As a coda to this book, we bring together various definitions of the curve with which so much of it has been concerned, the conic.

This curve has been considered from numerous points of view, and definitions abound. Indeed, a former Cambridge lecturer used to give the advice that, when faced with proving some property of a conic, the art is to select a definition tactfully close to what is required to be proved.

**1. The conic section.** As the name suggests, the conic section is a section of a cone, and it was as such that the Greeks first considered it. Let $V$ be the vertex of a cone which is such that all the lines of the cone (called *generators*) pass through $V$ and make an angle $\alpha$ with the axis $VW$ (Fig. 92). This surface is called a right circular cone and any section of it is called a conic section.

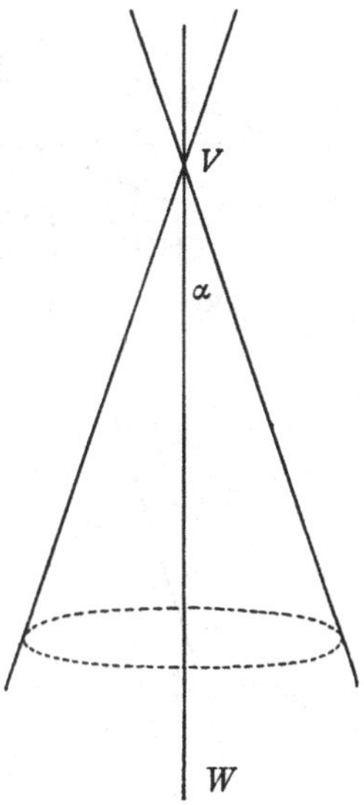

Fig. 92

Let $\pi$ be a plane such that the angle between it and the axis is $\beta$. When $\pi$ passes through $V$, it is clear that the section of the cone is (i) a line pair if $\beta < \alpha$, (ii) a line if $\beta = \alpha$, (iii) a point, $V$, if $\beta > \alpha$. When $\pi$ does not pass through $V$, the section by $\pi$ when $\beta = \pi/2$ is evidently a circle. It remains to consider this section for other values of $\beta$.

Let $S$ be a sphere touching the cone in a circle $C$ (Fig. 93), and touching $\pi$ at a point $F$. Let the plane of the circle $C$ be $\omega$, meeting $\pi$ in a line $d$. Let $P$ be any point on the curve which is the section of the cone by $\pi$, and let $VP$ meet $C$ in $Q$. Let $PM$ be perpendicular to $d$, and let the foot of the perpendicular from $P$ to $\omega$ be $D$, where $PD = h$. Then $\angle QPD = \alpha$, so $h = PQ \cos\alpha$, and $\angle PMD = \frac{1}{2}\pi - \beta$, so

$$h = PM \cos\beta.$$

Also $PF = PQ$ (tangents to a sphere), so $PF = PM \cos\beta \sec\alpha$.

It follows that for any position of $P$ on the curve,

$$PF = e.PM,$$

so that these conic sections have a focus directrix property. Moreover,

$$e = \cos\beta \sec\alpha,$$

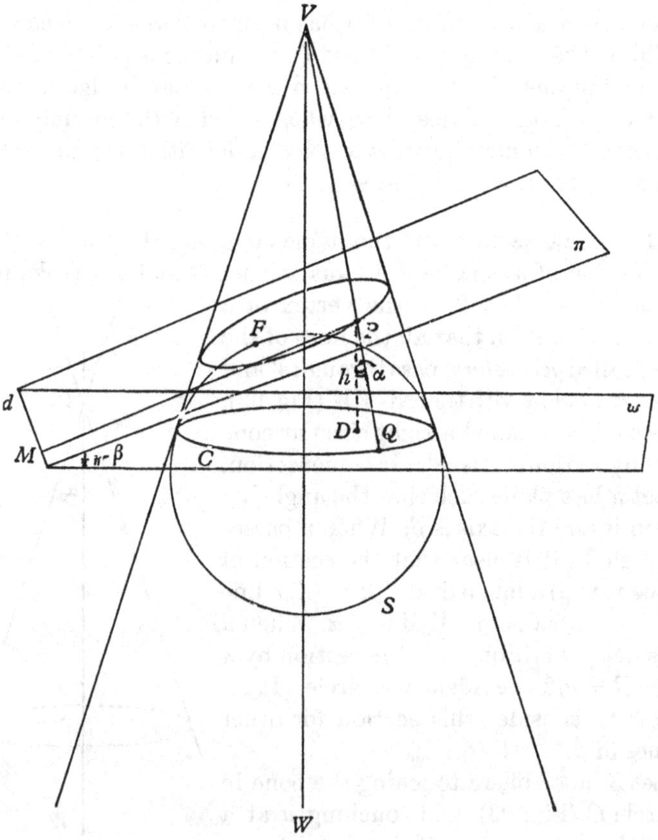


Fig. 93

so that (i) when $\beta > \alpha$, $e < 1$, (ii) when $\beta = \alpha$, $e = 1$, (iii) when $\beta < \alpha$, $e > 1$; thus the curves are ellipses, parabolas and hyperbolas, respectively.

We notice that the conic which is a pair of parallel lines never emerges from this definition.

**2. Focus directrix definition.** It is interesting to consider more closely the curves which arise when we *define* a conic as the locus of a point $P$ such that

$$SP = e.PM,$$

where $S$ is a fixed point and $M$ is the foot of the perpendicular from $P$ to a fixed line $d$. As we already know, if $e < 1$, the conic is an ellipse, if $e = 1$ it is a parabola, and if $e > 1$ it is a hyperbola. Also if $e = 0$, the locus is the point $S$ alone.

If the point $S$ is *on* $d$, the loci which arise are

if $e < 1$, nothing;

if $e = 1$, the line $p$ through $S$ perpendicular to $d$;

if $e > 1$, a pair of lines meeting at $S$ at an angle $2\sin^{-1}(1/e)$.

When $e = 0$, the locus is again the point $S$ only.

With this definition, again the pair of parallel lines never appears, also in this case no choice of $S$ or of $e$ gives a circle.

**3. $SP \pm S'P = 2a$.** This property of the constancy of the sum (for the ellipse) or the difference (for the hyperbola) of the focal distances may be used to define these conics. In the exceptional cases, when $SS' = 2a$, the first conic becomes the segment $SS'$, and the second becomes the whole line *except* $SS'$. When $S$ and $S'$ coincide, the first locus becomes a circle.

Line pairs do not count as conics under this definition.

**4. Curve given by a quadratic equation.** This was a much later development in thought, and arose after the idea of doing geometry by algebra had evolved. Equivalent to this is the definition of the conic as a curve which meets a general line of the plane in at most two points.

A full analysis of the conic from this point of view was given in Chapters VII and VIII; summarizing, we have that the equation $S = 0$ represents:

if $\Delta \neq 0$ and $ab - h^2 > 0$, an ellipse (which is a circle if $a = b, h = 0$);
$ab - h^2 = 0$, a parabola;
$ab - h^2 < 0$, a hyperbola;

if $\Delta = 0$ and $ab - h^2 < 0$, intersecting lines;
$ab - h^2 = 0$, parallel lines (or coincident lines);
$ab - h^2 > 0$, a point.

**5. A locus definition.** Let $A$ and $B$ be any two points in the plane, and take $AB$ as the $x$-axis. Let $a$, $b$ be lines through $A$, $B$ respectively, with gradients $m$, $m'$, and let $a$ and $b$ meet in $P$. If the gradients $m$ and $m'$ are connected by any relation of the form

$$pmm' + qm + rm' + s = 0,$$

the locus of $P$ is a conic through $A$ and $B$. The reader may like to verify this, taking $A$, $B$ as $(\pm 1, 0)$; the case $q = r = 0$, $p = s$ will be seen to give a result which is familiar.

The next significant step in geometry which awaits a specialist is probably twofold. The insistence on real numbers (instead of complex) is relinquished, and by a slight extension of the plane the distinction between intersecting and parallel lines goes too. The former step, while removing the need for 'reality conditions' as in Chapter VII, § 6, has a devastating effect on the ideas of distance and angle, and in the new 'complex projective' geometry which emerges, the definitions of §§ 1, 2 and 3 have no place, though the definition of § 4 still holds. Less obviously, the locus definition just given can be generalized so that the gradients of the lines $a$ and $b$ need not be mentioned, and in the next stage this definition of a conic is fundamental. But that is another story.

# FURTHER MISCELLANEOUS EXAMPLES

The examples below are arranged roughly in increasing order of difficulty. In the first section are examples on the more elementary parts of the work; these are taken from O level 'Additional Mathematics' papers of the G.C.E. Then follow problems using the bookwork of the first six chapters, and lastly some using the topics treated in the final chapters.

1. The points $A(48, 26)$, $B(20, 30)$, $C(0, -10)$ are three of the vertices of a parallelogram $ABCD$. Find (i) the coordinates of $D$, the fourth vertex of the parallelogram; (ii) the coordinates of two points $P$, $Q$ which are such that the area of the rhombus $APCQ$ is equal to the area of the parallelogram $ABCD$.

2. (i) Show that the three lines $3x-2y = 17$, $2x-5y = 4$ and $5x+y = 37$ pass through one point, and find its coordinates. (ii) Show that the three points $(4, 9)$, $(10, 11)$ and $(-5, 6)$ lie on one line, and find its equation. (iii) The line through $(3, 4)$ perpendicular to $3x-2y = 1$ cuts the axes at $A$, $B$ and $AB$ is one diagonal of a rectangle with the axes as sides. Find the equation of the other diagonal.

3. Find and identify the locus of a point which moves so that its distance from the point $(9, 12)$ is twice its distance from the origin.

4. In the triangle $ABC$, $B$ is the point $(2, -3)$, $C$ is the point $(5, 3)$ and the equations of the sides $AB$, $AC$ are $x+y+1 = 0$, $2x-7y+11 = 0$ respectively. Find (i) the coordinates of $A$; (ii) the equation of the line through $A$ at right angles to $BC$; (iii) the equation of the locus of a point which moves so that its distance from $B$ is equal to its distance from $C$.

5. The line joining the points $(4, 0)$ and $(3, 2)$ meets the $y$-axis at the point $(0, b)$; find $b$. Through $(0, 8)$ is drawn a line perpendicular to the line joining the origin to $(3, 2)$; find the coordinates of the point where these two lines meet, and the area of the triangle whose corners are $(0, 0)$, $(0, 8)$ and this meeting-point.

6. A point moves so that the sum of the squares of its distances from the three points $(0, 4)$, $(0, -4)$, $(6, 3)$ is 362. Find the equation of its locus. Show that this locus is a circle through the point $(8, 9)$ and find the coordinates of its centre and the length of its radius.

7. Three points $A$, $B$, $C$ have coordinates $(8, 1)$, $(1, 2)$ and $(4, -2)$ respectively. Lines are drawn through $A$ and $C$ parallel respectively to $BC$ and $BA$, and these lines meet at $D$. Find the equations of $AD$ and $CD$, and hence find the coordinates of $D$, the fourth vertex of the parallelogram $ABCD$. Verify by calculation that the angle $ACB$ is a right angle and find the area of the parallelogram.

8. Find the coordinates of the centroid of the triangle with vertices $A(2, 1)$, $B(4, -5)$ and $C(-2, 3)$. Find also the equation of the locus of the mid-point of the line joining $A$ to a varying point on $BC$.

9. Through $P$, whose coordinates are $(7, 12)$, a line is drawn with gradient $\frac{3}{4}$. Find its equation, and also the equation of the line perpendicular to it through $P$. Through the origin $O$ a line is drawn parallel to the first line to meet the second line at $Q$. Find the length of $OQ$, and the coordinates of $R$ on $OQ$ produced, where $OR = 2OQ$.

10. Three given points are $A(-3, 1)$, $B(1, -2)$ and $C(10, 10)$. Prove that the angle $ABC$ is a right angle, and find the coordinates of the fourth vertex, $D$, of the rectangle $ABCD$. Find also the equation of the diagonal $BD$.

11. A line whose gradient is $\frac{3}{4}$ passes through the point $(5, 2)$ and meets the $x$-axis at $A$ and the $y$-axis at $B$. Find its equation, and also the length of $AB$ and the coordinates of the mid-point of $AB$. Find also the coordinates of the point where this line meets the perpendicular to it through the origin.

12. Find the mid-point of the line joining the points $(-4, 2)$ and $(2, -6)$, and show that the equation of the perpendicular bisector of this line is $3x-4y = 5$. Prove that the point $(3, 1)$ is one vertex of a square of which the points $(-4, 2)$ and $(2, -6)$ are opposite vertices, and find the coordinates of the fourth vertex.

13. $ABCD$ is a quadrilateral in which $A$ is the point $(4, 4)$, $B$ is $(-2, 2)$, $C$ is $(-1, -1)$ and $D$ is $(1, -3)$. Calculate the lengths of the sides of this quadrilateral and of $AC$. Hence or otherwise prove that the angle $ABC$ is a right angle and calculate the area of the quadrilateral.

14. Find the equations of (i) the locus of a point which is equidistant from $(-1, 1)$ and $(7, 5)$; (ii) the locus of a point which is distant 5 units from the point $(-1, 1)$. If $(-1, 1)$ and $(7, 5)$ are opposite vertices of a rhombus of side 5 units, find the coordinates of the two remaining vertices.

15. The line $ax+by = 1$ meets the $x$-axis at $A$ and the $y$-axis at $B$. Find $OA$ and $OB$ in terms of $a$ and $b$ where $O$ is the origin. If the line passes through the point $(2, 2)$, find the coordinates of the mid-point of $AB$ in terms of $a$. Hence find the equation of the locus of this mid-point, for varying values of $a$.

16. Find the equation of the tangent to the circle

$$x^2+y^2+10x-12y+11 = 0$$

at the point $(2, 7)$. Prove that $y = x+1$ is a tangent to this circle, and find the coordinates of the point of contact.

17. Find the centre and radius of the circle $x^2+y^2-16x-12y+75 = 0$. Prove that the point $(11, 10)$ lies on this circle, and find the equation of a

circle of radius $2\frac{1}{2}$ units which touches the above circle internally at the point (11, 10).

18. Find the equations of the tangents to the circle $x^2+y^2+5x-12y=0$ at the two points where the circle cuts the $y$-axis, and find the point of intersection of these tangents. Find also the equation of the circle which has its centre at the origin and which is touched by the above circle internally.

19. Find the equation of a circle which touches the $x$-axis at the point (7, 0) and passes through the point (16, 9). State, without proof, how many circles could be drawn to touch this circle externally and *both* coordinate axes, and calculate the radius of the smallest of these circles.

20. Find the centre and radius of the circle $x^2+y^2+6x-8y+21=0$. Find the length of the tangent $OT$ drawn from the origin $O$ to touch the circle at $T$, and hence find the equation of the circle with $O$ as centre and $OT$ as radius.

21. Find (i) the equation of a circle with centre $(-2, 0)$ and radius 2; (ii) the centre and radius of the circle whose equation is $x^2+y^2-6x-16=0$. Prove that the point $(-5\frac{1}{3}, 0)$ is the point of intersection of the common tangents to these two circles, and hence find the equation of that common tangent which has a positive gradient.

22. Prove that the circles $x^2+y^2-12x-24y+80=0$ and $4x^2+4y^2+12x+48y+149=0$ lie entirely outside each other, and find the shortest distance between them.

23. Prove that the line joining the points $(-1, 3)$ and (5, 7) subtends a right angle at the point (0, 8). Find the coordinates of the centre and the radius of the circle through the three given points, and hence find the equation of this circle. Find the equation of the tangent to this circle at the point (0, 8).

24. A circle has its centre at the point (5, 2), and passes through the point (8, 6). Find its equation, and also the equation of the tangent to it at the point (8, 6). Prove that the chord joining the points (8, 6) and $(2, -2)$ is a diameter.

25. Two circles have the same centre, and the radius of one is five times the radius of the other. The equation of the greater circle is $x^2+y^2-48x-14y=0$; find the equation of the smaller circle. Find also the equation of the circle which has for a diameter the line joining the origin to the common centre of the two circles.

26. Find the coordinates of the foot of the perpendicular from the point (3, 2) to the line $3x+4y-2=0$. Hence find the equation of the circle with centre at (3, 2) touching the line. Prove that the line $3x+4y=32$ is also a tangent to the circle.

27. A point $P$ moves so that its distance from the origin is five times its distance from the point (12, 0). Prove that the locus of $P$ is the circle

$x^2+y^2-25x+150=0$, and find the centre and radius of this circle. Find also the equation of the tangent to this circle at the point (11, 2).

28. Write down the equation of the circle on the line joining (3, 2), (7, $-1$) as diameter. Prove that the tangents to this circle at the points ($3\frac{1}{2}$, $2\frac{1}{2}$) and (3, $-1$) are perpendicular and intersect at the point ($1\frac{1}{2}$, 1). Find also the length of one of the tangents from this point to the circle.

29. Find the coordinates of the centres, and the lengths of the radii, of the circles $x^2+y^2+8x+6y-11=0$ and $x^2+y^2-22x-10y+82=0$. Hence find the equation of their line of centres, and the shortest distance between the two circumferences.

30. Find the equation of the tangent to the curve $y=12/x$ at the point $P(3, 4)$. If the tangent at $P$ cuts the coordinate axes at $M$, $N$, and lines are drawn through $M$, $N$ parallel to the coordinate axes to cut the curve at $Q$, $R$, prove that the chord $QR$ is parallel to the tangent at $P$.

31. Find the equation of the tangent to the curve $y=36/x$ at the point (4, 9). Prove that the point (4, 9) is the mid-point of that part of this tangent cut off by the axes. If $(6c, 6/c)$ is the point on this curve at which the slope of the tangent is $-4$, find a positive value for $c$, and find also the equation of this tangent.

32. Find the equation of the tangent to the curve $y=x^2-4x+5$ at the point (5, 10) and also the equation of the line through (5, 10) which is perpendicular to the tangent. Find the point on the curve at which the tangent is parallel to the line $y=2x$.

33. The coordinates of a point $P$ on the curve $y=x^2$ are (3, 9). Find the equation of the tangent at $P$. If this tangent intersects the $y$-axis at $T$, prove that the $x$-axis bisects $PT$. Prove that the same result follows for the tangent at a point $(c, c^2)$, for any value of $c$.

34. A point moves so as to be always at a distance from the $x$-axis equal to its distance from the point (3, 2); find the equation of its locus. The curve $4y=x^2-6x+13$ meets the line $x=5$ at $A$ and the line $x=9$ at $B$. Find the equation of the chord $AB$.

35. Find the equations of the tangents to the curve $2y=x(x-1)^2$ at the points where $x=0$ and $x=1$. Find the coordinates of the point where the tangent at the origin again meets the curve.

---

36. Find the diagonals of the quadrilateral whose sides are given by $x=3$, $x-y+3=0$, $6x-y+8=0$ and $2x+5y+24=0$. Where do these lines meet? Calculate the acute angle between the lines, and the area of the quadrilateral.

37. Find the equations of the bisectors of the angles between the line joining the points ($-3$, $-9$), (2, 3) and the line cutting intercepts 3, 4 on the positive $x$- and $y$-axes respectively.

38. $ABCD$ is a parallelogram. The equations of $AB$, $AD$ are $2x-3y+1 = 0$, $4x-y-3 = 0$, and the coordinates of the centre of the parallelogram are (3, 4). Find (i) the equations of $BC$, $CD$, (ii) the angle $ABC$, (iii) the area of the parallelogram.

39. The vertices of a triangle are (4, 1), (6, 2), $(\frac{8}{3}, \frac{11}{3})$. Find the equations of the sides of the triangle and the coordinates of its incentre.

40. Write down the equations of the two bisectors of the angles between the pair of lines $2x+y-1 = 0$, $x+2y-2 = 0$, stating which bisector bisects the angle containing the origin. Prove that the pair of lines $ax+y-1 = 0$, $bx+y-1 = 0$, has the same point of intersection as this pair, and find the relation between $a$ and $b$ so that it should also have the same bisectors.

41. Given a point $P$ and a line $l$, the reflection $P'$ of $P$ in $l$ is a point such that $PP'$ is perpendicular to $l$ and $l$ bisects $PP'$. $P$ is the point (1, 1) and $l$, $m$ are the lines $y = 3x$, $3y = x$ respectively. Find the coordinates of the reflections of $P$ in each of these two lines. If $Q$ is a point on $l$ and $R$ is a point on $m$, by considering these two reflections find the minimum value of $PQ+QR+RP$.

42. Show that the points (1, $-1$), (0, $-2$), ($-3$, 1) and ($-2$, $-2$) lie one in each of the four regions of the plane defined by the lines $x+2y+3 = 0$ and $3x-2y+1 = 0$.

43. $A$, $B$, $C$ are the points ($-1$, $-1$), (2, 0), (0, 3) respectively, and $a'$, $b'$, $c'$ are the lines with equations

$$x+y+4 = 0,\ x-y-2 = 0,\ x-2y+1 = 0$$

respectively. If $A'$ is the point of intersection of $b'$ and $c'$ and $B'$, $C'$ are defined similarly, prove that the three lines through $A'$, $B'$, $C'$ perpendicular to $BC$, $CA$, $AB$ respectively are concurrent.

44. Two lines have equations $y = m_1x+n_1$, $y = m_2x+n_2$, where $m_1$, $m_2$ are the roots of the equation $m^2-bm+c = 0$. The bisectors of the angles between the lines have equations $y = m_3x+n_3$, $y = m_4x+n_4$. Prove that $m_3$, $m_4$ are the roots of $bm^2+2(1-c)m-b = 0$.

45. A rectangle is formed by the four points $RSTU$ whose coordinates are (2, 1), ($-2$, 1), ($-2$, $-1$), (2, $-1$) respectively. If $P$ is any point in the plane, find the equations of the perpendicular to $PR$ at $R$, and of the perpendicular to $PT$ at $T$. Show that the intersection $Q$ of these perpendiculars describes a hyperbola when $P$ moves along the line $SU$.

46. Show that the coordinates of any point on the line through $(a, b)$ making an angle $\alpha$ with the $x$-axis can be written in the form $x = a+r\cos\alpha$, $y = b+r\sin\alpha$. A line which makes an acute angle $\theta$ with the positive $x$-axis is drawn through the point $P(3, 4)$ to cut the curve $y^2 = 4x$ at $Q$ and $R$. Show that the lengths of the segments $PQ$ and $PR$ are the numerical values of the roots of the equation

$$r^2\sin^2\theta+4r(2\sin\theta-\cos\theta)+4 = 0.$$

Hence find the gradients of the tangents from $P$ to the curve.

47. Find the equation of the locus of mid-points of chords of the parabola $y^2 = 4ax$ which pass through the point $(h, k)$. Show that this locus is a parabola with the line $y = k/2$ as axis and the point $(h - k^2/8a, k/2)$ as vertex.

48. Prove that the parabola $y^2 = 4ax$ and the hyperbola $xy = c^2$ have only one real point of intersection. Calculate the angle of intersection of these two curves when $a = 1$ and $c = 4$.

49. $P$ is the point $(ap^2, 2ap)$ on the parabola $y^2 = 4ax$. Show that if the line through the focus perpendicular to the tangent at $P$ meets the directrix in $Q$, then $PQ$ is parallel to the axis of the parabola. A circle touches the parabola at the points where the line $x = c$ cuts it. Show that the centre of the circle is at the point $(c+2a, 0)$, and find its radius.

50. The tangents at points $P$ and $Q$ on a parabola meet at $T$. If the mid-point of $PQ$ lies on a fixed line perpendicular to the axis, prove that $T$ lies on a fixed parabola.

51. If the point of intersection of the tangents to the parabola $y^2 = 4ax$ at $P$, $Q$ lies on the parabola $y^2 + 4ax = 0$, show that the mid-point of $PQ$ lies on the parabola $3y^2 = 4ax$.

52. A parabola is given parametrically by the equations $x = at^2$, $y = 2at$. Show that the line through the point $A(\alpha, \beta)$ parallel to the line $y = mx$ is a tangent to the parabola if and only if $\alpha m^2 - \beta m + a = 0$. If the roots of this equation are $m$, $m'$, verify that

$$\frac{\beta - m'(\alpha - a)}{(\alpha - a) + \beta m'} = m,$$

and deduce that the angle between one of the tangents from $A$ and the axis is equal to the angle between the other tangent and the line $AF$, where $F$ is the focus.

53. Show that the parabolas $x^2 + 4y = 4$ and $x^2 + 2y = 2x$ touch one another and find the equation of their common tangent. Find the coordinates of the vertex of the inner parabola.

54. The point $P$ on the parabola $y^2 = 4ax$ has coordinates $(at^2, 2at)$. Find the equation of the line through $P$ perpendicular to $OP$, where $O$ is the vertex. A point $Q$, not on the parabola, has coordinates $(\alpha, \beta)$. Prove that there are three points $P_1$, $P_2$, $P_3$ on the parabola at which $OQ$ subtends a right angle. Prove also that the normals to the parabola at these three points are concurrent in a point $R$, and express the coordinates of $R$ in terms of $\alpha$ and $\beta$.

55. The tangents to a parabola at $P$ and $Q$ meet in $R$. If the chord $PQ$ subtends a right angle at the focus of the parabola, show that the locus of $R$ is a rectangular hyperbola. Show that the centre of this hyperbola is the reflection of the focus of the parabola in the directrix of the parabola.

56. The constants $h$ and $c$ are chosen so that the curve $y(x-h) = c^2$ meets the parabola $y^2 = 4ax$ at right angles at a given point $P(at^2, 2at)$. The curve meets the parabola again at $Q$ and $R$. Prove that (i) the tangents at $Q$ and $R$ to the parabola meet on the line $x = 2a$; (ii) the normals to the parabola at $Q$ and $R$ meet at $P$.

57. Find the equation of the locus of a point from which the two tangents to the parabola $y^2 = 4ax$ are inclined at a given angle $\alpha$. Prove that an infinity of equilateral triangles may be drawn with their vertices on the curve $y^2 = 3x^2 + 10ax + 3a^2$ and their sides touching the parabola $y^2 = 4ax$.

58. $ABC$ is an isosceles triangle in which $AB = AC$; $A$ is a fixed point, $B$ varies on a fixed line through $A$, and the mid-point of $BC$ varies on a fixed line which is perpendicular to $AB$ and does not pass through $A$. Find an equation for the locus of $C$ and prove that $BC$ touches the curve at $C$. If $Q$ is the point on the inward normal at $C$ such that $QC = BC$, and $P$ is the mid-point of $BQ$, give a description of the locus of $P$.

59. Show that if the normals to a parabola $y^2 = 4ax$ at $P$, $Q$, $R$ meet in a point $S$ then there is a rectangular hyperbola of the form $(x-b)\,y = c^2$ through $P$, $Q$, $R$, $S$.

60. Prove that the equation

$$2yt = x + (t^2-1)\,a, \tag{1}$$

where $a$ and $t$ are parameters, represents a family of parallel lines when $t$ is fixed and $a$ is variable, and write down the equation of the line which passes through the origin. When $a$ is fixed and $t$ is variable, prove that equation (1) represents a family of tangents to the parabola $y^2 = a(x-a)$. When $a$ and $t$ both vary, find the relation which must hold between $a$ and $t$ in order that equation (1) shall represent the family of tangents to the circle $x^2 + y^2 = 1$.

61. A point $P$ moves so that its distance from the line $x = 0$ is equal to the length of the tangent from it to the circle whose centre is $(p, 0)$, where $p$ is not zero, and whose radius is $a$. Prove that the locus of $P$ is a parabola, and find the equation of its directrix.

62. Two equal parabolas $S_1$ and $S_2$ have a common vertex and axes at right angles. Consider a set of lines parallel to a given direction. The locus of mid-points of chords of $S_1$ meets the locus of mid-points of chords of $S_2$ in $P$. Show that the locus of $P$ as the direction of these chords varies is a rectangular hyperbola and identify its centre and asymptotes.

63. A fixed line $l$ is parallel to the axis of a parabola but not coincident with it. From a variable point on $l$ three normals are drawn to the parabola. Prove that the triangle formed by the tangents at the feet of these normals is inscribed in a fixed rectangular hyperbola.

64. A line $l$ is drawn perpendicular to the axis of a parabola whose focus is $S$. A circle $C$ whose centre is a point $P$ of $l$ meets the parabola in

four points $Q_1, Q_2, Q_3, Q_4$. Show that $SQ_1+SQ_2+SQ_3+SQ_4$ is independent of the position of $P$ on $l$ and the radius of $C$. Prove further that $SQ_1^2+SQ_2^2+SQ_3^2+SQ_4^2$ is also independent of the position of $P$ on $l$ provided $C$ intercepts a constant length on the axis.

65. The circle having as extremities of a diameter the points $P(ap^2, 2ap)$, $Q(aq^2, 2aq)$ of the parabola $y^2 = 4ax$ meets the parabola again in $U(au^2, 2au)$, $V(av^2, 2av)$. Prove that $u$, $v$ are the roots of the equation $t^2+(p+q)t+pq+4 = 0$, and that

$$\frac{UV^2}{PQ^2} = 1-\frac{16}{(p-q)^2}.$$

66. A fixed point $O$ lies on a given line $OY$. Prove that a point $P$ whose distance from $O$ exceeds its perpendicular distance from $OY$ by $2a$ (where $a$ is constant) lies on one or other of two fixed parabolas. Determine whether *every* point of each parabola gives a possible position for $P$.

67. $P_2(x_2, y_2)$ is a point on the polar of the point $P_1(x_1, y_1)$ with respect to the parabola $y^2 = 4ax$. If $P_3(x_3, y_3)$ is the point of intersection of the polars of $P_1$ and $P_2$ show that $P_1P_2$ is the polar of $P_3$. Prove also that the lines joining the mid-points of the sides of the triangle $P_1P_2P_3$ touch the parabola.

68. A variable triangle is inscribed in the parabola $y^2 = 4ax$ and two of its sides touch the parabola $y^2 = 4bx$; prove that the third side touches the parabola $y^2 = 4cx$ where $(2a-b)^2c = ab^2$.

69. $P$ and $Q$ are variable points of the parabola $x = t^2$, $y = t$ such that $PA$ and $QA$ are perpendicular, where $A$ is the fixed point $(\alpha^2, \alpha)$. Prove that the chords $PQ$ pass through a fixed point $H$. Show also that, for two positions of $A$ on the parabola, $H$ is the mid-point of the normal chord at $A$.

70. A variable chord $QR$ of a parabola subtends a right angle at a point $P$ of the curve. Prove that $QR$ meets the normal at $P$ in a fixed point $F$, and that $PF$ is bisected by the axis. Prove that, as $P$ varies on the parabola, the locus of $F$ is an equal parabola.

71. Prove that, if the points $P(p, 1/p)$, $Q(q, 1/q)$, $R(r, 1/r)$, $S(s, 1/s)$ on the rectangular hyperbola $xy = 1$ are concyclic, then $pqrs = 1$. Deduce that, if another circle through $P$, $S$ cuts the hyperbola in $Q'$, $R'$ then $Q'R'$ is parallel to $QR$.

72. The point $P$ is taken on the auxiliary circle $x^2+y^2 = 4a^2$ of the rectangular hyperbola $xy = 2a^2$; $Q$ is the mirror image of $P$ in the major axis $x-y = 0$ of the hyperbola. The tangents to the circle at $P$, $Q$ meet in $R$. Prove that the triangle $PQR$ is self-polar with respect to the hyperbola (that is, each vertex of the triangle is the pole of the opposite side).

73. Prove that the lines joining any point of a rectangular hyperbola to the ends of a diameter are equally inclined to either asymptote. Prove

also that the tangent at any point of a rectangular hyperbola and the diameter through this point are equally inclined to either asymptote.

74. A point $P$ on a rectangular hyperbola has coordinates $(ct, c/t)$. Find the equation of the tangent at $P$ and show that it forms with the asymptotes a triangle of constant area. A similar hyperbola is obtained by rotating the above about the origin through an angle $\theta$ in its plane. If any tangent common to the two hyperbolas is inclined at an angle $\phi$ to the $x$-axis, prove that $\cos(\theta - 2\phi) = 0$.

75. The normal to the hyperbola $xy = c^2$ at the point $P(ct, c/t)$ meets the curve again at $Q$, and the tangents at $P$ and $Q$ meet at $T$. Prove that the locus of $T$ as $P$ moves on the hyperbola is given by the equation $(x^2 - y^2)^2 + 4c^2xy = 0$.

76. Show that the tangent at $P(8t^2, 8t^3)$ to the curve $8y^2 = x^3$ meets the curve again at the point $Q$ whose coordinates are $(2t^2, -t^3)$. If $S$ is the foot of the perpendicular from $P$ to the $x$-axis, $T$ is the point where the tangent at $P$ cuts the $y$-axis, and $O$ is the origin, prove that $OQ$ and $TS$ are equally inclined to the $x$-axis.

77. Sketch the curve whose parametric equations are $x = t^2$, $y = 4/t$. Show that the locus of the point of intersection of two real perpendicular tangents to the curve is a circle which touches the curve at two points.

78. A circle with centre $P(2, 3)$ and radius 15 meets a circle centre $Q$, radius 8 in two points $A$, $B$. Find the equation of the locus of $Q$ if the points $P$, $A$, $Q$, $B$ are concylic.

79. The pair of lines $x^2 + 3xy + y^2 = 0$ and the line $x + y = 1$ intersect in the points $P$ and $Q$. Show that the circle on $PQ$ as diameter passes through the points $(2, 0)$ and $(0, 2)$.

80. A coaxal system of circles is given by the equation

$$x^2 + y^2 + 2\lambda x + d = 0,$$

where $d > 0$. Show that the locus of poles of the line $y = mx + c$ $(m \neq 0)$ with respect to the circles of this system is a hyperbola passing through the limiting points.

81. Prove that the end points of a set of parallel diameters of a coaxal system of circles lie on a rectangular hyperbola.

82. $AB$ is a fixed diameter of a circle and $PQ$ is a variable chord whose direction is fixed. Prove that the locus of the point of intersection of $AP$ and $BQ$ is a rectangular hyperbola which is also the locus of the point of intersection of $AQ$ and $BP$.

83. The tangent at a point $P$ on the hyperbola $b^2x^2 - a^2y^2 = a^2b^2$ meets the asymptotes in $U$, $V$. Prove that $P$ is the mid-point of $UV$. Prove that $OP^2 - PU^2 = a^2 - b^2$ and that $OP^2 + PU^2 \geqslant a^2 + b^2$.

84. Prove that the feet of the normals from $P(h, k)$ to the ellipse $S$, $x^2/a^2+y^2/b^2 = 1$, are the intersections of $S$ with the rectangular hyperbola $H$, $(a^2-b^2)\,xy+b^2kx-a^2/hy = 0$, and deduce that in general, if $P$ lies on $S$, three normals (apart from the normal at $P$) can be drawn from $P$ to $S$. The feet of these normals from a point $P$ of $S$ are $L$, $M$, $N$ and $L$ is the point $(a\cos\theta, b\sin\theta)$. By using the fact that the pair of lines $PL$, $MN$ pass through the common points of $S$ and $H$, or otherwise, show that the equation of $MN$ is

$$\frac{x}{a^3}\cos\theta-\frac{y}{b^3}\sin\theta+\frac{1}{a^2-b^2} = 0.$$

85. A rectangular hyperbola of centre $O$ is cut in four points $A, B, C, D$ by a circle whose centre $P$ does not lie on either of the axes of symmetry of the hyperbola. Prove that two parabolas can be drawn through $A$, $B$, $C$, $D$, and that the axes of the parabolas meet at the mid-point of $OP$.

86. Find the condition that the lines

$$lx+my+n = 0, \quad l'x+m'y+n' = 0$$

should be conjugate with respect to the ellipse $x^2/a^2+y^2/b^2 = 1$, that is, that one should pass through the pole of the other. Hence show that, if the lines joining a variable point $P$ to two fixed points are conjugate, then the locus of $P$ is a conic. Deduce from this result that the locus of mid-points of the ellipse through a fixed point $A$ is a similar ellipse with its centre at the mid-point of $OA$ (where $O$ is the origin).

87. An ellipse $S$ has centre $O$. Prove that, if $R$ is any point other than $O$ inside $S$, there is just one chord of $S$ which has $R$ as mid-point. A chord $PQ$ of $S$ varies so that $AP = AQ$, where $A$ is a fixed point of $S$. Prove that the mid-points of $PQ$ lie on a rectangular hyperbola.

88. The point $A$ and an ellipse $S$ with centre $C$ are given. Prove that the locus of points $P$ such that $AP$ is parallel to the diameter conjugate to $PC$ is a conic $S'$ through $C$ and $A$. Prove also that the tangents to $S'$ at $C$ and $A$ are parallel.

89. If the polar of a point $P$ with respect to the ellipse

$$(x+a)^2/a^2+y^2/b^2 = 1$$

touches the parabola $y^2 = 4ax$, show that the locus of $P$ is the hyperbola $x(x+a)/a^4-y^2/b^4 = 0$. Find the asymptotes of this hyperbola, and show that they intercept on the $y$-axis a length equal to the semi latus rectum of the ellipse.

90. Show that the inverse of the curve $(x-y)(x^2+y^2)+xy = 0$ with respect to the circle $x^2+y^2 = 1$ is a hyperbola, and find the equations of the asymptotes of this hyperbola.

91. The circle having three-point contact with a parabola $S$ at a variable point $P$ meets $S$ again in $Q$. Prove that the locus of the mid-point of $PQ$ is another parabola whose latus rectum is one-fifth of that of $S$.

92. A variable circle through two fixed points $A$, $B$ meets a fixed line $h$ in $P$ and $Q$. Prove that the locus of the point of intersection of the lines $AP$ and $BQ$ is a rectangular hyperbola through $A$ and $B$ having the line $AB$ as diameter. Show that the tangents to the hyperbola at $A$ and $B$ are parallel to $h$, and that the asymptotes are parallel to the angle-bisectors of the lines $AB$ and $h$.

93. The tangent at a point $P$ of an ellipse meets the major axis $AA'$ (produced) at $Q$. Prove that $P$, $Q$ are conjugate with respect to the auxiliary circle of the ellipse. A system of ellipses is drawn, the extremities of whose major axes are the fixed points $A$, $A'$, and $Q$ is a fixed point on $AA'$ produced. Prove that the normals to the ellipses at points where the tangent passes through $Q$ touch a parabola.

94. Two ellipses congruent with the ellipse $ax^2+by^2 = 1$ are coplanar and concentric; prove that their common diameters bisect the angles between their major axes, and are of lengths

$$2(a\cos^2\alpha+b\sin^2\alpha)^{-\frac{1}{2}}, \quad 2(a\sin^2\alpha+b\cos^2\alpha)^{-\frac{1}{2}},$$

where $2\alpha$ is the angle between the major axes of the ellipses.

95. $D$ and $E$ are two points on an ellipse $x^2/a^2+y^2/b^2 = 1$, and the tangents at these points meet in $P$. If $P$ is the point $(h, k)$, obtain the combined equation of the diameters $CD$ and $CE$. Find the locus of $P$ (i) when the diameters are perpendicular, (ii) when they are conjugate. Show that the two loci meet in four points, which are the vertices of a rectangle.

96. A variable line through the fixed point $(p, q)$ cuts the fixed lines $ax^2+2hxy+by^2 = 0$ in the points $A$, $B$ and the parallelogram $OAPB$ is completed; prove that the locus of $P$ is the hyperbola

$$ax^2+2hxy+by^2-2(ap+hq)\,x-2(hp+bq)\,y = 0.$$

97. Show that the line $y = mx+c$ touches the ellipse $x^2/a^2+y^2/b^2 = 1$ if $c^2 = a^2m^2+b^2$. Hence, or otherwise, show that the equation of the pair of tangents from the point $P(\alpha, \beta)$ may be expressed in the form

$$(\beta x-\alpha y)^2 = a^2(y-\beta)^2+b^2(x-\alpha)^2.$$

These tangents cut the $x$-axis at the points $A$ and $B$. (i) If $PA$ and $PB$ are perpendicular, find the locus of $P$. (ii) If the mid-point of $AB$ is the fixed point $(k, 0)$ show that the locus of $P$ is the parabola $ky^2 = b^2(k-x)$.

98. Prove that if $ax^2+2hxy+by^2+2gx+2fy+c = 0$ represents a pair of lines $L$, $L'$, then the slopes $m$, $m'$ of the two lines are the roots of the equation in $m$, $a+2hm+bm^2 = 0$. If, further, one line of the pair

$$Ax^2+2Hxy+By^2+2Gx+2Fy+C = 0$$

is perpendicular to $L$, prove that

$$2(hA+bH)\,m = bB-aA, \quad (hA+bH)\,m^2 = -(hB+aH),$$

and deduce that

$$(aA-bB)^2+4(aH+hB)(hA+bH) = 0.$$

99. Find the equation of the pair of lines which are the perpendiculars through $(p, q)$ to the two lines of the pair $ax^2+2hxy+by^2 = 0$. The line $x-3y+16 = 0$ intersects the line pair $3x^2-4xy+y^2 = 0$ at $A$ and $B$. Find the equation of the circumcircle of the triangle $OAB$, where $O$ is the origin. If $R$ is the point diametrically opposite $O$ on this circle, find the equation of the line pair $RA$, $RB$.

100. The ellipse $b^2x^2+a^2y^2 = a^2b^2$ meets the rectangular hyperbola $xy = c^2$ in four real points. Prove that the equation of the four sides of the parallelogram defined by these four points can be expressed in the form

$$(b^2x^2+a^2y^2-a^2b^2)^2 = A(xy-c^2)^2,$$

and find the value of the constant $A$.

# FORMULAE FOR REFERENCE

Below is a list of the most important of the formulae obtained in this book, and with each the number of the page on which it first appears. Practice with examples will serve to make these formulae familiar, and this is the best way of learning them.

## I. THE POINT

1. The distance between points $P_1(x_1, y_1)$ and $P_2(x_2, y_2)$ is

$$\sqrt{\{(x_2-x_1)^2+(y_2-y_1)^2\}}. \qquad \text{(page 13)}$$

2. The coordinates $(X, Y)$ of the point which divides $P_1P_2$ in the ratio $k_2 : k_1$ are

$$X = \frac{k_1x_1+k_2x_2}{k_1+k_2}, \quad Y = \frac{k_1y_1+k_2y_2}{k_1+k_2}. \qquad \text{(page 16)}$$

## II. THE LINE

1. Forms of the equation of a line are as follows:

| | | |
|---|---|---|
| General form | $ax+by+c=0$ | (page 27) |
| Gradient form | $y-k=m(x-h)$ | (page 27) |
| Gradient-intercept form | $y=mx+c$ | (page 30) |
| Joining two points | $y-y_1=\frac{y_2-y_1}{x_2-x_1}(x-x_1)$ | (page 30) |
| Intercept form | $\frac{x}{a}+\frac{y}{b}=1$ | (page 30) |
| Normal form | $x\cos\alpha+y\sin\alpha=p$ | (page 31) |
| Parametric form | $x=h+r\cos\alpha,\ y=k+r\sin\alpha$ | (page 32) |
| Determinant form | $\begin{vmatrix} x & y & 1 \\ x_1 & y_1 & 1 \\ x_2 & y_2 & 1 \end{vmatrix}=0.$ | (page 32) |

2. Lines with gradients $m$, $m'$ are

parallel if and only if $\mathbf{m = m'}$,

perpendicular if and only if $\mathbf{mm' = -1}$. (page 26)

3. Lines $ax+by+c = 0$, $a'x+b'y+c' = 0$ are

parallel if and only if $\mathbf{ab' - a'b = 0}$

perpendicular if and only if $\mathbf{aa' + bb' = 0}$. (page 28)

4. If $L = 0$, $L' = 0$ are the equations of two lines, then any line through their intersection is

$$\mathbf{L + \lambda L' = 0}.$$ (page 42)

5. The acute angle between lines with gradients $m$, $m'$ is

$$\tan^{-1}\left|\frac{\mathbf{m - m'}}{\mathbf{1 + mm'}}\right|.$$ (page 46)

6. The acute angle between lines

$$ax+by+c = 0, \quad a'x+b'y+c' = 0$$

is

$$\tan^{-1}\left|\frac{\mathbf{ab' - a'b}}{\mathbf{aa' + bb'}}\right|.$$

7. The perpendicular distance from $P(h, k)$ to

$ax+by+c = 0$ is

$$\left|\frac{\mathbf{ah + bk + c}}{\sqrt{(\mathbf{a^2 + b^2})}}\right|.$$ (page 49)

8. The area of a triangle

$P_1(x_1, y_1)$, $P_2(x_2, y_2)$, $P_3(x_3, y_3)$ is

$$\frac{1}{2}\begin{vmatrix} \mathbf{x_1} & \mathbf{y_1} & \mathbf{1} \\ \mathbf{x_2} & \mathbf{y_2} & \mathbf{1} \\ \mathbf{x_3} & \mathbf{y_3} & \mathbf{1} \end{vmatrix}.$$ (page 52)

## III. THE PARABOLA

1. If the focus $S$ is $(a, 0)$ and the directrix $x = -a$, the equation is

$$\mathbf{y^2 = 4ax}.$$ (page 63)

2. The point $P(ap^2, 2ap)$ lies on the parabola for all values of $p$. (page 66)

3. The equation of the chord $PQ$ is

$$x - \tfrac{1}{2}(p+q)y + apq = 0. \qquad \text{(page 67)}$$

4. The equation of the tangent at $P$ is

$$x - py + ap^2 = 0. \qquad \text{(page 67)}$$

The equation of the tangent at $(x_1, y_1)$ is

$$yy_1 = 2a(x + x_1). \qquad \text{(page 67)}$$

5. The equation of the normal at $P$ is

$$px + y = ap^3 + 2ap. \qquad \text{(page 68)}$$

6. Tangents at $P$, $Q$ meet at

$$\{apq, a(p+q)\}. \qquad \text{(page 70)}$$

7. The chord of contact of tangents drawn through $(h, k)$ is

$$ky = 2a(x + h). \qquad \text{(page 70)}$$

8. The line $lx + my + n = 0$ is a tangent if and only if

$$am^2 = nl. \qquad \text{(page 71)}$$

9. The normals at $P$, $Q$, $R$ are concurrent if and only if

$$p + q + r = 0. \qquad \text{(page 89)}$$

## IV. THE RECTANGULAR HYPERBOLA

1. Referred to the asymptotes as axes, the equation is

$$xy = c^2, \qquad \text{(page 99)}$$

with general point $P(cp, c/p)$.

2. The equation of the chord $PQ$ is

$$x + pqy - c(p+q) = 0. \qquad \text{(page 101)}$$

3. The equation of the tangent at $P$ is

$$x + p^2y - 2cp = 0. \qquad \text{(page 101)}$$

4. The equation of the normal at $P$ is

$$p^3x - py = c(p^4 - 1). \qquad \text{(page 101)}$$

## V. THE CIRCLE

1. The equation of a circle, centre (0, 0), radius $r$, is

$$\mathbf{x^2+y^2=r^2.} \qquad \text{(page 133)}$$

2. The equation of a circle, centre $(-g,\ -f)$, radius $\sqrt{(g^2+f^2-c)}$, is

$$\mathbf{x^2+y^2+2gx+2fy+c=0.} \qquad \text{(page 134)}$$

3. The equation of the circle on

$$A(h,k),\ A'(h',k')$$

as diameter is

$$\mathbf{(x-h)(x-h')+(y-k)(y-k')=0.} \qquad \text{(page 141)}$$

4. Two circles $S \equiv x^2+y^2+2gx+2fy+c=0$

and $$S' \equiv x^2+y^2+2g'x+2f'y+c'=0$$

cut orthogonally if and only if

$$\mathbf{2gg'+2ff'=c+c'.} \qquad \text{(page 143)}$$

5. The radical axis of $S$ and $S'$ is

$$\mathbf{S-S'=0.} \qquad \text{(page 152)}$$

This is also the equation of the common chord, if the circles intersect.

6. The coaxal system of circles with $x=0$ as radical axis and $y=0$ as line of centres is

$$\mathbf{x^2+y^2+2\lambda x+c=0.} \qquad \text{(page 154)}$$

7. The orthogonal system of circles is

$$\mathbf{x^2+y^2+2\mu y-c=0.} \qquad \text{(page 155)}$$

8. The general coaxal system of circles determined by $S$ and $S'$ is

$$\mathbf{S+\lambda S'=0.} \qquad \text{(page 157)}$$

## VI. THE ELLIPSE AND HYPERBOLA

[The formulae are listed together as far as possible; alternatives for the hyperbola appear in square brackets.]

1. The ellipse [hyperbola] has equation

$$\frac{x^2}{a^2}+\frac{y^2}{b^2}=1, \quad \left[\frac{x^2}{a^2}-\frac{y^2}{b^2}=1\right],$$ (page 174) [page 191]

where the foci are $(\pm ae, 0)$, the directrices are $x = \pm a/e$, the eccentricity is $e$, and $b^2 = a^2(1-e^2)$, $[b^2 = a^2(e^2-1)]$.

2. The hyperbola has asymptotes

$$y = \pm bx/a.$$ (page 194)

3. If $S$, $S'$ are the foci and $P$ is a point on the ellipse, [hyperbola], then

$$SP+S'P = 2a,$$ (page 175)

$$[SP \sim S'P = 2a].$$ [page 192]

4. The point $(a\cos\theta, b\sin\theta)$ lies on the ellipse for all $\theta$. (page 176)

The point $(a\sec\theta,\ b\tan\theta)$ lies on the hyperbola for all $\theta$. [page 192]

5. The equation of a chord of the ellipse is

$$\frac{x}{a}\cos\tfrac{1}{2}(\alpha+\beta)+\frac{y}{b}\sin\tfrac{1}{2}(\alpha+\beta)=\cos\tfrac{1}{2}(\alpha-\beta).$$ (page 178)

6. The equation of the tangent at $(x_1, y_1)$ is

$$\frac{xx_1}{a^2}+\frac{yy_1}{b^2}=1, \quad \left[\frac{xx_1}{a^2}-\frac{yy_1}{b^2}=1\right].$$ (page 178)

7. The equation of the normal at $(x_1, y_1)$ is

$$\frac{xa^2}{x_1}-\frac{yb^2}{y_1}=a^2-b^2, \quad \left[\frac{xa^2}{x_1}+\frac{yb^2}{y_1}=a^2+b^2\right].$$ (page 178)

8. The equation of the auxiliary circle is

$$x^2+y^2=a^2.$$ (page 176)

9. The equation of the director circle is

$$\mathbf{x^2 + y^2 = a^2 + b^2}, \quad [\mathbf{x^2 + y^2 = a^2 - b^2}]. \qquad \text{(page 187)}$$

10. The line $lx + my + n = 0$ touches the ellipse, [hyperbola] if

$$\mathbf{a^2l^2 + b^2m^2 = n^2}, \quad [\mathbf{a^2l^2 - b^2m^2 = n^2}]. \qquad \text{(page 198)}$$

11. Diameters with gradients $m, m'$ are conjugate if and only if

$$\mathbf{mm' = -b^2/a^2}, \quad \mathbf{mm' = b^2/a^2}. \qquad \text{(page 205)}$$

## VII. THE LINE PAIR

1. The equation of a line pair through the origin is

$$\mathbf{ax^2 + 2hxy + by^2 = 0}. \qquad \text{(page 221)}$$

2. The angle between these lines is

$$\tan^{-1}\left|\frac{2\sqrt{(h^2 - ab)}}{a + b}\right|. \qquad \text{(page 222)}$$

3. Bisectors of the angle between these lines have equation

$$\frac{x^2 - y^2}{a - b} = \frac{xy}{h}. \qquad \text{(page 222)}$$

4. The equation $ax^2 + 2hxy + by^2 + 2gx + 2fy + c = 0$ represents a line pair if and only if

$$\Delta = \begin{vmatrix} a & h & g \\ h & b & f \\ g & f & c \end{vmatrix} = 0. \qquad \text{(page 226)}$$

## VIII. THE GENERAL CONIC

1. If the origin is moved to $(\lambda, \mu)$ the new coordinates $(x^*, y^*)$ are

$$x^* = x - \lambda,$$
$$y^* = y - \mu. \qquad \text{(page 22)}$$

2. If the axes are turned through an angle $\theta$, the new coordinates are $(x^*, y^*)$ where

$$x = x^* \cos\alpha - y^* \sin\alpha,$$
$$y = x^* \sin\alpha + y^* \cos\alpha,$$

and
$$\mathbf{x^* = x\cos x + y\sin\alpha},$$
$$\mathbf{y^* = -x\sin\alpha + y\cos\alpha}. \qquad \text{(page 54)}$$

3. If the equation of the curve

$$S = ax^2 + 2hxy + by^2 + 2gx + 2fy + c = 0$$

is transformed by either of the above substitutions, the expressions
$$\mathbf{ab - h^2} \quad \text{and} \quad \mathbf{a + b}$$
are invariant. (page 237)

4. $S$ has centre $(\lambda, \mu)$ if $ab - h^2 \neq 0$, and then

$$\mathbf{a\lambda + h\mu + g = 0},$$
$$\mathbf{h\lambda + b\mu + f = 0}. \qquad \text{(page 238)}$$

5. $S$ is an ellipse if $\mathbf{ab - h^2 > 0}$ and $\Delta < 0$, (page 240)

a parabola if $\mathbf{ab - h^2 = 0}$, (page 247)

a hyperbola if $\mathbf{ab - h^2 < 0}$. (page 240)

6. The principal axes of a central conic are parallel to
$$\mathbf{\frac{x^2 - y^2}{a - b} = \frac{xy}{h}}. \qquad \text{(page 242)}$$

7. The asymptotes of a hyperbola $S = 0$ have equation
$$\mathbf{S + p = 0},$$
where $p$ is a constant. (page 246)

8. The line $P_1(x_1, y_1)$ $P_2(x_2, y_2)$ meets $S$ in

$$\left(\frac{x_1 + \lambda x_2}{1 + \lambda}, \quad \frac{y_1 + \lambda y_2}{1 + \lambda}\right),$$

where
$$\mathbf{S_{11} + 2\lambda S_{12} + \lambda^2 S_{22} = 0},$$
where
$$S_{11} = ax_1^2 + 2hx_1y_1 + by_1^2 + 2gx_1 + 2fy_1 + c,$$
$$S_{12} = ax_1x_2 + h(x_1y_2 + x_2y_1) + by_1y_2 + g(x_1 + x_2) + f(y_1 + y_2) + c,$$
$$S_{22} = ax_2^2 + 2hx_2y_2 + by_2^2 + 2gx_2 + 2fy_2 + c. \qquad \text{(page 250)}$$

9. The polar line of $P_1$ with respect to $S$ is $S_1 = 0$, where

$$S_1 = ax_1x + h(x_1y + y_1x) + by_1y + g(x_1 + x) + f(y_1 + y) + c.$$

This is also the equation of the tangent at $P_1$, if $P_1$ lies on $S$. (page 252)

10. The equation of the pair of tangents drawn from $P_1$ to $S$ is

$$S_1^2 - S_{11}S = 0. \qquad \text{(page 252)}$$

11. The equations of any conic through the intersections of $S = 0$, $S' = 0$ is

$$S + \lambda S' = 0. \qquad \text{(page 256)}$$

12. The equation of any conic with the same foci as $x^2/a^2 + y^2/b^2 = 1$ is

$$\frac{x^2}{a + \lambda} + \frac{y^2}{b^2 + \lambda} = 1. \qquad \text{(page 258)}$$

# HINTS ON THE SOLUTION OF EXAMPLES

In some of the cases where a problem is not an immediate application of the preceding bookwork, or there is some other difficulty, a hint about how it may be solved is given below. These hints are mainly designed to assist those readers who are working with little or no help; it is emphasized that in each case they indicate merely one possible approach to the problem. Most geometrical problems can be tackled in several ways, and it is not claimed that the method indicated here is the only, or even the best, one. It would be a valuable exercise to look for alternatives that improve on the suggestions given.

## EXAMPLES I C (page 17)

11. Take $A$ as $(\alpha, \alpha')$, $B(\beta, \beta')$, $C(\gamma, \gamma')$, $D(\delta, \delta')$, and find $P$ as $\{\frac{1}{2}(\beta+\gamma), \frac{1}{2}(\beta'+\gamma')\}$, and so on.

## EXAMPLES I D (page 21)

1. Let $P$ be $(x, y)$ and find $AP^2$, $PB^2$; express the fact that $AP^2 = PB^2$.

3. $AP^2 = 4PB^2$.

4. $AP = \sqrt{\{(x-1)^2+y^2\}}$, $PB = \sqrt{\{(x+1)^2+y^2\}}$; $AP = 3-PB$; square both sides, take the surd alone to one side of the equation and square again.

## MISCELLANEOUS EXAMPLES I (page 23)

6. If the foot of the perpendicular from $C$ to $BD$ is $E$, the triangles $BEC$, $DAB$ are similar and so $CE/BC = AB/BD$, and $BD = AC$.

7. Prove that $OA/OB = \frac{5}{3}$, and use the converse of the theorem that the bisectors of the angle $AOB$ divide $AB$ in the ratio $OA/OB$.

12. If $O$ is the origin, the square of the length of a tangent from $P$ to a circle centre $O$, radius 4, is $OP^2-16$.

13. Take $A$ as $(x_1, y_1)$, and so on (as in the next example).

## EXAMPLES II A (page 28)

19. Use the fact that the tangent of a circle is perpendicular to the radius.

## EXAMPLES II C (page 42)

6. Find the vertices of the triangle, and so the length of each side.

7. Let the lines be $BC$, $CA$, $AB$ and show that $A$ is $(1, -3)$, $B(-\frac{5}{9}, 2\frac{1}{3})$, $C(-3\frac{4}{9}, -6\frac{1}{3})$. Show that $AB = AC$ and that the mid-point of $BC$ is $(-2, -2)$. Find the length of the altitude through $A$, and hence the other altitudes by expressing the area of the triangle in an alternative way.

## MISCELLANEOUS EXAMPLES II (page 54)

5. Factorize the equation.

21. Find the vertices $A(2, 3)$, $B(-1, 1)$, $C(3, -2)$ and the equations of the altitudes.

23. Find the equation of an altitude (a line through the intersection of two of the sides perpendicular to the third). Find the $x$-coordinate of the intersection of two of these altitudes and show that it is $-a$.

24. Factorize the equation.

26. Factorize the equation.

30. The diagonals of a rhombus are perpendicular.

32. The vertex $(-1, 10)$ of the two triangles is the mid-point of their common altitude.

36. Notice that $C$ lies on $A'B$.

43. If $P$ is $(h, k)$ we have $h^2+k^2-8h-4k = 0$ (product of gradients of $AP$, $BP$ is $-1$), and $3h^2+3k^2+16h-32k = 0$ ($PB^2 = 4PA^2$). Subtraction gives $k = 2h$, and the result follows by substitution. (Alternatively, the first solution may be spotted to be 0, and then $OP$ is perpendicular to $AB$ and bisected by it, since $AB$ is a diameter of a circle through $O$ and $P$).

56. The expressions $x+y-1$, $x-y-1$, $x-3y+3$ are positive, negative, positive when $x = 1\frac{1}{3}$, $y = 1$. The centroid is inside the triangle, and so any other point which gives the same distribution of signs will also be inside.

61. If $P$ is $(2\alpha, 0)$ and $Q$ is $(0, 2\beta)$, the mid-point is $(\alpha, \beta)$. Express the fact that $P$, $Q$ and $(h, k)$ are collinear.

64. Prove that the four lines form a cyclic quadrilateral, by considering the sum of the opposite angles.

68. $BD$ is parallel to $AC$.

70. The required lines must be parallel to the bisectors of the given lines.

## EXAMPLES III A (page 68)

2. Eliminate $t$.

12. Use the facts that points on a circle are a constant distance from the centre, and that the tangent is perpendicular to the radius.

## EXAMPLES III B (page 73)

5. Prove that $qr = p^2$, and that the $x$-coordinate of $M$ is $ap^2$.

9. If $Q$ has parameter $u$, prove that $\tan\theta = -t$, $u+t = 2\cot\theta$, $u-t = 2\cot\theta\sec^2\theta$.

## EXAMPLES III C (page 78)

1. (ii) $SO.SP = AS^2$ (page 77 (8)).
   (iii) Congruent triangles $OAT$, $OAN$.
   (iv) $PA = AT$.
   (v) Alternate segment theorem.
   (vi) $PA = AT$.
   (vii) Square (vi) and use $AS^2 = SO.SP$.
   (viii) Congruent triangles $SAR$, $SAK$.
   (x) Prove that the angles $FPG$, $PGN$ are equal, and use $SL = NG$.

2. (iii) $SMRU$ is a parallelogram.
   (iv) $\angle MRP = \angle PQR$.
   (v) The line through $R$ parallel to the axis is a radius of the circle $PQR$.
   (vii) A circle centre $R$ goes through $M$, $V$, $S$.
   (viii) $SGHV$ is a parallelogram.
   (ix) $SQ = VQ$.
   (x) Alternate segment theorem.

## EXAMPLES III D (page 84)

6. Show that the equation of the locus of $R$ is $x^2+y^2-4ax = 0$; to find the centre use the method of p. 21.

9. If the parameters of $P$, $Q$ are $p$, $q$, show by finding the condition that $PQ$ touches $y^2 = 4bx$ (see § 6) that $b(p+q)^2 = 4apq$.

21. Put $p+q = \theta$, $pq = \phi$, then $2b-\theta c+2a\phi = 0$ and

$$X = a(\phi^2+2-\theta), \quad Y = -a\theta\phi.$$

Eliminate $\theta$, and find expressions in terms of $X$, $Y$ for $\phi$ and $\phi^2$.

25. For definition of polar line, see p. 70.

## EXAMPLES III E (page 90)

5 (iii). See Example III D, 21.

## MISCELLANEOUS EXAMPLES IV (page 127)

15. One root of $2x^3 - 19x^2 + 225 = 0$ is $x = 5$.

26. The tangents are

$$x \sin\theta + y\cos\theta = a\sin 3\theta \quad \text{and} \quad x\cos\theta - y\sin\theta = -a\cos 3\theta.$$

To find the equation of the curve on which the tangents meet, square and add these equations.

29. If the curve is $xy = c^2$ and $A$ is $(h, k)$, the locus is $kx + hy = c^2 + hk$.

## EXAMPLES V A (page 139)

12. Use the geometry of the figure.

## EXAMPLES V C (page 158)

9. Prove, by considering gradients of lines, that $\angle CAD = 90$.

12. Use conditions obtained from the quartic equation of p. 150 by applying [A 7].

## MISCELLANEOUS EXAMPLES V (page 163)

28. Prove that there is a circle through the common points of the given circles which also contains the two given points.

56. Notice that two of the lines are perpendicular, so the third line is a diameter.

61. Prove geometrically that the points in question are equidistant from (0, 0), and in each case find the distance by using the geometry of the figure.

74. Find the equations of the radical axis and the line of centres of the given system.

80. If $(p, q)$ and $(x, y)$ are ends of a diameter of the required circle, the first part gives that

$$p(x+g) + q(y+f) + (gx + fy + c) = 0,$$

with two similar equations. Elimination of $p : q : 1$ [A 3] gives the result.

## EXAMPLES VI D (page 201)

14. Take $P$ as $\{a(1-p^2)/(1+p^2), 2bp/(1+p^2)\}$ and use Example VI A, 12 and Example VI D, 16.

## MISCELLANEOUS EXAMPLES VI (page 208)

23. $CQ.CR$ is the square of the length of a tangent from $C$ to the circle on $PQ$ as diameter.

47. Write down the equation of $PQ$ in terms of the mid-point $(h, k)$ of $PQ$; then $W$ lies on the line perpendicular to this through $PQ$.

57. Notice that $Q$ lies on the normal at $P$, and that $PM = MQ \tan\alpha$, where $\alpha$ is the angle between $PA$ and the tangent at $P$.

## EXAMPLES VII B (page 230)

9. Since the line pair obtained is the line $2x-y=0$ repeated, the given locus meets the given line in coincident points, for all values of $a$ and $b$.

## MISCELLANEOUS EXAMPLES VII (page 231)

1. Take $S$ as origin and use the method of § 5.

14. Notice that each pair of lines includes a right angle, and find the ends of the diameter $x+2y-5=0$.

## FURTHER MISCELLANEOUS EXAMPLES

1. $P$ and $Q$ lie on the perpendicular bisector of $AC$ and on lines parallel to $AC$ through $B$, $D$.

8. The locus is a line parallel to $BC$ through the mid-points of $AB$ and $AC$.

17. The required circle has (11, 10), (8, 6) as ends of a diameter.

18. The radius of the required circle has radius twice that of the given circle.

19. A circle touching the axes will have as centre either $(\alpha, \alpha)$ or $(\beta, -\beta)$; $\alpha, \beta$ must be chosen to make such a circle touch the given one.

21. Use the geometry of the figure; if the centres are $C_1$, $C_2$ and the tangents meet in $T$, then $TC_1/TC_2 = \frac{2}{5}$.

41. The minimum occurs when $Q$ and $R$ are collinear with the two reflections of $P$.

44. Find the combined equation of the pair of bisectors.

45. When $P$ is on $SU$, $\alpha+2\beta = 0$. Find the locus of $Q$ without finding the coordinates of $Q$; eliminate $\beta$ by adding and subtracting the equations of the lines.

49. Use the geometry of the figure.

58. Take the fixed lines as axes, $A$ as $(a, 0)$, and the mid-point of $BC$ as $(0, at)$.

60. To obtain the last relation, find the distance from the origin to the line.

62. Prove that $P$ is $(2am, 2a/m)$.

63. Show that $pq(p+q)$ is a constant, where $p$, $q$ are parameters of two of the feet.

64. $\Sigma SQ_i^2 = \Sigma a^2(p^4+2p^2+1)$, where $Q_1$ has parameter $p_1$, and so on. Use the fact that

$$\Sigma p^2 = \sigma_1^2 - 2\sigma_2, \quad \Sigma p^4 = \sigma_1^4 - 4\sigma_1^2\sigma_2 + 2\sigma_2^2 + 4\sigma_1\sigma_3 - 7\sigma_4,$$

where $\sigma_1 = \Sigma p_i$, $\sigma_2 = \Sigma p_i p_j$, $\sigma_3 = \Sigma p_i p_j p_k$, $\sigma_4 = p_1 p_2 p_3 p_4$.

69. $H$ is $(1+\alpha^2, -\alpha)$.

72. Prove that $Q$, $R$ are conjugate points with respect to the hyperbola (see p. 251) and also $R$, $P$ and $P$, $Q$.

# ANSWERS

## EXAMPLES I A (page 12)

1. On $x$-axis, (3, 0); on $y$-axis, (0, $-1$), (0, 5); in first quadrant, (2, 7), (2, 2); in second quadrant, ($-5$, 3), ($-\frac{3}{2}$, 2); in third quadrant, ($-4$, $-\frac{1}{2}$); in fourth quadrant, (5, $-1$), (2, $-6$).

2. Lie on a line; $t = 2$, $t = -3$, $-3 < t < 2$.

3. (i) Lie on a line; $t = -4$, $t = 1$, $t > 1$.
(ii) Lie on a curve; $t = 0$, $t = 0$, $t > 0$.
(iii) Lie on a curve; $t = 0$, $t = \pm 1$, $0 < t < 1$.

## EXAMPLES I B (page 15)

1. $\sqrt{89}$, $\sqrt{10}$, $8\sqrt{2}$, $\frac{1}{6}\sqrt{2}$, $\frac{1}{2}\sqrt{5}$, $\sqrt{13}$, $\sqrt{(a^2-4ab+5b^2)}$.

2. 1, 3.

5. 5, 5, 5, 5, 5, 5; lie on a circle.

8. (19, $-14$).

9. 5.

10. $x^2+y^2-2x-4y-20 = 0$, $(1\pm\sqrt{21}, 0)$, $(0, 2\pm 2\sqrt{6})$.

## EXAMPLES I C (page 17)

1. (i) (4, 8), (ii) ($-4\frac{1}{2}$, $1\frac{1}{2}$), (iii) $\{\frac{1}{2}(a+b), \frac{1}{2}(a+b)\}$.

3. ($-1\frac{1}{2}$, $-1$).

4. (i) ($\frac{2}{3}$, 0), (ii) ($\frac{2}{3}$, 0); yes.

6. (2, $-7$).

7. (1, 1).

8. (i) (0, 4), (ii) ($-3$, 6), (iii) (12, $-4$).

9. $24\sqrt{2}$.

10. (i) $1 : -3$, (ii) $7 : -3$.

## EXAMPLES I D (page 21)

1. $x = 0$.

2. $2x^2+2y^2 = 1$.

3. $3x^2+3y^2+10x+3 = 0$.

4. $20x^2+36y^2 = 45$.

5. $12x^2-4y^2 = 3$.

6. $x+y = 2$.

7. $2x^2+2y^2-14x+6y+27 = 0$.

8. $3x^2+3y^2-26x+4y+55 = 0$.

9. $32x^2+32y^2-8xy-236x+124y+443 = 0$.

10. $8xy+12x-28y-43 = 0$.

11. (i) Circle centre (0, 0), radius 3, (ii) circle centre ($-1$, 0), radius 2.

13. Line $x = 0$.

## EXAMPLES I E (page 23)

**1.** $x^{*2}+y^{*2} = 36$. **2.** $x^{*2}+y^{*2} = 4$.

**3.** $y^{*2} = 4ax^*$. **4.** $x^{*2}+2y^{*2} = 2$.

**5.** $y^* = x^{*3}$.

## MISCELLANEOUS EXAMPLES I (page 23)

**2.** $(\frac{2}{3}, \frac{2}{3})$. **3.** $(2\frac{1}{2}, 2\frac{1}{6})$.

**4.** $(2-\frac{1}{2}\sqrt{34}, 3+\frac{1}{2}\sqrt{34}), (2+\frac{1}{2}\sqrt{34}, 3-\frac{1}{2}\sqrt{34})$.

**5.** $(-14, 20)$. **6.** $\frac{2}{5}\sqrt{65}$.

**8.** $8x^2-y^2+4x-4 = 0$. **9.** $2x+3y = 8$.

**10.** $3x-y = 6$.

**11.** (i) $x+2y = 8$, (ii) $x^2+y^2-4x-6y+8 = 0$, (iii) $x^2+y^2-4x-6y-12 = 0$.

**12.** $4x+6y = 29$.

**14.** $\{\frac{1}{4}(x_1+x_2+x_3+x_4), \frac{1}{4}(y_1+y_2+y_3+y_4)\}$.

## EXAMPLES II A (page 28)

**1.** $-5/7, -2, 0, -1/9, -1$. **3.** $(2, 0), (0, 3), (-2, 6)$.

**4.** (i) $(0, 2)$, (ii) $(1, 6)$, (iii) $(-3, -10)$, (iv) $(1, 6)$, (v) $(-2\frac{3}{4}, -9)$, (vi) $(\frac{3}{4}, 5)$, (vii) $(-\frac{3}{10}, \frac{4}{5})$.

**5.** $c = 0$. **6.** $-a/b$.

**7.** (i) $x-2y+8 = 0$, (ii) $3x-y-17 = 0$, (iii) $2x-y+8 = 0$, (iv) $x+y-1 = 0$, (v) $x+3y-4 = 0$, (vi) $x-y-a = 0$.

**8.** (i) $8x+y-13 = 0$, (ii) $3x-8y-29 = 0$, (iii) $2x-y+5 = 0$, (iv) $y = 11$, (v) $x = -1$, (vi) $x+y = a+b$.

**9.** $y = 1$, $3x-y+1 = 0$.

**10.** $2x+3y-11 = 0$; $(5\frac{1}{2}, 0), (0, 3\frac{2}{3})$.

**11.** $\frac{4}{3}$; $4x-3y-19 = 0$, $3x+4y-8 = 0$.

**12.** $x+3y-6 = 0$, $8x+3y-13 = 0$, $x = 1$.

**13.** $2x-y+4 = 0$. **14.** $(2a+b, c)$.

**15.** $x+y = 0$. **16.** $4x-y-8 = 0$, $x-4y-2 = 0$.

**17.** $-3\frac{1}{2}$. **19.** $3x-4y = 25$.

## EXAMPLES II B (page 37)

**1.** (i) $-\frac{3}{2}, 2, 3,$ (ii) $\frac{2}{7}, -\frac{3}{2}, \frac{3}{7},$ (iii) $1, -1, 1,$
(iv) $-\frac{1}{3}, k, \frac{1}{3}k,$ (v) $\frac{3}{5}, \frac{2}{3}, -\frac{2}{5},$ (vi) $-1, -3, -3.$

**2.** $\frac{1}{3}, -1.$

**3.** (i) $a, c, h$; $b, g$; $d, e$; $f, k$; (ii) $a, c, h$ perpendicular to $d, e$; $b, g$ perpendicular to $f, k$.

**5.** $2x+4y = 3$, $x-2y-2 = 0$, $x+y+1 = 0$, $4x-9y+12 = 0$, $lx+my = 1$; $-\frac{1}{2}, \frac{1}{2}, -1, \frac{4}{9}, -l/m.$

**6.** $x-y = 5.$ **7.** $x+11y = 0.$

**8.** $3x+4y-12 = 0$, $4x-3y-16 = 0$, $3x+4y-37 = 0$, $4x-3y+9=0.$

**9.** (i) Parallel, through the origin, (ii) perpendicular, through the origin, (iii) parallel to the image of the given line in the $x$-axis, through the origin.

**10.** $3x-y-2 = 0$, $x+3y-4 = 0.$

**11.** $3x+4y = 15$; $4x-3y = 1.$ **12.** $3x-7y+2 = 0.$

**14.** (i) $x\sqrt{3}+y-6 = 0,$ (ii) $x\sqrt{3}+y+4 = 0,$
(iii) $x-y-4\sqrt{2} = 0,$ (iv) $x-y+\sqrt{2} = 0,$
(v) $12x-16y = \pm 15,$
(vi) $2x+y = \pm 2\sqrt{5}$; $-\sqrt{3}, -\sqrt{3}, 1, 1, \frac{3}{4}, -2.$

**15.** (i) $\frac{3}{5}x+\frac{4}{5}y = 1,$ (ii) $\frac{2}{5}\sqrt{5}x-\frac{1}{5}\sqrt{5}y = \frac{4}{5}\sqrt{5},$
(iii) $-\frac{5}{74}\sqrt{74}x+\frac{7}{74}\sqrt{74}y = \frac{6}{37}\sqrt{74},$
(iv) $-\frac{1}{5}\sqrt{5}x-\frac{2}{5}\sqrt{5}y = \frac{3}{5}\sqrt{5},$ (v) $-\frac{1}{2}\sqrt{2}x-\frac{1}{2}\sqrt{2}y = \frac{1}{2}\sqrt{2},$

(vi) $$-\frac{ax}{\sqrt{(a^2+b^2)}}-\frac{by}{\sqrt{(a^2+b^2)}}=\frac{+c}{\sqrt{(a^2+b^2)}};$$

$1, \frac{4}{5}\sqrt{5}, \frac{6}{37}\sqrt{74}, \frac{3}{5}\sqrt{5}, \frac{1}{2}\sqrt{2}, |c/\sqrt{(a^2+b^2)}|.$

**16.** $x+2y = 5$; $-\frac{1}{2}.$ **17.** $(a, 2a)$, $(25a, -10a).$

**19.** $(1+\frac{4}{5}r, -3+\frac{3}{5}r)$; (i) 5, $(5, 0)$, (ii) $-\frac{5}{4}$, $(0, -3\frac{3}{4})$, (iii) $-20$, $(-15, -15).$

**20.** $x\sqrt{3}+y = 4+3\sqrt{3}$; $(3, 4)$, $(\frac{3}{2}+2\sqrt{3}, \frac{3}{2}\sqrt{3}-2).$

## EXAMPLES II C (page 42)

**1.** (i) $(-3, -1)$, (ii) $(1, 2)$, (iii) $\{ab/(a+b), ab/(a+b)\}.$

**2.** (i) Any line through $(-1, -1)$, (ii) any line through $(\frac{1}{17}, -\frac{22}{17})$, (iii) any line parallel to $x+2y = 0.$

**3.** (i) $13x+11y = 0,$ (ii) $x+5y+6 = 0,$
(iii) $5x-2y-9 = 0,$ (iv) $7x-10y-23 = 0,$
(v) $9x+9y+2 = 0.$

4. $(-6, -1), (-3, -2), (2, 3); (-1, -6)$.

5. 5. 6. $3\sqrt{3}$. 7. $\sqrt{10}, 5\frac{1}{5}, 5\frac{1}{5}$.

8. (i) $161y+8 = 0$, (ii) $23x+14 = 0$,
(iii) $169x-420y+82 = 0$, (iv) $322x-161y+188 = 0$.

9. (i) $2x+3y+5 = 0$, $4x-y+3 = 0$,
(ii) $8x-3y+5 = 0$,
(iii) $64x-9y+25 = 0$.

10. (i) $(-1, -3), (3, 1), (-3, 1)$.
(ii) $4x-y+1 = 0$, $2x-5y-1 = 0$, $x+2y+1 = 0$,
(iii) $(-\frac{1}{3}, -\frac{1}{3})$,
(iv) $x = -1$, $x-2y-1 = 0$, $x+y+2 = 0$,
(v) $(-1, -1)$,
(vi) $x = 1$, $x-2y+1 = 0$, $x+y-2 = 0$,
(vii) $(1, 1)$; $(-1, -3), (3, 1), (-3, 1), (1, -3), (2\frac{3}{5}, 1\frac{4}{5}), (-1, 3), (1, 3), (-3, -1), (3, -1)$ are all a distance $\sqrt{10}$ from $N(0, 0)$.

11. $19x+5y-14 = 0$.

12. (i) 2, 1, (ii) $x-y+4 = 0$, $2x-y+8 = 0$; (iii) $(-4, 0)$, (iv) $(-\frac{1}{2}, 3)$.

13. $9x-46y = 0$, $\sqrt{13}$. 14. $x+4y = 9$.

15. $x-y = 0$, $x+y = 6$; $(2, 4), (4, 2)$; $2\sqrt{5}$.

16. $(2, 2)$. 18. $(5, 6), (-2, 5)$, 24.

## EXAMPLES II D (page 45)

2. $\sqrt{5}$. 5. $-5$. 6. $-1$.

## EXAMPLES II E (page 50)

1. (i) $71° 34'$, (ii) $78° 41'$, (iii) $45°$, (iv) $15° 57'$, (v) $81° 52'$.

2. $2x-y+1 = 0$, $2x+11y-35 = 0$.

3. (i) 3, (ii) $17/\sqrt{73}$, (iii) $\frac{1}{5}\sqrt{10}$, (iv) $2ab/\sqrt{(a^2+b^2)}$, (v) $\frac{2}{5}$.

4. $6x-2y+1 = 0$, $2x+6y+15 = 0$. 6. $2, \frac{1}{2}\sqrt{2}, \frac{2}{5}\sqrt{5}$.

7. $7x+y-19 = 0$, $x-7y+33 = 0$; $x+3y+3 = 0$, $3x-y-61 = 0$; $x-2y-7 = 0$, $2x+y+21 = 0$; $(3, -2), (8, -37), (-12, 3), (23, 8)$.

8. $x-7y+4 = 0$, $7x+y-22 = 0$.

9. $(\frac{9}{2}, -\frac{13}{2}), (\frac{21}{10}, \frac{7}{10}), (-\frac{43}{10}, \frac{39}{10}), (\frac{47}{46}, \frac{21}{46})$.

10. $(\frac{1}{2}, \frac{1}{2})$, (inside the triangle); $(\frac{1}{2}, -\frac{9}{2}), (-\frac{3}{2}, \frac{3}{2}), (\frac{7}{2}, \frac{3}{2})$.

11. (i) $-c(ha+kb)/2ab$, (ii) $c^2/2ab+cd\sqrt{(a^2+b^2)}/2ab$.

**12.** (i) $\{-(bk+c)/a,\ k\}$, (ii) $(ah+bk+c)/a$.

**13.** $\{(b^2h-abk-ac)/(a^2+b^2),\ (a^2k-abh-bc)/(a^2+b^2)\}$.

**14.** $-(ah+bk+c)/(a\cos\alpha+b\sin\alpha)$.

**15.** $219x-23y-35=0$, $141x-127y+95=0$.

**16.** $5x-y-21=0$, $x+5y-25=0$.

## EXAMPLES II F (page 52)

**1.** (i) $22\frac{1}{2}$, (ii) 7, (iii) 9, (iv) $\frac{1}{2}(a^2-b^2)$, (v) 23.

**3.** 6. **4.** $\frac{13}{16}$.

**5.** $(0, -3)$; $x-y-2=0$, $x+y+3=0$; 7.

**7.** 3, $(8, -15)$. **8.** 2.

## MISCELLANEOUS EXAMPLES II (page 54)

**1.** $3x-y-6=0$; 40. **2.** $x+2y-12=0$, $(6, 3)$, $3:2$.

**3.** $x-y-2=0$, $3x-5y+10=0$, $(7\frac{1}{2}, 5\frac{1}{2})$.

**4.** $3x+y-14=0$, $3x+5y-28=0$, $(3\frac{1}{2}, 3\frac{1}{2})$.

**5.** $2x-3y=0$, $x+y=0$, $78^\circ\ 41'$.

**6.** $2x+y=0$, $\frac{3}{5}\sqrt{5}$. **7.** 4; $x-y-2=0$, $x+y-1=0$.

**8.** $4x-5y=0$, $y=2$. **9.** $2x+3y-31=0$.

**10.** $(3, 5)$, $(4, 2)$, $(\frac{3}{2}, 2)$; $y=5$, $6x+2y-13=0$.

**11.** $4x+6y-5=0$. **12.** $7:18$, $5x-6y+15=0$, $15/\sqrt{61}$.

**13.** $(2\frac{2}{5}, 5\frac{4}{5})$, $4x+3y-27=0$, $157^\circ\ 50'$.

**14.** $2x+y-5=0$, $2x+y+12=0$; $17/\sqrt{5}$; $(-4\frac{24}{29}, 0)$.

**15.** $(1\frac{1}{3}, \frac{2}{3})$.

**16.** $(-3, \frac{1}{3})$, $(2, 2)$, $(2, 2\frac{1}{3})$, $(-3, \frac{2}{3})$; $(-\frac{1}{2}, 1\frac{1}{3})$.

**17.** $x-y-1=0$, $x-3y-1=0$; $(1, 0)$; 5.

**18.** 10, $(2\frac{1}{3}, 3)$; $\frac{24}{7}$. **19.** $9\frac{1}{2}$; $(\frac{6}{19}, 1\frac{7}{19})$; $2\frac{55}{57}$.

**20.** $2x+y-10=0$, $x-2y+5=0$; $15\frac{3}{5}$.

**21.** $(\frac{13}{17}, 1\frac{6}{17})$. **22.** $6x-5y+1=0$, $2x-y-1=0$.

**24.** $21^\circ\ 48'$.

**25.** $(\frac{9}{10}, 2\frac{7}{10})$, $(1\frac{1}{5}, \frac{3}{5})$, $(\frac{4}{5}, 3\frac{2}{5})$; $7x+y-9=0$.

**26.** $53^\circ\ 8'$; $\frac{9}{20}$.

**27.** $(2, 6), (-7, 8)$; 40.

**28.** $7x-2y-7=0,\ 3x+y-2=0$; $(\frac{11}{13}, -\frac{7}{13})$; 13.

**29.** $(-1\frac{1}{2}, 4\frac{1}{2}), (2\frac{1}{2}, 1\frac{1}{2})$; $12\frac{1}{2}$.

**30.** $\{3/2(a+b), 3/2(a+b)\}$.

**31.** 20; $y=0,\ 10x-y-5=0$.

**32.** $x-2y+41=0$.

**33.** $8x-8y+1=0,\ x+2y-2=0,\ 9x-6y-1=0$; $(\frac{7}{12}, \frac{17}{24})$.

**34.** $2x+11y-5=0,\ 22x-4y-15=0$.

**35.** $48^\circ\ 22'$; $17x+5y-26=0$; $2x-y+7=0$.

**36.** $(0, -3)$; $(-1, 2)$; $45^\circ$.

**37.** $x+4y+10=0,\ 3x-5y-4=0$.

**38.** $4x+3y-12=0,\ 3x-2y+8=0,\ x+2y=0$; $(-2\frac{2}{3}, 0), (4\frac{4}{5}, -2\frac{2}{5})$.

**39.** $x+y-5=0,\ x-2y+1=0$.

**40.** 3 sq.cm., 27 sq.cm., 9 sq.cm., 9 sq.cm.

**41.** $x-y-1=0,\ x-y-3=0,\ x+y-5=0$; 3; $x-3y-1=0$,
$x=3$, $(3, \frac{2}{3})$.

**42.** $(0, 6\frac{1}{4}), (8, \frac{1}{4})$; $3:3:2$.

**43.** $(0, 0)$, $x=0$, $y=0$; $(3\frac{1}{5}, 6\frac{2}{5})$, $3x-4y+16=0,\ 4x+3y-32=0$.

**44.** $3x^2+2xy+12y^2-16x-32y-12=0$.

**45.** $4\frac{1}{12}$, $2x=3y$.

**46.** $13:5$.

**47.** $(6, 4)$, $1:1$.

**48.** $ax-by=\frac{1}{2}(a^2-b^2)$; $b(3a^2-b^2)/8a$.

**49.** $4x-5y-16=0,\ y=3,\ 4x-2y-25=0$; $x-4y-4=0$,
$7x+8y-10=0,\ 4x+2y-7=0$; $(7\frac{3}{4}, 3), (2, -\frac{1}{2})$.

**50.** $(21, 13)$, 165, 55/76.

**51.** $x=2$.

**52.** $(7, 3), (2, -2), (-5, -1)$; 20.

**53.** 26.

**54.** $(5, -4)$, $4x-7y-48=0$.

**55.** $(\frac{3}{5}, 1\frac{3}{5}), (-\frac{2}{5}, \frac{3}{5}), (1\frac{2}{5}, -\frac{3}{5}), (2\frac{2}{5}, \frac{2}{5})$; $x-y-2=0,\ 2x+3y-1=0$.

**56.** $(1\frac{1}{3}, 1)$.

**57.** $2x+y-2=0$.

**58.** $x-y=0,\ 3x+2y-3=0$.

**59.** $3x-y-9=0,\ x-3y+5=0$; $6x+22y-45=0$.

**60.** $(3, 2), (-5, -4)$; $7x-y-19=0,\ x+7y-17=0,\ 7x-y+31=0$,
$x+7y+33=0$.

**61.** $(15, -6)$.

62. (i) $2x-3 = \pm k\sqrt{2}$, $2y+1 = \pm k\sqrt{2}$,
(ii) $2x^2+2y^2-6x+2y+5-2p^2 = 0$; $k = \pm p\sqrt{2}$.

63. $(2+m-n)x-(m+n)(y-1) = 0$, $(2-m+n)x-(m+n)(y+1) = 0$.

67. $\{am/(m-1),\ am/(m-1)\}$, $\{am/(m+1),\ -am/(m+1)\}$;
$\{2am^2/(m^2-1),\ 2am/(m^2-1)\}$.

68. $(6\frac{3}{5}, 5\frac{1}{5})$, $19\frac{1}{2}$.

69. $(-\frac{1}{2}, -5)$.

70. $x-y+1 = 0$, $x+y-9 = 0$.

71. $(1, 0)$, $(8\frac{17}{25}, 5\frac{19}{25})$.

73. $QP$; $(2\frac{1}{5}, 2\frac{3}{5})$.

74. $(x_1+x_2, y_1+y_2)$, $|x_1y_2-x_2y_1|$;
(i) $x_1x_2+y_1y_2 = 0$, (ii) $x_1^2+y_1^2 = x_2^2+y_2^2$.

76. $x-ty+at^2 = 0$, $tx+y = at^3+2at$; $(0, at)$, $(0, at^3+2at)$.

77. $\frac{9}{16}$.

78. $\frac{5}{11}$.

79. $(4a+3b)x+(3a-4b)y = 2a^2+3ab-2b^2$; $(3x+y)(x-3y) = 0$.

80. A line parallel to the given lines; $x-2y-7 = 0$, $7x+3y-15 = 0$.

## EXAMPLES III A (page 68)

1. (i) $y^2 = 4a(x+a)$, (ii) $x^2 = 4b(b-y)$,
(iii) $16x^2-24xy+9y^2-100x-50y+125 = 0$.

2. $gy^2+2v^2x-2uvy = 0$.

3. $1, -1, 0, -\frac{1}{2}, 4$.

6. $2x-y-12 = 0$, $x-3y+9 = 0$, $3x+y-33 = 0$, $x+2y+4 = 0$,
$2x-y-12 = 0$.

7. $x+y-3 = 0$, $(9, -6)$.

8. $(2, -4)$.

9. $x+8y-33 = 0$, $68\frac{1}{16}$.

10. $x-2y+4 = 0$, $2x+y-12 = 0$; 20.

11. $(\frac{1}{4}a, -a)$, $(a, 2a)$; $4x+2y+a = 0$, $x-y+a = 0$, $(-\frac{1}{2}a, \frac{1}{2}a)$.

13. $3x-2y+4 = 0$.

14. $(\frac{1}{9}, \frac{2}{3})$, $9x+27y = 19$.

15. $x+y+1 = 0$, $x+5y+25 = 0$.

17. $x-ty+3t^2 = 0$.

18. $8a\sqrt{2}$, $x-y+a = 0$, $x+3y+9a = 0$, $32a^2$.

## EXAMPLES III B (page 73)

2. $3x-4y-3a = 0$, $x-3y+9a = 0$, $9x+3y+a = 0$.

6. $\frac{1}{2}$, $(4a, 4a)$; $(-a, \frac{3}{2}a)$.

7. $90°$; $2a$.

10. $25x+10y+4a = 0$.

12. $(4a, 4a)$.

## EXAMPLES III D (page 84)

**1.** $(y+2a)^2 = 8ax$.
**2.** $\{ap^2, ap(1-p^2)\}$.
**4.** $\{-ap^2/(1+p^2), ap^3/(1+p^2)\}$.
**5.** $9y^2 = 4a(3x-2a)$.
**6.** $(4a, 0), (2a, 0)$.
**7.** $y^2 = a(x-3a)$.
**8.** $y^2+ax = 0$.
**10.** $\{apq, a(p+q)\}$.
**12.** $\{apq, a(p+q)\}$.
**13.** $ly+2am = 0$.
**14.** $x+hy+a(h^2+4) = 0$, where $h$ is the parameter of $H$.
**16.** (i) $y = k_1x$, (ii) $y^2 = k_2x^2+2ax$.
**18.** $\{a(p^2+q^2+1), a(p+q)\}$, $y^2 = a(x-3a)$.
**19.** $y^2 = a(x-3a)$.
**20.** $y^2 = 16a(x-6a)$.
**22.** $\{a(2m^{-2}-1), 2am^{-1}\}$.

## EXAMPLES III E (page 90)

**1.** (i) $y^2 = 2a(x-a)$, (ii) $(y^2-4ax)(y^2+4a^2)+d^2a^2 = 0$, (iii) $y^2 = 2a(x+a)$, (iv) $y^4-2axy^2+4a^2y^2+8a^4 = 0$, (v) $y^2-y\mu = 2a(x-\lambda)$.
**2.** Normal at $(ak^2, -2ak)$.
**3.** $y^2 = 16a(x-6a)$.
**4.** $4(x-2a)^3 = 27ay^2$; yes, from $(2a, 0)$.
**5.** (i) $y^2 = 4ax$, (ii) $y^2 = a(x-3a)$, (iii) $2(kx+2ay-ak)^2 = (4a^2+k^2)(2ax-ky-2a^2)$, where $k$ is the ordinate of the fixed point, (iv) $y^2 = a(x-3a)$.
**7.** $27ay^2 = 2(x-2a)^3$.
**10.** $y^2(x+2a)+4a^3 = 0$.
**11.** $(\lambda-6a, -\frac{1}{2}\mu)$, $\{\frac{2}{3}(\lambda-2a), 0\}$.
**12.** (i) $y^2 = a(x+6a)$, (ii) $y = 0$.

## MISCELLANEOUS EXAMPLES III (page 91)

**3.** $\sqrt{2}$.
**6.** $(4, -4), (-8, 2)$.
**7.** $4x+2y+a = 0$, $(\frac{1}{4}a, -a)$, $(-\frac{1}{2}a, \frac{1}{2}a)$.
**9.** $y^2 = 4ax$ where $(a, 0)$ is the fixed point and $x+a = 0$ is the fixed line; $(\frac{1}{2}a, 0)$, $2x+a = 0$.
**10.** $y^2 = 4ax$ where $A$ is $(a, 0)$ and $l$ is $x+a = 0$, $(a, 0)$, $x = 0$.
**12.** $y^2 = 4ax$, $y^2 = 4aq^2(aq^2+a-x)$.

**13.** $y^2 = 4a(x+a)$; $45°$.

**20.** $(-ap^2, 0)$, $(0, ap)$, $(2a+ap^2, 0)$; $2px+(1-p^2)y-2ap = 0$,
$px+(2+p^2)y = ap(2+p^2)$.

**22.** $2a^2$.

**25.** (i) $\{\frac{1}{3}a(qr+rp+pq), \frac{2}{3}a(p+q+r)\}$,
(ii) $\{-a, a(pqr+p+q+r)\}$,
$\{\frac{1}{2}a(qr+rp+pq+1), \frac{1}{2}a(p+q+r-pqr)\}$.

**26.** $2/(p+q)$, $p^{-1}$, $q^{-1}$.

**27.** $\{l^{-2}(4am^2-nl+2al^2), 2mnl^{-2}\}$, $y^2 = a(x-3a)$.

**30.** $k(x+ky-2a)^2+ay = 0$.

**31.** $2at^2x+(b+c)ty+2bc = 0$; when $b+c = 0$, $QR$ is parallel to $x = 0$ for all positions of $P$.

**39.** $xy = y(h-2a)+2ak$; the directrix and the axis.

**40.** $\{-a(pq+2), a(p+q)(pq+2)\}$.

**46.** $y^2 = 2a(x-a)$, $y^2 = a(x-3a)$.

**49.** $3y^2 = 4ax$.

**50.** A horizontal line at a height $V^2/g$ above $O$.

## EXAMPLES IV A (page 101)

**4.** $2(x-y) = 3c$, $x+4y = 4c$, $4x+y+4c = 0$, $8x-2y = 15c$,
$2x-8y = 15c$.

**5.** $4x-y = 15$, $(-\frac{1}{4}, -16)$.

**6.** $0{\cdot}02$.

**7.** $(3c, \frac{1}{3}c)$, $(-2c, -\frac{1}{2}c)$; $x+9y = 6c$, $x+4y+4c = 0$, $(-12c, 2c)$.

**10.** $(-c/p^3, -cp^3)$.

## EXAMPLES IV B (page 106)

**3.** $\frac{1}{2}c^2(p^2+p^{-2})^2$.

**5.** $(c, -3c)$.

**6.** $xy = 4c^2$.

**7.** $c^2(l^2-m^2)^2+lm = 0$.

**8.** $2c^2$, $\frac{1}{3}c^2$.

**9.** $xy(x^2+y^2)^2+4c^2(x^2-y^2)^2 = 0$.

**10.** $-1, -1$; $\frac{3}{2}\sqrt{2}$.

**15.** $(-c, -c)$, $(-\frac{1}{2}c, -2c)$.

## EXAMPLES IV C (page 116)

**1.** $(a^2-3p^2)x+2py = (a^2-p^2)^2$.

**2.** $(\frac{1}{4}t^2, -\frac{1}{8}t^3)$.

**4.** $3x-2y = 1$, $2x+3y = 5$.

**5.** $x-ty+t^3 = 0$.

**6.** $(\frac{1}{4}at^2, -\frac{1}{8}at^3)$.

## MISCELLANEOUS EXAMPLES IV (page 127)

**5.** $8xy = 9c^2$. **7.** $2xy = kx + hy$.

**8.** $(xy - c^2)(x^2 + y^2) = a^2xy$. **10.** 0.

**11.** $x - 2y + 2 = 0$, $2x + y = 11$.

**12.** $3a^2x - y + 1 - 2a^3 = 0$, $x + 3a^2y = 3a^5 + 3a^2 + a$.

**13.** 1, (2, 0). **14.** $-\frac{1}{2}$, $\frac{3}{8}\sqrt{13}$; $(2-x)^3 = (y-3)^2$.

**15.** $(6\frac{1}{3}, 225)$.

**16.** $(1-t^2)x - (1+t^2)^2y + 2t^3 = 0$, 0, $\pm\sqrt{3}$.

**17.** 1, $-\frac{3}{2}$, $-6$, 10.

**18.** $x = \frac{1}{2}a(3t^2-1)$, $y = \frac{1}{2}at(3t^2-1)$; $4x - 3y - a = 0$, coincident points at $(a, a)$.

**19.** $\frac{4}{9}$, $-3$, 6, $(-\frac{3}{2}, 9)$.

**20.** $x = 2at^2/(1+t^2)$, $y = 2at^3/(1+t^2)$.

**21.** $(\frac{1}{3}, 0)$, $(-\frac{1}{9}, -\frac{4}{81})$, $16\sqrt{2}/81$.

**22.** $x\sin^2\theta + y\cos^2\theta = a\cos^2\theta\sin^2\theta$.

**23.** $x\sin\theta + y\cos\theta = a\cos\theta\sin\theta$, $x\cos\theta - y\sin\theta = a\cos 2\theta$.

**24.** $x\sin t(2 - 3\sin^2 t) + y\cos t(2 - 3\cos^2 t) - a\sin^2 t\cos^2 t = 0$,
$2x\cos t(2 - 3\cos^2 t) - 2y\sin t(2 - 3\sin^2 t) + a\sin 2t\cos 2t = 0$.

**25.** 8. **28.** $x\sin\frac{1}{2}\theta - y\cos\frac{1}{2}\theta = a\theta\sin\frac{1}{2}\theta$.

**31.** $\theta^4 - (t+\lambda)\theta^3 + (t^{-1} + \lambda t^2)\theta - 1 = 0$.

**34.** $x - 3p^2y + 2ap^3 = 0$; if $R$ is the origin the tangents at $P$ and $Q$ are parallel.

**36.** $\lambda_1\lambda_2\lambda_3 = -1$, $\lambda x(\lambda^3 - 2) - y(2\lambda^3 - 1) + \lambda^2 = 0$.

**37.** $ct^4 + t^3(a+b) + t(b-a) - c = 0$.

**38.** $bc + ca + ab = 3$; (4, 3), $(-4, 3)$.

**39.** Another line.

**40.** $x = at/(1+t^3)$, $y = at^2/(1+t^3)$, $t_1t_2t_3 = -1$.

**41.** $(\frac{27}{4}, -\frac{3}{2})$, $2x + y - 8 = 0$.

**44.** $b^2x^2 + a^2y^2 + 2abxy\sin\alpha = a^2b^2\cos^2\alpha$, $bx + ay\sin\alpha \pm ab\cos\alpha = 0$,
$bx\sin\alpha + ay \pm ab\cos\alpha = 0$.

## EXAMPLES V A (page 139)

1. (i) (0, 0), 5, (ii) (2, −1), 4, (iii) $(-1\frac{1}{2}, \frac{1}{2})$, $\frac{1}{2}\sqrt{14}$,
(iv) $(-\frac{1}{4}, -\frac{3}{4})$, $\frac{1}{4}\sqrt{10}$, (v) (5, −7), 0.

2. (i) $x^2+y^2-8x-2y+8=0$,
(ii) $x^2+y^2-6x+8y=0$,
(iii) $x^2+y^2+2ax-2by+b^2=0$,
(iv) $x^2+y^2-2xat^2-4yat+a^2(2t^2-1)=0$.

3. ± 6.

4. $3x^2+3y^2-14x-10y+3=0$.

5. $4x^2+4y^2-20x-13y+24=0$.

6. (0, −2), (−7, 5), (−3, 7); $3x-4y-8=0$, $4x-3y+43=0$, $y=7$.

7. (11, −1), (−14, −1), (2, 11); (2, 11).

8. (−10, 0), (6, 8).

9. $\{\frac{1}{25}(-2 \pm 16\sqrt{34}), \frac{1}{25}(86 \pm 12\sqrt{34})\}$.

10. $(-3, -\frac{1}{3})$, (−3, 3), $(-\frac{1}{2}, 3)$.

12. $x^2-y^2+a^2=0$.

13. $x^2+y^2+10x+20y+25=0$, $x^2+y^2-6x+4y+9=0$;
(−5, −10), 10; (3, −2), 2.

14. 12, −38.

15. (3, 2), 5; 8; (−1, 5).

16. (3, 2), 5.

17. (3, 0), 4.

18. $5x^2+5y^2-16x-12y=0$.

19. $3x^2+3y^2-10y+3=0$, $x^2+y^2-30y+81=0$; $(\frac{4}{5}, \frac{3}{5})$, $(7\frac{1}{5}, 5\frac{2}{5})$.

20. $x^2+y^2-8x-6y+16=0$, $x^2+y^2-14x-12y+76=0$.

21. $7x-4y=3$, $12x-5y=27{\cdot}8$.

22. $8x-15y=0$, $(4\frac{7}{17}, 2\frac{6}{17})$.

23. $\{4/(1+m^2), 4m/(1+m^2)\}$, $x^2+y^2-4x=0$.

24. $\{a, (c^2-a^2)/b\}$, $\{0, (a^2+b^2-c^2)/2b\}$,
$bx^2+by^2-y(a^2+b^2-c^2)-bc^2=0$.

25. $x^2+y^2-2x=0$, $x^2+y^2-10x+8y+16=0$.

## EXAMPLES V B (page 146)

1. (i) $x^2+y^2-7x+3y-28=0$,
(ii) $x^2+y^2-x-y-2=0$,
(iii) $x^2+y^2+2x-y-66=0$,
(iv) $x^2+y^2+x-6=0$.

2. (i) $5(x^2+y^2)+11x+2y+5=0$,
(ii) $5(x^2+y^2)-x+18y-85=0$,
(iii) $25(x^2+y^2)+49x+18y-20=0$.

3. $3x-4y=40$, $3x+4y=8$.

4. $x^2+y^2-6x-7y=0$, $x^2+y^2-3x-2y-18=0$,
$x^2+y^2+3x-4y=0$, $3(x^2+y^2)-11x-32y+60=0$.

5. $x^2+y^2-12x-8y+34=0$, (3, 1).

6. (3, 5), $2x+3y=21$; (6, 3), (0, 7).

7. $2x+y=4$, $x^2+y^2-2x-4y=0$; (2, 0), (0, 4).

8. $x^2+y^2-5x-4y+4=0$.

9. $x^2+y^2-bx-cy+a^2=0$, $b^2+c^2-4a^2\geqslant 0$; $b$, $a^2$.

10. $x^2+y^2-10x+26y+25=0$, 13, (5, −13); $65x-72y=0$.

11. $x^2+y^2-5x-3y+8=0$, $x^2+y^2+3x-11y=0$,
$x^2+y^2+2x-10y+1=0$, $x^2+y^2-6x-2y+9=0$.

13. $x^2+y^2-8x-8y+27=0$, (5, 6), (3, 2).

14. $(l^2+m^2)a^2=n^2$, $x\pm\sqrt{3}y=2a$, $x^2+y^2=4a^2$.

15. (0, 1), (3, 0), (4, 3), $x^2+y^2-4x-4y+3=0$.

16. $\frac{1}{2}$, 3.

17. $x^2+y^2-8x+6y=0$, $x^2+y^2+6x-8y=0$; 4.

18. $\{(c+\lambda a\cos\theta)/(1+\lambda),\ \lambda a\sin\theta/(1+\lambda)\}$, $\{c/(1+\lambda),\ 0\}$, $\lambda a/(1+\lambda)$.

20. $x^2+y^2-2x-2y=0$, $x^2+y^2-2x+14y=0$; (2, 2), (6, −2).

21. $x^2+y^2-4x-4y+4=0$, $x^2+y^2-20x-20y+100=0$;
71° 30′, $x+y=6$, (4, 2).

22. $x^2+y^2-10x-5y+25=0$, $4x-3y=0$.

23. $x^2+y^2-2ay-a^2=0$. 24. (1, 2), 5; 4.

25. $(x-9)^2+(y-12)^2=81$, $(3\frac{3}{5}, 4\frac{4}{5})$.

## EXAMPLES V C (page 150)

4. $2y^2+ax=0$.

6. $\{\frac{1}{2}(4a+3m^2al^{-2}-nl^{-1}),\ \frac{1}{2}(mnl^{-2}-m^3al^{-3})\}$.

## EXAMPLES V D (page 158)

1. (i) $4x-3y+13=0$, (ii) $x-4y+5=0$, (iii) $ax+by=0$.

7. $4\sqrt{3}$.

8. $(3, 4)$, $x^2+y^2-22x-16y+65 = 0$.

9. $x+y = 1$.

10. $(2, 0)$, $3$; $(-3, 1)$, $2$.

11. $4x+2y-1 = 0$, $x-2y+1 = 0$; $x^2+y^2+6x+2y = 0$, $x^2+y^2+2x+1 = 0$, $x^2+y^2-2x-2y+2 = 0$, $(-1, 0)$, $(1, 1)$.

12. $(2, 2\frac{1}{2})$.

13. Diameter $x-my = 0$, where the gradient of the parallel chords is $m$.

14. $(-1, 1)$, $(\frac{1}{5}, 1\frac{3}{5})$; $x^2+y^2-2x-4y+3 = 0$, $x^2+y^2+102x+48y-23 = 0$.

15. $3x+y = 0$, $4x^2+4y^2-9x-3y+4 = 0$.

16. $(1, 1)$, $(3, 3)$, $x^2+y^2-4x-4y+6\pm\sqrt{14}(x-y) = 0$.

17. $2gx+c-f^2 = 0$, $\{(f^2-c)/g, -f\}$.

18. $ax-by+1 = 0$, $bx+ay+2ab = 0$. $2ab(x^2+y^2)+(2+\mu+c)\,bx+(\mu+c-2)\,ay+2ab\mu = 0$.

## MISCELLANEOUS EXAMPLES V (page 163)

1. $2x^2+2y^2-16x-11y+24 = 0$, $\{4\pm k, \frac{1}{4}(11\pm k)\}$ where $17k^2 = 185$.

2. $(3, \frac{1}{2})$, $\frac{1}{2}\sqrt{37}$.

3. $\frac{3}{4}$.

4. $x+2y = 5$, $x^2+y^2-4x-3y-25 = 0$, $2x-y+10 = 0$.

5. $x^2+y^2-8x-6y+16 = 0$, $4x+3y = 10$.

6. $x\cos\frac{1}{2}(\theta-\phi) = a\cos\frac{1}{2}(\theta+\phi)$, $y\cos\frac{1}{2}(\theta-\phi) = a\sin\frac{1}{2}(\theta+\phi)$; $x+y\tan\alpha = 0$.

7. $4(x-a)^2+9y^2 = 4a^2$.

8. $x^2+y^2-4ax-2a^2y+4a^2 = 0$, $x^2 = 4y$.

9. $(6, 3)$, $(7, 1)$.

10. $x^2+y^2-14x-8y+20 = 0$.

11. $c^2 = a^2(1+m^2)$, $16\sqrt{2}$.

12. 12.

13. $(-1, 3)$, $\frac{1}{2}\sqrt{10}$; $x-3y+15 = 0$; $\sqrt{2}$.

14. $3x+4y = 8$, $3x+4y = 28$; $(\frac{4}{5}, \frac{7}{5})$, $(\frac{16}{5}, \frac{23}{5})$, $3x+4y = 18$.

15. $(5, 3)$, $(1, 5)$, $2\sqrt{5}$; $\frac{5}{2}(4-\pi)$.

16. $3x+4y = 17$, $x = 3$, $7x-24y+123 = 0$.

17. $x^2+y^2-4y-1 = 0$, $(0, -\frac{1}{2})$, $2x^2+2y^2-3y-2 = 0$.

18. $x^2+y^2+4x-8y = 0$, $126°\,53'$, $14{\cdot}15$, $47{\cdot}69$.

19. $(7\frac{1}{5}, \pm 9\frac{3}{5})$.

20. $3(x^2+y^2)-5(x+y) = 0$.

**21.** $4x+3y+7=0$, $(-3\frac{2}{5}, 2\frac{1}{5})$; $5x^2+5y^2+2x-46y+26=0$.

**22.** $x^2+y^2-4\sqrt{3}x-4y+12=0$.

**23.** $x^2+y^2-16x-20y+64=0$, $x^2+y^2+24x-40y+144=0$;
$4x+3y=112$.

**24.** $(1, 0)$, $\frac{1}{2}\sqrt{3}$.

**25.** $x^2+y^2-4ax+a^2=0$.

**26.** $2\frac{2}{5}$, $17\frac{1}{3}$.

**27.** (i) $-1$, (ii) $27/8$, (iii) $-113/123$, (iv) $\frac{1}{4}$;
$x-2y+4=0$, $(0, 2)$, $(-2\frac{2}{5}, \frac{4}{5})$.

**29.** $x^2+y^2-2x-4y+3=0$, $x^2+y^2-8x-6y+17=0$,
$7(x^2+y^2)-5x-25y=0$.

**30.** $x^2+y^2-3x+4y+2=0$, $x^2+y^2-12x-4y+8=0$.

**31.** $19x^2+19y^2+84x-88y=0$, $21x-22y=0$, $2x-5y+16=0$.

**32.** $a$.

**33.** $\{(a+b), -\sqrt{3}(a^2+2ab-3b^2)/6b\}$.

**34.** $x=0$, $3x \pm 4y=8$.

**35.** $4x^2+4y^2-20x-20y+25=0$.

**36.** Circle concentric with the given circle; the given circle.

**37.** $(3, 1)$; $(6, 2)$, $(-1, 3)$, $(2, -6)$.

**38.** $(-5, 0)$, $(1, 2)$, $30$.

**39.** $4x+3y-8=0$; $4x^2+4y^2-12x-7y+12=0$,
$x^2+y^2-2x-2y+1=0$.

**40.** $x^2+y^2+x-2y+1=0$, $5x^2+5y^2-x+2y-1=0$.

**42.** $x^2+y^2+ax(t^2-1)-a^2t^2=0$.

**43.** $x^2+y^2-ax(4+3t^2)+at^3y=0$.

**45.** $2\sqrt{\{a(a+b)\}}$.

**46.** $x^2+y^2-ax(pq+1)-ay(p+q)+a^2pq=0$.

**47.** $\{0, (a^2-b^2)/2b\}$, $(a^2+b^2)/2b$.

**50.** $x^2+y^2+4x-6y=0$.

**51.** $x^2+y^2-6x-4y+9=0$.

**52.** $x^2+y^2=25$, $(3, 4)$.

**56.** $x^2+y^2-8x-6y=0$.

**59.** $(0, \frac{1}{2}\lambda)$, $\frac{1}{2}\lambda$, $x^2+y^2-2\alpha x-\lambda' y+\alpha^2=0$.

**60.** $x^2=2y$.

**61.** $x^2+y^2=a^2-bc$.

**62.** $(-2\frac{2}{5}, 5\frac{4}{5})$.

**65.** $g^2+2ga-a^2t^2=0$, $a\{-1+\sqrt{(1+t^2)}\}$.

**74.** $(a^2+b^2)(x^2+y^2)+acx+bcy+\mu(bx-ay) = 0$.

**75.** $\alpha = 0$, $\gamma = -c$.

**77.** $(0, 0)$, $\{-ln/(l^2+m^2),\ -mn/(l^2+m^2)\}$,
$(l^2+m^2)(x^2+y^2)+x(ln+\mu m)+y(mn-\mu l) = 0$.

**78.** When the three given circles belong to one coaxal system.

## EXAMPLES VI A (page 180)

**1.** $x^2(1-e^2)+y^2+2e^2dx-e^2d^2 = 0$.

**3.** $\frac{4}{5}$, 8.

**4.** $x^2/32+y^2/30 = 1$.

**5.** $x^2/169+y^2/25 = 1$, $\frac{12}{13}$, $x = \pm 169/12$.

**7.** (i) $\frac{1}{3}\sqrt{5}$, $(\pm 2\sqrt{5}, 0)$, $x = \pm\frac{18}{5}\sqrt{5}$,
(ii) $\frac{1}{3}\sqrt{6}$, $(\pm\sqrt{6}, 0)$, $x = \pm\frac{3}{2}\sqrt{6}$,
(iii) $\frac{1}{4}\sqrt{7}$, $(1\pm\sqrt{7},\ -4)$, $x = 1\pm 16/\sqrt{7}$.

**8.** (i) $\frac{1}{2}\sqrt{2}$, (ii) $\frac{1}{2}\sqrt{3}$.

**9.** $e^4+e^2-1 = 0$.

**11.** $x-2y\pm 5 = 0$.

**12.** $(1-p^2)x/a+2py/b = 1+p^2$.

**13.** $x^2/25+y^2/9 = 1$, $9x+20y-75 = 0$; $(\frac{25}{4}, \frac{15}{16})$.

**16.** $(\frac{1}{2}a\sec\alpha,\ \frac{1}{2}b\operatorname{cosec}\alpha)$.

**17.** $\frac{7}{17}$.

**20.** $\{(a+b)\cos\theta,\ (a+b)\sin\theta\}$.

**21.** $(e^2a\cos\theta,\ 0)$, $(0, b\operatorname{cosec}\theta)$.

**22.** $(ka\cos\theta,\ -kb\sin\theta)$ where $k(a^2+b^2) = a^2-b^2$.

## EXAMPLES VI C (page 196)

**1.** $x^2/4-y^2/12 = 1$, $2x-y = 2$, $x+2y = 16$.

**2.** $2xy = 1$.

**3.** $\frac{2}{3}\sqrt{3}$.

**4.** $(0, 0)$, $(\pm 6\frac{1}{4}, 0)$, $\frac{5}{3}$.

**6.** $x+y = 6$, $x-y = 12$, $(4\frac{1}{2}, 1\frac{1}{2})$.

**7.** $(1\frac{1}{2}, \frac{1}{2})$, $63^\circ\,26'$.

**8.** $x^2/9-y^2/7 = 1$, $x^2+y^2 = 2$.

**13.** $\{a(\sec\theta\pm\tan\theta),\ \pm b(\sec\theta\pm\tan\theta)\}$.

**14.** $(x^2-y^2)^2 = a^2(2x^2-y^2)$.

**15.** $x\cos\frac{1}{2}(\theta-\phi)-y\sin\frac{1}{2}(\theta+\phi) = a\cos\frac{1}{2}(\theta+\phi)$.

## EXAMPLES VI D (page 201)

**1.** $(\alpha m^2+\beta)x^2+2\alpha mcx+\alpha(c^2-\beta) = 0$.

**2.** $(3\frac{1}{5}, 1\frac{4}{5})$.

**3.** (i) $3x-2y = 6$, (ii) $x+y+2 = 0$, (iii) $x/a+y/b = 1$; (ii).

4. (i) $x^2+y^2 = 7$, (ii) $x^2+y^2 = 4$, (iii) $x^2+y^2 = a^2+b^2$.

5. $x^2+y^2-10x+6y+9 = 0$.

6. $(-a^2ln^{-1}, -b^2mn^{-1})$, $(\pm\frac{2}{5}, \mp\frac{1}{5})$.

13. $x^2/a^2+y^2/b^2 = 4$.

16. $bx-ay = 0$.

17. $(a^2p/r, -b^2q/r)$.

18. $(x^2+y^2)^2 = a^2x^2-b^2y^2$.

19. $x^2\sec^2\frac{1}{2}\alpha = y^2+a^2$; $x = 0$.

## EXAMPLES VI E (page 207)

1. $b^2x+a^2my = 0$.

2. $\beta x+m\alpha y = 0$.

9. $3x+4y = 7$, 5.

13. $b$.

14. $\alpha b^2(x-\alpha)-\beta a^2(y-\beta) = 0$; tangent at $(\alpha, \beta)$.

15. $\alpha b^2(x-\alpha)+\beta a^2(y-\beta) = 0$; $xb^2m = ya^2l$,
$(b^2x^2+a^2y^2)^2 = b^2(b^4x^2+a^4y^2)$.

## MISCELLANEOUS EXAMPLES VI (page 208)

1. $b^4x^2+a^4y^2 = a^2b^2(a^2+b^2)$.

3. $\frac{9}{5}, \frac{5}{3}$; $3x+5y = 25$, $25x-15y = 27$.

4. $\frac{1}{4}x^2+\frac{1}{3}y^2 = 1$, $12x^2-4y^2 = 3$, $(\pm 1, \pm 1\frac{1}{2})$, $(\pm 1, \mp 1\frac{1}{2})$.

7. 2, $2x\pm y-2 = 0$.

8. Yes, if the first hyperbola is rectangular.

9. $(h, k)$ where $(a^2-b^2)h^2 = a^2(a^2-2b^2)$, $(a^2-b^2)k^2 = b^4$.

12. $\frac{1}{2}\sqrt{2}$.

18. $x\pm y+a = 0$.

26. $\pm x\sqrt{(c^2-b^2)}\pm y\sqrt{(a^2-c^2)} = c\sqrt{(a^2-b^2)}$.

36. $e > \frac{1}{2}\sqrt{2}$.

38. $4ab\operatorname{cosec}(\alpha-\beta)$.

41. $nxy(b^2A^2-a^2B^2) = mx(a^2-A^2)b^2B^2+ly(b^2-B^2)a^2A^2$.

43. $(a^2+b^2)(b^2x^2+a^2y^2)^2 = a^4b^4(x^2+y^2)$.

44. $4(a^2x^2+b^2y^2)^3 = (a^2x^2-b^2y^2)^2(a^2-b^2)^2$.

49. $bx = \pm ay$.

## EXAMPLES VII A (page 224)

1. (i) $\tan^{-1}\frac{7}{11}$, (ii) $\alpha$, (iii) $\tan^{-1}\frac{1}{2}\sqrt{37}$.

2. $-\frac{1}{2}$, $-\frac{1}{11}(5-\sqrt{3})$, 0, $1+\sqrt{3}$, $-1$.

3. (i) $4x^2+xy-4y^2=0$, (ii) $x^2+42xy-y^2=0$, (iii) $7x^2+26xy-7y^2=0$.

4. $m=-1$.

5. $\sqrt{(h^2-ab)}/(a+b-2h)$.

6. 2.

7. $\frac{1}{35}(5+10\sqrt{2}+7\sqrt{5})$.

## EXAMPLES VII B (page 230)

1. (i) $\pm 2$, (ii) $-16/49$, (iii) $\frac{1}{3}$, (iv) 1, $1\frac{2}{3}$, (v) $-2\frac{2}{3}$.

2. (i) $(0, -2)$, (iv) $(0, -2)$, (v) $(1, -2)$.

3. 45°.

5. $-10$, $\tan^{-1} 11{\cdot}5$.

6. $0, \{gf \pm \sqrt{(g^2f^2-g^2c-f^2c+c^2)}\}/(g^2+f^2-c)$; $g=f=0$.

7. (i) 45°, (ii) 60°, (iii) $\tan^{-1}\frac{3}{5}$.

8. 45°.

9. $4x^2-4xy+y^2=0$; the given locus is a line pair with vertex $(1, 2)$.

10. (i) $2x^2-xy-2y^2-5x-3y+1=0$,
(ii) $3x^2-2xy-3y^2=0$,
(iii) $3x^2+xy-3y^2+4x+13y-11=0$.

11. $2a=b+c$.

12. $x^2+y^2=a^2$.

## MISCELLANEOUS EXAMPLES VII (page 231)

1. If $S$ is $(0, 0)$ and $C$ is $(c, 0)$, $c > 4a$ or $c < 0$, where $4a$ is the latus rectum.

2. $2\sqrt{5}$; $x^2+y^2 \pm 4x \mp 2y=0$.

3. $ab(x^2-y^2)=(a^2-b^2)xy$.

4. $5x+3y+14=0$.

7. $b(x-x')^2-2h(x-x')(y-y')+a(y-y')^2=0$,
$(a+b)(x^2+y^2)-2x(bx'-hy')-2y(ay'-hx')+bx'^2-2hx'y'+ay'^2=0$.

8. $(2, -1)$, $3x^2+xy-3y^2+11x+8y+7=0$, $11x^2+30xy+16y^2=0$, $28/\sqrt{37}$.

9. $0, -2, -3$; $3x-y+3=0$, $x-4y-1=0$, $x+y+1=0$,
$x=1$, $y=0$, $4x-y+2=0$.

11. $\{\frac{1}{2}(m^{-2}-1), \frac{1}{2}(m+m^{-1})\}$.

12. $2x \pm 7y+12a=0$.

13. $3x-y-1=0$, $3x+y-14=0$.

14. $3x^2-8xy-3y^2=0$, $x^2+y^2-2x-4y=0$.

15. $x^2(a-2h\sqrt{3}+3b)+2xy(a\sqrt{3}-2h-b\sqrt{3})+y^2(3a+2h\sqrt{3}+b)=0$,
$\sqrt{3}xy-y^2=0$.

**16.** $(\frac{23}{40}, \frac{19}{30})$, $\tan^{-1}\frac{30}{19}$.

**17.** $x-2y-1=0$, $x-y-2=0$; 2.

**18.** $3x+2y+4=0$, $2x-3y+7=0$.

**20.** $bx^2-2hxy+ay^2+2x(hq-bp)+2y(hp-aq)$
$+(aq^2-2hpq+bp^2)=0$,
$x^2+y^2-x-2y-5=0$.

**21.** $2gx+2fy+c=0$, $(af-gh)x=(bg-hf)y$.

**23.** $\{k(hm-bl), k(hl-am)\}$, where $k=2n/(am^2-2hlm+bl^2)$.

**27.** $\frac{1}{2}a^2(2\phi-\sin 2\phi)$.

**29.** $2(ap+hq)x+2(hp+bq)y=3(ap^2+2hpq+bq^2)$.

## EXAMPLES VIII A (page 248)

**2.** $x-y+1=0$.

**3.** $3x+4y+1=0$, $4x-3y-7=0$, 5, 3.

**4.** (i) $2\sqrt{2}$, $2\sqrt{3}$; $\tan^{-1}\frac{2}{5}$, $\tan^{-1}-\frac{5}{2}$,
(ii) $2\sqrt{5}$, $\frac{2}{3}\sqrt{15}$; $\tan^{-1}\frac{1}{4}$, $\tan^{-1}-4$,
(iii) $\sqrt{10}$, $\frac{2}{3}\sqrt{15}$; $\tan^{-1}2$, $\tan^{-1}-\frac{1}{2}$.

**5.** $4x+7y=0$, $8x+y=0$, $3x+2y=0$, $2x-3y=0$, $2x=3y$.

**6.** (2, 3), 8, 2.

**7.** (1, 1), $\frac{1}{2}\sqrt{2}$.

**8.** 3.

**9.** $-22$, $-29$.

**10.** $(-\frac{1}{4}, -\frac{1}{4})$.

**11.** $4x-3y=0$, $3x+4y=0$; 4, 2; $\frac{1}{2}\sqrt{3}$.

**12.** $(-7, -4)$.

**13.** $(2, -1)$, $(-2, 1)$.

**15.** $12/\sqrt{13}$.

**16.** $\alpha=1$, $\beta=2$, $\gamma=-6$.

## EXAMPLES VIII B (page 254)

**1.** (1, 1).

**2.** $(ab+h^2)(x^2+y^2)-2bcx-2acy=0$.

**4.** $(0, -1)$.

**5.** $hx+ky+\lambda(x+h)+c=0$, $hx+ky+\lambda(y+k)-c=0$, $x^2-y^2=2c$.

**6.** $cx=4a^2$, $2acx-c^2y+8a^3=0$.

## EXAMPLES VIII C (page 260)

**1.** $(x^2-y^2-2y-1)+\lambda(y^2+y-6) = 0$, $x^2 = y+7$.

**2.** $h'(a-b) = h(a'-b')$.

**4.** $x^2+2x-15 = 0$, $x^2+y^2-25 = 0$, $(x^2+2x-15)+\lambda(x^2+y^2-25)=0$, $y^2 = 2x+10$, $(x-2)^2+3y^2 = 49$.

**5.** $x^2+y^2-2ak^{-1}x+2aky-c^2(k^2+k^{-2}) \pm (xy-c^2) = 0$.

**6.** (i) $2-\sqrt{\cdot 8} < \lambda < 2+\sqrt{\cdot 8}$, (ii) $\lambda = 2 \pm \sqrt{\cdot 8}$,
(iii) $\lambda < 2-\sqrt{\cdot 8}$, $\lambda > 2+\sqrt{\cdot 8}$, (iv) $\lambda = -3, 0, 1$,
$y = 0$, $4x-y+2 = 0$, $x = 1$, $x+y+1 = 0$, $3x-y+3 = 0$,
$x-4y-1 = 0$.

## MISCELLANEOUS EXAMPLES VIII (page 261)

**1.** $4x+3y = 0$, $3x-4y = 0$; $\frac{2}{5}$, $\frac{1}{5}$.

**2.** $x-3y = 0$, $3x+y+1 = 0$, 2. **3.** $16x-12y+13 = 0$, $(0, \frac{1}{4})$.

**4.** $5x+3y = 8$, $3x+5y = 8$, $x-y = 0$, $x+y = 2$; $32X^2-2Y^2 = 1$.

**5.** $3x+4y = 1$, $(\frac{1}{5}, \frac{2}{5})$, $(\frac{16}{125}, \frac{38}{125})$.

**6.** $(2, -1)$, 6, 2; $x+y = 1$, $x-y = 3$.

**7.** $\tan^{-1}\frac{2}{3}$; hyperbola, $\frac{1}{2}\sqrt{26}$.

**8.** $X^2+24XY-6Y^2+1 = 0$, $12X^2-7XY-12Y^2 = 0$; $\dfrac{1}{\sqrt{10}}, \dfrac{1}{\sqrt{15}}$.

**9.** $3x-2y = 1$, $2x+3y = 5$. **10.** $6a$, $4a$.

**11.** $2x+y+8 = 0$, $2x+11y-2 = 0$.

**12.** $5x^2+5xy-7x-16y = 0$, $5x = 16$, $5x+5y+9 = 0$.

**13.** $y = 1$, $4x-3y-1 = 0$, $2x+y-3 = 0$, $x-2y+1 = 0$;
$4xy-3y^2-4x+2y = 0$.

**14.** $(2, -2)$, $(-1, 1)$, $(\frac{1}{2} \pm \frac{3}{2}\sqrt{2}, -\frac{1}{2} \mp \frac{3}{2}\sqrt{2})$.

**15.** $x-y(2 \pm \sqrt{5})+7 \pm 2\sqrt{5} = 0$. **16.** $(-\frac{7}{4}, \frac{3}{4})$, $2x-2y+7 = 0$.

**17.** $4\sqrt{2}$, $2\sqrt{2}$, $x+y = 0$, $x-y = 0$.

**18.** $\frac{1}{2}(13 \pm \sqrt{61})$, $(5 \pm \sqrt{61})x-6y = 0$.

**19.** $7x^2+48xy-7y^2-62x-34y-577 = 0$.

**20.** $2x^2-2xy+y^2 = 1$. **21.** $(0, -\frac{36}{23})$.

**22.** $\{ab^2/(a^2+b^2), a^2b/(a^2+b^2)\}$.

**23.** $2x-y = 0$, $x+2y = 0$, $6(x^2+y^2) = 11$.

**25.** (i) $h = 0$, (ii) $(a+c\alpha)f^2+(b+c\beta)g^2 = 2fgh$,
(iii) $g^2(\alpha+\beta)^2 = (a\beta-b\alpha)(c\alpha+c\beta+a+b)$.

**27.** $ax^2+2hxy+by^2+2fy = 0$.

**29.** $cdx^2+acxy+aby^2-cd(a+b)x-ab(c+d)y+abcd = 0$.

**30.** $5x-12y+1 = 0$, $12x+5y+2 = 0$, $(-\frac{29}{169}, \frac{2}{169})$, $\frac{1}{13}$.

**34.** 185/147. **35.** $a\beta x^2+hxy(\alpha+\beta)+b\alpha y^2 = 0$.

## MISCELLANEOUS EXAMPLES IX (page 268)

**1.** $\frac{3}{8}\pi$, $\frac{7}{8}\pi$. **2.** $r\cos(\theta-\frac{2}{3}\pi) = \frac{1}{2}$.

**3.** $x\sin\frac{3}{2}t-y\cos\frac{3}{2}t = 3a\sin\frac{1}{2}t$.

**4.** $er\cos(\theta-\alpha) = l$, $e^2\sin^2\alpha(1-r^2/4a^2) = \left(1-\frac{l}{r}-\frac{er\cos\alpha}{2a}\right)^2$.

## FURTHER MISCELLANEOUS EXAMPLES (page 273)

**1.** (i) $(28, -14)$, (ii) $(12, 24)$, $(36, -8)$.

**2.** (i) $(7, 2)$, (ii) $x-3y+23 = 0$, (iii) $2x = 3y$.

**3.** $x^2+y^2+6x+8y-75 = 0$, circle centre $(-3, -4)$, radius 10.

**4.** (i) $(-2, 1)$, (ii) $x+2y = 0$, (iii) $2x+4y = 7$.

**5.** 8, $(\frac{48}{13}, \frac{32}{13})$, $14\frac{10}{13}$.

**6.** $x^2+y^2-4x-2y-95 = 0$, $(2, 1)$, 10.

**7.** $4x+3y-35 = 0$, $x+7y+10 = 0$, $(11, -3)$, 25.

**8.** $(1\frac{1}{3}, -\frac{1}{3})$, $4x+3y = 6$.

**9.** $3x-4y+27 = 0$, $4x+3y-64 = 0$, $12\frac{4}{5}$, $(\frac{512}{25}, \frac{384}{25})$.

**10.** $(6, 13)$, $3x-y = 5$.

**11.** $3x-4y-7 = 0$, $2\frac{11}{12}$, $(1\frac{1}{6}, -\frac{7}{8})$, $(\frac{21}{25}, -1\frac{3}{25})$.

**12.** $(-1, -2)$, $(-5, -5)$. **13.** $\sqrt{40}$, $\sqrt{10}$, $\sqrt{8}$, $\sqrt{58}$, $\sqrt{50}$; 20.

**14.** (i) $2x+y-9 = 0$, (ii) $x^2+y^2+2x-2y-23 = 0$; $(2, 5)$, $(4, 1)$.

**15.** $1/a$, $1/b$, $\{1/2a, 1/(1-2a)\}$, $1/x+1/y = 1$.

**16.** $7x+y-21 = 0$, $(0, 1)$.

**17.** $(8, 6)$, 5, $x^2+y^2-19x-16y+148 = 0$.

**18.** $5x-12y = 0$, $5x+12y = 144$, $(14\cdot4, 6)$, $x^2+y^2 = 169$.

**19.** $x^2+y^2-14x-18y+49 = 0$; three, 1.

**20.** $(-3, 4)$, 2, $\sqrt{21}$, $x^2+y^2 = 21$.

**21.** (i) $x^2+y^2+4x = 0$, (ii) $(3, 0)$, 5; $3x-4y+16 = 0$.

**22.** $8\frac{1}{2}$.

**23.** $(2, 5)$, $\sqrt{13}$, $x^2+y^2-4x-10y+16 = 0$, $2x-3y+24 = 0$.

**24.** $x^2+y^2-10x-4y+4 = 0$, $3x+4y-48 = 0$.

**25.** $x^2+y^2-48x-14y+600 = 0$, $x^2+y^2-24x-7y = 0$.

**26.** $(1\frac{1}{5}, -\frac{2}{5})$, $x^2+y^2-6x-4y+4 = 0$.

**27.** $(12\frac{1}{2}, 0)$, $2\frac{1}{2}$, $3x-4y-25 = 0$.

**28.** $x^2+y^2-10x+19 = 0$, $2\frac{1}{2}$.

**29.** $(-4, -3)$, $(11, 5)$, 6, 8, $8x-15y-13 = 0$, 3.

**30.** $4x+3y = 24$.

**31.** $9x+4y = 72$, $\frac{1}{2}$, $4x+y = 24$.

**32.** $6x-y-20 = 0$, $x+6y-65 = 0$, $(3, 2)$.

**33.** $6x-y-9 = 0$.

**34.** $x^2-6x-4y+13 = 0$.

**35.** $x-2y = 0$, $y = 0$, $(2, 1)$.

**36.** $2x-y = 0$, $2x+y = 0$, $(0, 0)$, $2\tan^{-1}2$, 40.

**37.** $112x+14y-201 = 0$, $8x-64y+111 = 0$.

**38.** (i) $4x-y-13 = 0$, $2x-3y+11 = 0$, (ii) $137°\,44'$, (iii) 10.

**39.** $x+2y-10 = 0$, $2x+y-9 = 0$, $x-2y-2 = 0$; $(\frac{13}{3}, 2)$.

**40.** $x-y+1 = 0$, (bisecting the line containing the origin), and $x+y-1 = 0$; $ab = 1$.

**41.** $(-\frac{1}{5}, 1\frac{2}{5})$, $(1\frac{2}{5}, -\frac{1}{5})$; $\frac{8}{5}\sqrt{2}$.

**45.** $(\alpha-2)x+(\beta-1)y = 2\alpha+\beta-5$, $(\alpha+2)x+(\beta+1)y = -2\alpha-\beta-5$, where $P$ is $(\alpha, \beta)$.

**46.** $\frac{1}{3}$, 1.

**47.** $y(y-k) = 2a(x-h)$.

**48.** $71°\,34'$.

**49.** $2\sqrt{(a^2+ac)}$.

**53.** $x+y = 2$, $(1, \frac{1}{2})$.

**54.** $tx+2y = at^3+4at$, $R$ is $(\alpha-2a, 2\beta)$.

**57.** $y^2 = 4ax+(x+a)^2\tan^2\alpha$.

**58.** If $A$ is $(a, 0)$, where the fixed lines are axes, $C$ lies on $y^2 = 4ax$, $P$ lies on $x^2 = a(x-y)$, a parabola.

**60.** $x-2yt = 0$, $a^2(t^2-1)^2 = 1+4t^2$.

**61.** $x = -a^2/2p$.

**62.** $(0, 0)$, $x = 0$, $y = 0$.

**66.** Taking $OY$ as $y$-axis, the parabolas are $y^2 = 4a(x+a)$ and $y^2 = -4a(x-a)$, where the portions for which $|y| < 2a$ are removed.

**78.** $x^2+y^2-4x-6y-276 = 0$.

**86.** $a^2ll'+b^2mm' = nn'$.

**89.** $a^2y = \pm b^2(x+\frac{1}{2}a)$.

**90.** $x = 1, y = -1$.

**95.** $a^2b^2(b^2x^2+a^2y^2) = (b^2hx+a^2ky)^2$,
(i) $b^4x^2+a^4y^2 = a^2b^2(a^2+b^2)$, (ii) $b^2x^2+a^2y^2 = 2a^2b^2$.

**97.** (i) $x^2+y^2 = a^2+b^2$.

**99.** $b(x-p)^2-2h(x-p)(y-q)+a(y-q)^2 = 0$,
$x^2+y^2-14x-2y = 0$, $x^2+4xy+3y^2-36x-68y+320 = 0$.

**100.** $4a^2b^2$.

# INDEX